Third Edition

business communication

Kathryn Rentz
University of Cincinnati

Paula Lentz
University of Wisconsin—Eau Claire

Mc Graw Hill Education

business communication

SENIOR VICE PRESIDENT, PRODUCTS & MARKETS **KURT L. STRAND**

VICE PRESIDENT, CONTENT PRODUCTION & TECHNOLOGY SERVICES **KIMBERLY MERIWETHER DAVID**

MANAGING DIRECTOR **PAUL DUCHAM**

SENIOR BRAND MANAGER **ANKE BRAUN WEEKES**

EXECUTIVE DIRECTOR OF DEVELOPMENT **ANN TORBERT**

DEVELOPMENT EDITOR II **KELLY I. PEKELDER**

MARKETING MANAGER **MICHAEL GEDATUS**

DIRECTOR, CONTENT PRODUCTION **TERRI SCHIESL**

CONTENT PROJECT MANAGER **KATIE KLOCHAN**

CONTENT PROJECT MANAGER (OLC) **SUSAN LOMBARDI**

SENIOR BUYER **CAROL A. BIELSKI**

DESIGN **SRDJAN SAVANOVIC**

COVER IMAGE **MAN HOLDING GLOBE: © DIMITRI OTIS, GETTY IMAGES; BLOG KEYBOARD: © PETER DAZELEY, GETTY IMAGES; GUY SKYPING: © IMAGE SOURCE, GETTY IMAGES; GUY HOLDING POWERPOINT SLIDE: © COLIN ANDERSON, GETTY IMAGES**

BACK COVER IMAGE **© ALEJANDRO RIVERA/GETTY IMAGES**

CONTENT LICENSING SPECIALIST **JOANNE MENNEMEIER**

TYPEFACE **10/12 TIMES LT STD**

COMPOSITOR **MPS LIMITED**

PRINTER **R. R. DONNELLEY**

brief
contents

contents

part two MASTERING WRITING AND PRESENTATION BASICS

part three   WRITING EFFECTIVE MESSAGES

CHAPTER 5   Writing Good-News and Neutral Messages 98

CHAPTER 7 Writing Persuasive Messages and Proposals 156

part four   WRITING EFFECTIVE REPORTS

CHAPTER 8   Researching and Writing Reports 198

CHAPTER 9 Writing Short Reports 246

part five DEVELOPING ADDITIONAL BUSINESS COMMUNICATION SKILLS

bonus chapters
(ONLINE)

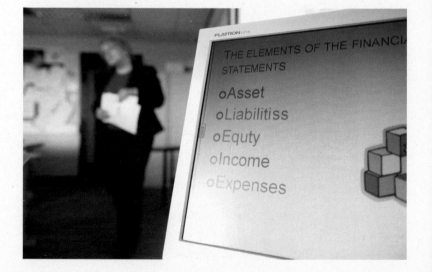

BONUS CHAPTER C Cross-Cultural Communication

chapter changes

CHAPTER 1
- Latest evidence of the importance of communication skills in business.
- Current research on the skills needed in the 21st-century workplace.
- New boxed features: "Demonstrating Your Value on a High-Profile Team," "This Just In: What You Can Do Is Even More Important than What You Know," "Why Companies Promote Workplace Diversity," "What's the Dominant Metaphor in *Your* Workplace?"
- Updated photos and exhibits.

CHAPTER 2
- Current advice on letter writing, particularly on avoiding the use of greetings such as "to whom it may concern" and other outdated expressions.
- Updated advice on current email practices in the workplace and on the role of email among other communication technologies such as text and instant messaging.
- Expanded information on text and instant messaging and social media communication as forms of business messages.
- New advice on preparing print vs. online documents and discussion of best practices for writing Web content.

CHAPTER 3
- Use of "visuals" rather than "graphics" to better reflect the wide range of options for visual communication.
- Emphasis on visuals as communication tools.
- Many new visuals to illustrate common types used in business communication.

CHAPTER 4
- New boxed features: "Writing with Clarity and Courtesy," "The Most Annoying Business Clichés," "Don't Be Hoodwinked by Homophones," "Understanding the Different Generations in the Workplace," "Beware the Vague or Illogical *This*," "Courtesy in the Age of Mobile Devices."
- A more logically organized section on selecting appropriate words.
- Clearer advice about using sentence structure (e.g., coordination and subordination) to manage emphasis.
- New sections on being courteous and on determining the right level of formality.

CHAPTER 5
- New "Workplace Scenario" that uses a running narrative of routine communication scenarios at White Label Industries. This provides an opportunity for instructors to simulate communication situations in a single company, creating a consistent scenario for addressing audience, context, and communication goals within an organization.
- New Case Illustration examples.
- New boxed features: "Choosing the Right Font," "A Workplace without Email? One Company's Strategy."
- Over 30 new or revised problem-solving cases (online).

CHAPTER 6
- Opportunity to continue use of the White Label Industries narrative from Chapter 5 for bad-news messages.
- New "Workplace Scenarios" throughout the chapter.
- New Case Illustrations of bad-news messages written in the indirect approach: a refused request to an external audience, a refused request to an internal audience, and a negative announcement.
- Over 30 revised or new problem-solving cases (online).

CHAPTER 7
- New boxed features: "Generating More Customers for Your Business," "Learn about e-Selling from Chief Marketer and MailChimp," "Are Sales Letters Becoming Extinct? Absolutely Not!," "What Type of Decision Maker Is Your Reader?," "Can Your Sales Message Pass This Test?," "Gaining—and Keeping—Readers' Attention on Facebook and Twitter," "Current Trends in Promotional Writing: A Q&A with a Young PR Professional," "CAN-SPAM: It's the Law," "Web Resources for Proposal Writing."
- New Case Illustrations and exhibits from Scotts Lawn Service, Skillpath Seminars, Delta Airlines, the American Society of Training and Development, and the state government of Vermont (an RFP).
- Incorporation of new media and use of visuals in the discussion of sales messages.
- Over 30 new problem-solving cases (online).

CHAPTER 8
- New boxed features: "How Far Should Your Report Go?," "Report-Writing Tools Help Businesses Succeed," "Managing Citations with Zotero."

- A completely reorganized and updated research section, including over 30 screenshots of online research tools.
- Addition of a wide variety of Web-based resources, a more helpful discussion of library research, a new table of useful library resources, and an updated list of resources organized by research question.
- Expanded discussion of designing a questionnaire.

CHAPTER 9
- New boxed features: "Are Tweets, Blog Comments, and Text Messages Undermining Your Report-Writing Skills?," "When Is a Report not a Report?," "The Monetary Value of a Good Report."
- Removal of audit reports; expanded discussion of progress reports, with a new Case Illustration.
- Over 30 new problem-solving cases, plus a list of 152 general report topics in different functional areas of business (online).

CHAPTER 10
- New boxed features: "Finding Your Professional Voice," "The Art of Negotiation," "What's in a Handshake?," "Virtual Presentations: The Real Thing," "Have You Met TED?," "Look Like a Pro with PowerPoint Keyboard Shortcuts."
- Updated discussion of phone etiquette.
- Current research on the relationship between "digital natives'" (e.g., millennials, Gen-Yers) technology use and the development of their nonverbal communication skills.
- Updated section on "Delivering Web-Based Presentations."

CHAPTER 11
- Discussion of the importance of internships.
- New boxed features: "The Where, What, and Whys of Hiring," "The Most Important Six Seconds in Your Job Search," "Developing a Professional Portfolio," "Answers to the 10 Toughest Interview Questions," "What's the Number One Interviewing Mistake?," "Make Your LinkedIn Profile Work for You," "Web Sites Offer Valuable Interview Advice."

- Discussion in various parts of the chapter on how employers and job seekers use social networking sites in the hiring or job-search process.
- Discussion of the features of print résumés and electronic résumés (e.g., email, scannable, Web-based).

BONUS CHAPTER A
- Discussion of document layout principles.
- Use of Word 2013 screenshots.

BONUS CHAPTER B
- Fifty new practice sentences to build students' skills in the use of pronoun case, pronoun-antecedent agreement, subject-verb agreement, punctuation, and the apostrophe.
- Additional guidelines on pronoun-antecedent agreement.
- New boxed features: "Can You Detect the Difference That Punctuation Makes?," "Good Grammar: Your Ticket to Getting and Keeping a Job."

BONUS CHAPTER C
- Updated discussion of the dimensions of cultural difference.
- Updated list of resources for effective cross-cultural communication.
- Addition of an exercise comparing Japanese and U.S. versions of an email message.

BONUS CHAPTER D
- Improved format for the sample long report.
- Use of Word 2013 screenshots in boxed features.

BONUS CHAPTER E
- Thoroughly updated examples of footnote and bibliography format for different types of sources.
- Use of Word 2013 screenshots in boxed features.

business
communication

Communicating in
the Workplace

chapter one

As Head of Learning & Development for Facebook, Stuart Crabb knows what it takes to be an attractive job candidate and a successful employee. He has over 20 years' experience helping companies hire the right people, develop their talent, and become more culturally diverse.

What does it take to succeed at Facebook? According to Crabb, the answers are "critical thinking," "problem solving," "creativity," and "performance." It also takes being "motivated," "individually accountable," and a "good fit" with the company culture.

These happen to be key traits of successful business communicators, too. They understand that communicating well takes analysis, judgment, and even ingenuity. It takes being attuned to people and to each communication situation. And it takes not only verbal skill but also technological and visual literacy.

Like business itself, business communication can be challenging. But the challenge can be fun, and solving communication problems can bring enormous rewards. This book will help prepare you for an exciting future as both a businessperson and a communicator. ■

LEARNING OBJECTIVES

LO 1-1 Explain the importance of communication to you and to business.

LO 1-2 Describe the main challenges facing business communicators today.

LO 1-3 Describe the three main categories of business communication.

LO 1-4 Describe the formal and informal communication networks of the business organization.

LO 1-5 Describe factors that affect the types and amount of communicating that a business does.

LO 1-6 Explain why business communication is a form of problem solving.

LO 1-7 Describe the contexts for each act of business communication.

LO 1-8 Describe the business communication process.

workplace scenario

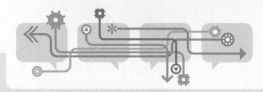

Demonstrating Your Value on a High-Profile Team

You were thrilled to be hired a few months ago as a customer service representative for OrgWare.com, a company that sells management software specially designed for professional associations. The software enables organizations like the American Marketing Association and the Association for Business Communication to manage their finances, keep track of their members, schedule events, and much more.

The company is doing well. In 12 years, it has grown from a five-person business into one that employs 120 people. There are now six regional sales teams located across the United States, and there's even a development team in Malaysia. But this growth has created a problem: The extensive face-to-face communication that helped make OrgWare.com a thriving business has, in many cases, become difficult or impossible. As a result, the sense of teamwork in the organization is weakening. And it is clear that phone calls, emails, and instant messaging are not sufficient to keep employees engaged and well informed.

The CEO has formed a task force to find an internal communication solution. Will it be an intranet? An electronic newsletter? A secure social networking site? Virtual meetings? A combination? Which would the employees be most likely to read and use? How should the solution be implemented, and what will it cost?

To your surprise, you were asked to help find the answers. The CEO felt that your familiarity with new media could be an asset to the team. You'll also be expected to represent the customer service area and the viewpoints of young employees like yourself.

Everyone on the team will need to research the pros and cons of different media, acquire employees' opinions, write progress reports, share ideas, and ultimately help present the team's recommendation to the top executives.

Are you ready?

> "Your work in business will involve communication—a lot of it—because communication is a major part of the work of business."

THE ROLE OF COMMUNICATION IN BUSINESS

Your work in business will involve communication—a lot of it—because communication is a major part of the work of business. The overview that follows will help you prepare for communication challenges like those described in the Workplace Scenario.

LO 1-1 Explain the importance of communication to you and to business.

The Importance of Communication Skills

Because communication is so important in business, businesses want and need people with good communication skills.

Evidence of the importance of communication in business is found in numerous surveys of executives, managers, and recruiters. Without exception, these surveys have found that communication ranks at or near the top of the business skills needed for success.

For example, the 431 managers and executives who participated in a survey about graduates' preparedness for the workforce named "oral communications," "teamwork/collaboration," "professionalism/work ethic," "written communications," and "critical thinking/problem solving" as the top "very important skills" job applicants should have.[1] The employers surveyed for the National Association of Colleges and Employers' *Job Outlook Survey* for 2011 rated "communication" as the most valuable soft skill, with "teamwork skills" and "analytical skills" following closely behind.[2] Why is communication ability so highly valued? As one professional trainer explains, "you will need to request information, discuss problems, give instructions, work in teams, and interact with colleagues and clients" to achieve cooperation and team efficiency. To advance, you'll also need to be able to "think for yourself," "take initiative,"

communication matters

This Just In: What You Can Do Is Even More Important than What You Know

In its latest annual survey of executives, the Association of American Colleges and Universities found that "cross-cutting capacities" like communication skills are now more valued than a particular choice of major. More specifically,

- Nearly all those surveyed (93%) agree that "a candidate's demonstrated capacity to think critically, communicate clearly, and solve complex problems is more important than their undergraduate major."
- More than nine in ten of those surveyed say it is important that those they hire demonstrate ethical judgment and integrity, intercultural skills, and the capacity for continued new learning.
- More than three in four employers say they want colleges to place more emphasis on helping students develop key learning outcomes, including critical thinking, complex problem solving, written and oral communication, and applied knowledge in real-world settings.

Source: "It Takes More than a Major: Employer Priorities for College Learning and Student Success," *Association of American Colleges and Universities*, AAC&U, 10 Apr. 2013, Web, 29 Apr. 2013.

and "solve problems."[3] On the managerial level, you'll find that communication skills are even more essential. In the words of an international business consultant, "nothing puts you in the 'poor leader' category more swiftly than inadequate communication skills."[4]

Unfortunately, businesses' need for employees with strong communication skills is all too often unfulfilled. When NFI Research asked senior executives and managers what areas of their companies they'd most like to see improved, they put "efficiency" and "communication" at the top of the list.[5] According to Solari Communications, "poor communication costs business millions of dollars every single day" in the form of wasted time, misunderstandings, eroded customer loyalty, and

Improving your communication skills improves your chances for success in business.

Why Business Depends upon Communication

Every business, even a one-person business, is actually an economic and social system. To produce and sell goods and services, any business must coordinate the activities of many groups of people: employees, suppliers, customers, legal advisors, community representatives, and government agencies that might be involved. These connections are achieved through communication.

[Whatever position you have in business, your performance will be judged largely on the basis of your ability to communicate.]

lost business.[6] SIS International Research found that poor communication is a problem for small and mid-sized businesses, not just for big corporations. Its data indicated that in 2009 a business with 100 employees spent an average downtime of 17 hours a week on clarifying its communications, which translated into an annual cost of $524,569.[7]

The communication shortcomings of employees and the importance of communication in business explain why you should work to improve your communication skills. Whatever position you have in business, your performance will be judged largely on the basis of your ability to communicate. If you perform and communicate well, you are likely to be rewarded with advancement. And the higher you advance, the more you will need your communication ability. The evidence is clear:

Consider, for example, the communications of a pharmaceutical manufacturer. Throughout the company, employees send and receive information about all aspects of the company's business:

- Salespeople receive instructions and information from the home office and submit orders and regular reports of their contact with customers.

- Executives use written and oral messages to conduct business with customers and other companies, manage company operations, and perform strategic planning.

- Production supervisors receive work orders, issue instructions, receive status reports, and submit production summaries.

- Shop floor supervisors deliver orders to the employees on the production line, communicate and enforce guidelines for safety and efficiency, troubleshoot problems that arise, and bring any concerns or suggestions to management.

- Marketing professionals gather market information, propose new directions for company production and sales efforts, coordinate with the research and development staff, and receive direction from the company's executives.

- Research specialists receive or propose problems to investigate, make detailed records of their research, monitor lab operations for compliance with government regulations, and communicate their findings to management.

- Public relations professionals use various media to build the company's brand and maintain the public's trust.

Numerous communication-related activities occur in every other niche of the company as well: finance and accounting, human resources, legal, information systems, and other departments. Everywhere, employees receive and send information as they conduct their work, and they may be doing so across or between continents as well as between buildings or offices.

Oral communication is a major part of this information flow. So, too, are various types of written communication—instant messaging, text messaging, online postings and comments, email, memos, letters, and reports, as well as forms and records.

All of this communicating goes on in business because communication is essential to the organized effort involved in business. Simply put, communication enables human beings to work together.

LO 1-2 Describe the main challenges facing business communicators today.

Current Challenges for Business Communicators

While communication has always been central to business, the nature of work today presents special communication challenges. Here we discuss four interrelated trends that are likely to influence how you will work and communicate.

the need for expanded media literacy When email arrived on the scene in the late 1980s, it created something of a revolution. Instead of being restricted to letters, memos, and printed reports and proposals, business writers could now correspond electronically. As a result, many tasks formerly conducted via printed documents—memos in particular—were performed through email instead, and email replaced many phone and face-to-face conversations as well. Email has also had the effect of speeding up communication and of enabling a communicator to

- Media literacy
- Social intelligence
- Cross-cultural competency
- Computational thinking
- Visual literacy
- Interpretive skills
- Ethical reasoning

reach many more readers simultaneously. It has increased what we can achieve—and are expected to achieve—each day.

Email is still the most heavily used medium in business, but many other media have appeared on the scene. In addition to instant messaging and text messaging, businesses are now using blogs, tweets, podcasts, social networking, virtual meetings, videos, animation, simulations, and even online games. Collectively referred to as **new media**, these forms of communication and the mobile devices with which people access them are causing another revolution.

The impacts of this change are many and far reaching. It is easy now to network with others, even on the other side of the world, and to tap the intelligence of those outside the boundaries of the organization. Obviously, these "new ways for groups to come together and collaborate" will require that employees be "highly conversant with digital networking and virtual collaboration."[8] But new media are also increasing the need for employees who have **social intelligence**—the ability "to quickly assess the emotions of those around them and adapt their words, tone, and gestures accordingly."[9]

With information coming in so fast and from so many sources, organizations are becoming less hierarchical and more brain-like, with each employee acting as a kind of sensor. As a result, front-line employees now have a higher level of decision-making power than ever before.[10] Performing well in such an environment takes "novel and adaptive thinking,"[11] a willingness to "embrace change," and "fierce problem-solving skills."[12] The approach to business communication that this book takes will help you develop these strengths.

increasing globalism and workplace diversity Countries and cultures continue to grow more interconnected as businesses expand their reach around the world. According to a panelist for a recent webinar on workplace trends, we are seeing "the emergence of the truly globally integrated enterprise," which means that the likelihood of working on a global team is increasing, as is the importance of "global social networks."[13]

Cross-cultural competency should thus be a part of your skillset.[14] You will need to be aware that your assumptions about business and communication are not shared by everyone everywhere. As Bonus Chapter C explains, businesspeople from other countries may have distinctly different attitudes about

communication matters

Why Companies Promote Workplace Diversity

Diversity programs are becoming widespread. Why? A Web article posted by American Express lists these benefits:

1. **It builds your employer brand.** You can attract better talent from around the world. Also, a company that has a strong diversity program will have a good reputation because it will be seen as having fair employment practices.

2. **It increases creativity.** When you bring a variety of different people from various backgrounds together, you'll end up getting better solutions to business problems.

3. **It encourages personal growth.** Employees, especially younger ones, are striving to use their corporate experience to learn and grow their careers. This is a major advantage to workplace diversity because it can help employees learn new ideas and perspectives and connect intellectually and personally to different people.

4. **It makes employees think more independently.** If you have similar people at a company, it will be harder to solve complex problems. One study by Katherine Phillips, a professor at Kellogg, shows that adding even one employee from a different background can get people out of their comfort zones and thinking differently about a situation.

How a company will define diversity will depend on the company. The visual to the left, from the Nissan Web site, incorporates 10 different types—and you might be able to think of others.

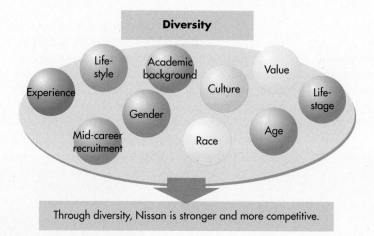

Through diversity, Nissan is stronger and more competitive.

Sources: Dan Schawbel, "Why Diversity Matters in the Workplace," *American Express Open Forum*, American Express Company, 8 Nov. 2012, Web, 28 Apr. 2013; "Diversity," *Nissan*, Nissan Motor Company, n.d., Web, 28 Apr. 2013.

---

punctuality and efficiency. They can also differ from you in their preference, or lack thereof, for directness and the show of emotion. And the core features of their culture—such as their preference for individualism or collectivism, their religious beliefs, their political environment, their ideas about social hierarchy, and their attitudes toward work itself—can make their view of how to do business quite different from yours.

You will encounter other kinds of diversity as well. To have adequate retirement income, the so-called Baby Boomers—those born soon after World War II—are extending their careers. This means that organizations are likely to have employees in their twenties, in their sixties and seventies, and every age in between.[15] The influx of women into the workplace has meant increased gender diversity. And according to a diversity officer for a major health care firm, each generation of U.S. workers has grown more ethnically diverse, with the so-called Generation Y cohort (those born after 1979) having the most ethnic diversity.[16] This trend is making organizations more innovative and productive,[17] and it means that "cultural agility" will need to figure into your workplace communications.[18] (See the Communication Matters feature above for more about the benefits and types of diversity in the workplace.)

an increased need for strong analytical skills Adapting to a quickly changing business landscape requires being able to assess information quickly, focus on what's relevant, and interpret information reliably and usefully. As data-gathering devices are built into more objects, there will be more numerical data for us to process. The need for **computational thinking**—the ability "to interact with data, see patterns in data, make data-based decisions, and use data to design for desired outcomes"[19]—will increase. So will the need for **visual literacy**, the ability to create and interpret graphics.[20]

The value of **interpretive skills** extends beyond interpreting numbers. As we've pointed out, being able to understand other people is critical. As "smart machines" automate many workplace tasks, employees will spend more time on tasks that require "sense-making," or "the ability to determine the deeper meaning or significance of what is being expressed."[21]

> ## "We've got to recognize that the real high-value work ... may actually have an *imaginative* component."

As one expert put it. "We've got to recognize that the real high-value work ... may actually have an *imaginative* component."[22] This quality is required to discern the key facts, to explore "what if," and to choose the best solution—all central components of successful business communication.

an increased focus on ethics and social responsibility
One more widespread trend under way in business will likely affect the goals of the organization you work for: an increased focus on ethical and socially responsible behavior.

While ethical scandals have plagued businesses throughout modern history, the Enron and WorldCom scandals of 2002, in which false reports of financial health cheated employees and shareholders alike, seemed to usher in a new era of concern. That concern was well founded: With 2008 came unprecedented discoveries of mismanagement and fraud on the part of some of the United States's largest financial institutions. Accounts of predatory lending, business espionage, and exploitative labor practices continue to shake the public's confidence in business. On a moral level, doing business in a way that harms others is wrong. On a practical level, doing so undermines trust, which is critical to the success of business. The more an organization builds trust among its employees, its shareholders, its business partners, and its community, the better for the business and for economic prosperity overall. A key way to build trust is through respectful, honest communication backed up by quality goods and services.

Lately, another important dimension of business ethics has developed: **corporate social responsibility**. The Internet has brought a new transparency to companies' business practices, with negative information traveling quickly and widely. Nongovernmental organizations (NGOs) such as Corp-Watch, Consumer Federation of America, and Greenpeace can exert a powerful influence on public opinion and even on

Nongovermental organizations (NGOs) such as CorpWatch attest to the growing importance of social responsibility in business.

Source: CorpWatch, Home page, CorpWatch, 1 May 2013, Web, 3 May 2013.

I WANT YOU TO MEET THE CLIENT, SHOW HIM OUR CATALOG, MAKE YOUR SALES PRESENTATION, LET HIM TEST THE DEMO, THEN CLOSE THE SALE, ARRANGE FOR SHIPPING, AND PROCESS THE INVOICE.

I ALREADY DID ALL THAT WITH MY PHONE WHILE YOU WERE TALKING!

Source: © Randy Glasbergen/glasbergen.com

governments. Businesses now operate in an age of social accountability, and their response has been the development of corporate social responsibility (CSR) departments and initiatives. While the business benefits of CSR have been debated, the public demand for such programs is strong. You may well find that social issues will influence how you do business and communicate in business.

LO 1-3 Describe the three main categories of business communication.

Main Categories of Business Communication

Such newer media as blogs and social networking have weakened the boundary between "inside" and "outside" the organization. One post on a company's blog, for example, could draw comments from its employees, from employees in a similar organization or industry, or from potential customers.

Even so, most communication on the job can still be categorized as either internal operational, external operational, or personal. These categories, while not completely distinct, can help you understand your purposes for communicating.

internal-operational communication All the communication that occurs in conducting work within a business is internal operational. This is the communication among

the business's employees that is done to perform the work of the business and track its success.

Internal-operational communication takes many forms. It includes the ongoing discussions that senior management undertakes to determine the goals and processes of the business. It includes the orders and instructions that supervisors give employees, as well as written and oral exchanges among employees about work matters. It includes reports that employees prepare concerning sales, production, inventories, finance, maintenance, and so on. It includes the messages that they write and speak in carrying out their assignments and contributing their ideas to the business.

Much of this internal-operational communication is performed on computer networks. Employees send email, chat online, and post information on company portals and blogs for others throughout the business, whether located down the hall, across the street, or in other countries. And today, much of this communication takes place via smartphones and other mobile devices.

external-operational communication The work-related communicating that a business does with people and groups outside the business is **external-operational communication**. This is the business's communication with suppliers, service companies, customers, government agencies, the general public, and others.

External-operational communication includes all of the business's efforts at selling—from sales letters, emails, and phone calls to Web and television ads, trade-show displays, the company Web site, and customer visits. Also in this category is all that a business does to gain positive publicity, such as promoting its community-service activities, preparing appealing materials for current and prospective investors, writing press releases for the media, and contributing expert insights at professional meetings and on webinars. In fact, every act of communication with an external audience can be regarded as a public-relations message, conveying a certain image of the company. For this reason, all such acts should be undertaken with careful attention to both content and tone.

The importance of these kinds of external-operational communication hardly needs explaining. Because the success of a business depends on its ability to attract and satisfy customers, it must communicate effectively with those customers.

But businesses also depend on one another in the production and distribution of goods and services. Coordinating with contractors, consultants, and suppliers requires skillful communication. In addition, every business must communicate to some extent with a variety of other external parties, such as government agencies and public-interest groups. Some external audiences for today's businesses are illustrated in Exhibit 1-2. Like internal communication, external communication is vital to business success.

Companies often use carefully designed portals or intranets, such as this one at Procter & Gamble, to communicate with employees and enable them to communicate with each other.

Source: Reprinted with permission.

personal communication

Not all the communication that occurs in business is operational. In fact, much of it is without apparent purpose as far as the operating plan of the business is concerned. This type of communication is personal. Do not underestimate its importance. **Personal communication** helps make and sustain the relationships upon which business depends, and it is more important than ever.

Personal communication is the exchange of information and feelings in which we human beings engage whenever we come together—or when we just feel like talking to each other. We are social animals, and we will communicate even when we have little or nothing to say.

EXHIBIT 1-2 Likely External Audiences for Today's Businesses

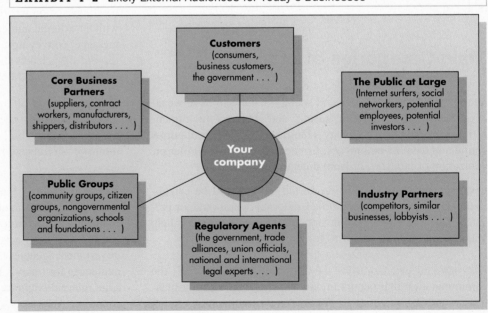

Although not an official part of the business's operations, personal communication can have a significant effect on their success. This effect is a result of the influence that personal communication can have on the attitudes of the employees and those with whom they communicate.

The employees' attitudes toward the business, one another, and their assignments directly affect their productivity. The nature and amount of personal talk at work affect those attitudes. In an environment where heated words and flaming tempers are often present, the employees are not likely to give their best efforts to their jobs. Likewise, a rollicking, jovial workplace can undermine business goals. Wise managers cultivate the optimum balance between employees' focus on job-related tasks and their freedom to engage with others on a personal level. Chat around the water cooler or in the break room encourages a team attitude and can often be the medium in which actual business issues get discussed. Even communication that is largely internal-operational will often include personal elements that relieve the tedium of daily routine and enable employees to build personal relationships.

Similarly, communication with external parties will naturally include personal remarks at some point. Sometimes you may find yourself writing a wholly personal message to a client, as when he or she has won a major award or experienced a loss of some kind. Other times, you may compose an external-operational message that also includes a brief personal note, perhaps thanking a client for a pleasant lunch or referring to a personal matter that came up in the course of a business meeting.

Using both online and face-to-face networking, you will also cultivate business-related friends. Your relationships with these contacts will not only help you do your current job; they will also be an important resource as you change jobs or even careers. Research shows that "the idea of the steady, permanent job is becoming a relic of another era."[23] Employees are now taking "an entrepreneurial approach" to their lives and skills, considering carefully where to work, what work to do, how much to work, and how long to work.[24] The personal connections you make in your current employment will contribute to your future success.

LO 1-4 Describe the formal and informal communication networks of the business organization.

Communication Networks of the Organization

Looking over all of a business's communication (internal, external, and personal), we see an extremely complex system of information flow and human interaction. We see dozens, hundreds, or even thousands of individuals engaging in

Personal communication in business is both inevitable and important.

untold numbers of communication events throughout each workday.

In fact, as Exhibit 1-3 shows, there are two complex networks of information in any organization—one formal and one informal. Both are critical to the success of the business.

the formal network In simplified form, information flow in a modern business is much like the network of arteries and veins in the body. Just as the body has blood vessels, the business has major, well-established channels for information exchange. This is the **formal network**—the main lines of operational communication. Through these channels flows the bulk of the communication that the business needs to operate. Specifically, the flow includes the upward, lateral, and downward movement of information in the form of reports, memos, email, and other media within the organization; the downward movement of orders, instructions, advisories, and announcements; and the broad dissemination of company information through the organization's newsletter, bulletin boards, email, intranet, or blogs.

As we have seen, information routinely flows outward as well. Order acknowledgments, invoices, receipts, correspondence with suppliers and consultants, and other standard external-operational communications can make external audiences part of the formal communication network.

Formal and Informal Communication Networks in a Division of a Small Business

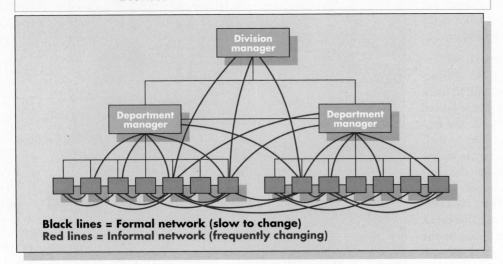

Black lines = Formal network (slow to change)
Red lines = Informal network (frequently changing)

These officially sanctioned lines of communication cause certain forms of communication, or **genres**, to exist within the organization. For example, it may be customary in one company for project leaders to require a weekly report from team members. In another company, the executives may hold monthly staff meetings. Whatever the established form, it will bring with it certain expectations about what can and cannot be said, who may and may not say it, and how the messages should be structured and worded. You will need to understand

Internal-operational communication enables employees to work together toward business goals.

these expectations in order to use the approved lines of communication to get things done.

the informal network Operating alongside the formal network is the **informal network**. It consists of the thousands upon thousands of personal communications that may or may not support the formal communication network of a business. Such communications follow no set pattern; they form an ever-changing and infinitely complex structure linking the members of the organization to each other and to many different external audiences.

The complexity of this informal network, especially in larger organizations, cannot be overemphasized. Typically, it is really not a single network but a complex relationship of smaller networks consisting of certain groups of people. The relationship is made even more complex by the fact that these people may belong to more than one group and that group memberships and the links between groups are continually changing. The department you belong to, the other employees with whom you come in contact in the course of your workday, and the many connections you make with those outside your organization can cause links in this network to form.

The informal network inside an organization is often referred to as the **grapevine**. This communication network is more valuable to the company's operations than a first impression might indicate. Certainly, it carries much gossip and rumor. Even so, the grapevine usually carries far more information than the formal communication system, and on many matters it is more effective in determining the course of an organization. Skillful managers recognize the presence of the grapevine, and they know that the powerful people in this network are often not those at the top of the formal organizational hierarchy. They find out who the talk leaders are and give them the information that will do the most good for the organization. They also make management decisions that will cultivate positive talk.

Employees' personal relations with external audiences add another dimension to a company's informal network. The widespread use of social media has dramatically increased employees' informal communication with outsiders. Such communication can either help or hurt the company. Here again, wise managers will be sensitive to the informal network and encourage talk that is beneficial to the company.

As an employee, you need to be careful about how you participate in the informal network. Unwise remarks can get you known as a troublemaker and even get you fired,

communication matters

What's the Dominant Metaphor in *Your* Workplace?

Prominent management scholar Gareth Morgan asserts that companies are shaped by powerful, yet often unconscious, metaphors. Below are the eight metaphors he discusses in his book *Images of Organization*. How do you think communication practices would vary across these different types of cultures?

The organization as a machine. An organization based on this way of seeing will be hierarchical and bureaucratic—strong on control but poor at adaptation.

The organization as an organism. This type of organization understands itself as a living organism that must pay attention to its various environments as well as foster healthy development internally.

The organization as a brain. Here the emphasis is on enabling quick adaptability through "organizational intelligence," which is achieved by establishing a minimal set of rules and then allowing employees at all levels to gather, share, and act on information.

The organization as a culture. This vantage point enables us to see organizations as meaning-making systems, with rituals, myths, heroes, values, and shared frames of reference that sustain an interpretive world, much like that of a tribe.

The organization as a political system. All organizations are "intrinsically political" because the people who work there will have diverse and conflicting interests. But conflict, coalition building, and the use of power will be more pronounced in some organizations than in others.

The organization as a psychic prison. "Organization *always* has unconscious significance," Morgan asserts: People bring their egos, anxieties, repressions, and many other psychic elements to the workplace, and the organization as a whole can develop tunnel vision or neuroses. These can block positive change and even threaten organizational survival.

The organization as flux and transformation. Organizations that embrace change (and understand that change is inevitable) are more willing than others to redefine the business they're in, question the traditional boundaries between themselves and other organizations, and let their identities continually evolve.

The organization as an instrument of domination. Organizations can and often do have a dark side, with the will to compete and expand taking precedence over regard for individuals, society, and the well-being of other countries.

Morgan's list is not exhaustive; an organization could be like a sports team, for example, or a family. And several different metaphors could be operating within the same company. But looking for your organization's dominant metaphor will help you interpret your place of employment and make more successful communication choices.

Source: Gareth Morgan, *Images of Organization, Executive Edition* (San Francisco: Berrett-Koehler, 1998), print.

---

whereas representing yourself and your company well can result not only in more pleasant relations but also in professional success.

LO 1-5 Describe factors that affect the types and amount of communicating that a business does.

Variation in Communication Activity by Business

Just how much and what kind of communicating a business does depends on several factors. The nature of the business is one. For example, insurance companies have a great need to communicate with their customers, especially through letters and other mailings, whereas housecleaning service companies have little such need. Another factor is the business's size and complexity. Relatively simple businesses, such as repair services, require far less communication than complex businesses, such as automobile manufacturers.

The business's relation to its environment also influences its communication practices. Businesses in a comparatively stable environment, such as textile manufacturing or food processing, will tend to depend on established types of formal communication in a set organizational hierarchy, whereas those in an unpredictable environment, such as software development or online commerce, will tend to improvise more in terms of their communications and company structure.

Yet another factor is the geographic dispersion of the operations of a business. Obviously, internal communication in a business with multiple locations differs from that of a one-location business. Enabling employees to work from home, requiring them to travel, or relying on outside contractors can also increase a company's geographical reach and thus affect its communication. Related to this factor is how culturally diverse the company is.

The communication of a multicultural organization will require more adaptation to participants' values, perspectives, and language skills than that of a relatively homogeneous organization.

Each business can also be said to possess a certain **organizational culture**, which has a strong effect upon, and is strongly affected by, the company's communication. The concept of organizational or corporate culture was popularized in the early 1980s, and it continues to be a central focus of management consultants and theorists.[25] You can think of a given company's culture as its customary, but often unstated, ways of perceiving and doing things. It is the medium of preferred values and practices in which the company's members do their work.

Recall places you've worked or businesses you've patronized. In some, the employees' demeanor suggests a coherent, healthy culture in which people seem to know what to do and be happy doing it. At the other extreme are companies where employees exhibit little affiliation with the business and may even be sabotaging it through poor customer service or lack of knowledge about their jobs. The content and quality of the company's communication have a great deal to do with employees' attitudes and behavior. (See the Communication Matters feature on page 13 for eight metaphors that often shape companies' cultures and communications.)

Take care to note that the official culture and the actual culture in a company are not necessarily the same. Officially, the company management may announce and try to promote a certain culture through formal communications such as mission statements and mottoes. But the actual culture of a company is a dynamic, living realm of meaning constructed daily through infinite behaviors and communications at all levels of the company. Having your antennae out for the assumptions that actually drive people's conduct in your or your client's workplace will help you become a more effective communicator.

> "Each business can also be said to possess a certain organizational culture, which has a strong effect upon, and is strongly affected by, the company's communication."

THE BUSINESS COMMUNICATION PROCESS

While business communication involves many different skills, from verbal and visual literacy to technological know-how, none are more important than problem-solving skills and people skills. These are central to the business communication process.

LO 1-6   Explain why business communication is a form of problem solving.

Business Communication as Problem Solving

Virtually every significant communication task that you will face will involve analyzing a unique set of factors that requires at least a somewhat unique solution. For this reason, it makes sense to think of business communication as **problem solving**.

Researchers in many fields—management, medicine, writing, psychology, and others—have studied problem solving. In general, they define *problem* as a gap between where you are now and where you want to be.[26] Within this framework, a problem isn't always something negative; it can also be an opportunity to improve a situation or do things in a better way. As a goal-focused enterprise, business is all about solving problems, and so, therefore, is business communication.

The problem-solving literature divides problems into two main types: *well defined* and *ill defined*. The former can be solved by following a formula, such as when you are computing how much money is left in your department's budget. But most real-world problems, including business communication problems, cannot be solved this way. They do not come to us in neat packages with the path to the best solution clearly implied. Instead, they require research, analysis, creativity, and judgment. One reason why this is the case in business communication is that, as in any communication situation, people are involved—and people are both complex and unique. But the business context itself is often complex, presenting you with multiple options for handling any given situation. For example, if a customer has complained, what will you do about it? Nothing? Apologize? Imply that the customer was at fault? Give a conciliatory discount? Refuse to adjust the bill? Even a "simple" problem like this one requires thinking through the likely short- and long-term effects of several possible solutions.

Solving ill-defined problems involves combining existing resources with innovation and good judgment. Although this book presents basic plans for several common types of business communication messages, you will not be able to solve particular communication problems by just filling in the blanks of these plans. The plans can be thought of as **heuristics**—"rules of thumb" that keep you from reinventing the wheel with each new problem. But the plans do not tell you all you need to do to solve each unique communication problem. You must decide how to adapt each plan to the given situation.

What this means is that successful business communication is both more challenging and more exciting than you may have thought. You will need to draw on your own powers of interpretation and decision making to succeed with your communication partners.

Of course, people will handle communication tasks somewhat differently depending on who they are, how they interpret the situation, and who they imagine their recipients to be. Does this mean that all communication solutions are equally valid? Not at all. While there is no perfect solution, there can be many bad ones that have been developed without enough analysis and effort. Focused thinking, research, and planning will not guarantee success in the shifting, complex world of business communication, but they will make your chances of success as high as possible. The next section will help you perform this kind of analysis.

and Receiver or Communicator and Audience. Certainly any communication event begins with someone deciding that communication is needed and initiating that communication, with an intended recipient on the other end. But in many situations, especially those involving real-time conversation, the two parties work together to reach a mutual understanding. Even in situations where a communicator is attempting to deliver a complete,

> ## Solving ill-defined problems involves combining existing resources with innovation and good judgment.

carefully prepared message—as in a letter, report, or oral presentation—the intended recipients have already participated in the construction of the message because the writer or presenter has kept them in mind when composing and designing the message. The labels in this model are thus intended to convey the cooperative effort behind every successful communication event.

A Model of Business Communication

Exhibit 1-4 shows the basic elements of a business communication event. Even though people can, and often do, communicate inadvertently, this communication model focuses on what happens when someone deliberately communicates with someone else to achieve particular business-related goals.

You'll notice that the two communicators in the figure are labeled simply Communicator 1 and Communicator 2 instead of Sender

LO 1-7 Describe the contexts for each act of business communication.

the contexts for communication Certain features of the communication situation are already in place as the communicators in our model begin to communicate.

The *larger context* includes the general business-economic climate; the language, values, and customs in the surrounding culture; and the historical moment in which the communication is taking place.

▼**EXHIBIT 1-4** The Business Communication Process

Communicator 1 ...
1. Senses a communication need
2. Defines the problem
3. Searches for possible solutions
4. Selects a course of action (message type, contents, style, format, channel)
5. Composes the message
6. Delivers the message

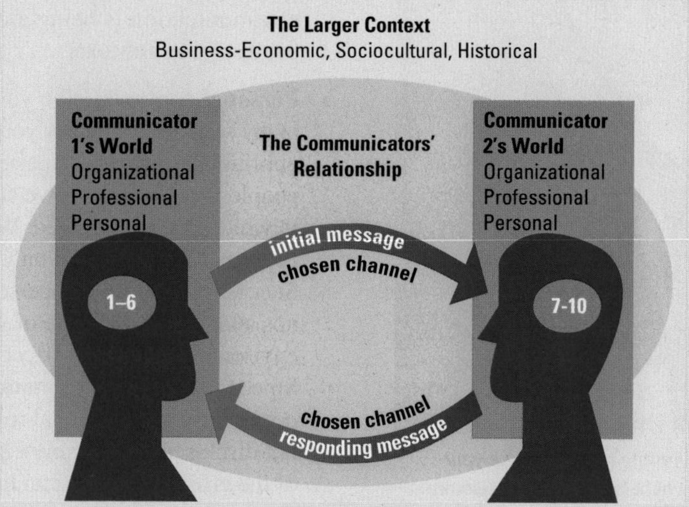

The Larger Context
Business-Economic, Sociocultural, Historical

Communicator 1's World
Organizational
Professional
Personal

The Communicators' Relationship

Communicator 2's World
Organizational
Professional
Personal

*initial message
chosen channel*

*chosen channel
responding message*

1–6

7–10

Communicator 2 ...
7. Receives the message
8. Interprets the message
9. Decides on a response
10. May send a responding message

Think about how these contexts might influence communication. For example, when the country's economy or a particular industry is flourishing, a communicator's message and the recipient's response may well be different from what they would be during an economic slump. The sociocultural context also affects how individuals communicate. Whether they are communicating in the context of U.S. urban culture, for instance, or the culture of a particular region or another country, or whether they are communicating across cultures, their communication choices will be affected. The particular historical context of their communication can also be a factor. Consider how recent financial scandals in the United States or the increased focus on the environment are influencing the language of business. The skillful communicator

The communicators' *particular contexts* exert perhaps the strongest influence on the act of communication. These interrelated contexts can be

- **Organizational contexts**. As we've discussed, the type and culture of the organization you represent will shape your communication choices in many ways, and the organizational contexts of your audiences will, in turn, shape theirs. In fact, in every act of business communication, at least one of the parties involved is likely to be representing an organization. What you communicate and how you do so will be strongly shaped by the organization for whom you speak. In turn, the organization

> ## "The relationship of the communicators also forms an important context for communication."

is sensitive to these larger contexts, which always exert an influence and, to some extent, are always changing.

The **relationship of the communicators** also forms an important context for communication. Certainly, communication is about moving information from point A to point B, but it is also about interaction between human beings. Your first correspondence with someone begins a relationship between the two of you, whether as individuals, people in certain business roles, or both. All future messages between you will continue to build this relationship.

Like this technician and manager, you will often need to adapt your communication when speaking to those whose areas of expertise are different from yours.

to which your audience belongs—its priorities, its current circumstances, even how fast or slow its pace of work—can strongly influence the way your message is received.

- **Professional contexts**. You know from school and experience that different professionals—whether physicians, social workers, managers, accountants, or those involved in other fields—possess different kinds of expertise, speak differently, and have different perspectives. What gets communicated and how can be heavily influenced by the communicators' professional roles. Be aware that internal audiences as well as external ones can occupy different professional roles and therefore favor different kinds of content and language. Employees in management and engineering, for example, have been demonstrated to have quite different priorities, with the former focusing on financial benefit and the latter on technological achievement.[27] Part of successful communication is being alert to your audiences' different professional contexts.

- **Personal contexts**. Who you are as a person comes from many sources: the genes you inherited, your family and upbringing, your life experiences, your schooling, the many people with whom you've come in contact, and the culture in which you were reared. Who you are as a person also depends to some extent on your current circumstances. Successes and failures, personal relationships, financial ups and downs, the state of your health, your physical environment—all can affect a particular communicative act. Since much business communication is between individuals occupying organizational roles, personal matters are usually not disclosed. But business professionals should be mindful of the effect that these can have on the communicators. If you're aware, for example, that the intended recipient of your message is under stress or having a bad day, you can adapt your communication accordingly.

LO 1-8   Describe the business communication process.

the process of communication No one can know exactly what occurs inside the minds of communicators when they undertake to create a message, but researchers generally agree that the process includes the following steps:

1. *Sensing a communication need.* A problem has come to your attention, or you have an idea about how to achieve a certain goal. You believe that some form of communication will help you achieve the desired state.

2. *Defining the situation.* To create a successful message or plan a communication event, you need to have a well-informed sense of the situation. What exactly is the problem? What further information might you need to acquire in order to understand the situation? How might your or your organization's goals be hindered or helped depending on your communication choices?

3. *Considering possible communication strategies.* As your definition of the situation takes shape, you will start considering different options for solving it. What kind of communication event will you initiate, and what will you want to achieve with it? What image of yourself, your company, and your communication partners might you project in your message?

4. *Selecting a course of action.* Considering the situation as you've defined it and looking at your communication options, you will consider the potential costs and benefits of each option and select the optimum one. Your decision will include preliminary choices about the message type, contents, structure, verbal style, and visual format, and about the channel you will use to deliver the message. (Read about a poor choice of channel in the Communication Matters feature on this page.)

5. *Composing the message.* Here is where you either craft your written message or plan your presentation or conversation. If you have decided to convey your message orally, you will make careful notes or perhaps even write out your whole message and also design any visuals you need. If you have decided to write your message, you will draft it and then revise it carefully so that it will get the job done and reflect well on you (see the next chapter for helpful writing and revising techniques).

6. *Sending the message.* When your message is prepared or carefully planned, you are ready to deliver it to your intended recipients in the channel you have chosen. You choose a good time to deliver it, realizing, for example, that Monday morning may not be the best time to make an important phone call to a busy executive. You also consider

communication matters

Channel Choice Affects Message Success

"Its official, you no longer work for JNI Traffic Control and u have forfided any arrangements made." Can you imagine getting such a text message? The Sydney employer was sued over this inappropriate choice of a communication channel for firing an employee. In settling the matter the commissioner ruled that email, text messages, and even answering machines were inappropriate for official business communication. Or what about being notified by text message of an overdue bill? While some might think of that as a service, others would regard it as invasive and inappropriate.

Historically, the importance of channel choice has been disputed, with some arguing that it is simply a means for transmitting words and others arguing that the chosen channel is, in itself, a message. However, today most people realize that the appropriate choice of communication channel contributes significantly, along with the words, to the success of the message. While research has provided guidelines for understanding when to use very lean (printed material) to very rich (face-to-face) channels, new technologies and laws have added new elements to consider. Not only are there no clear-cut rules or guidelines, but the smallest change in context may make one choice better than another.

In selecting a channel, a communicator needs to weigh several factors. These include the message content, the communicators' levels of competency with the channel, the recipient's access to the channel, and the assumptions associated with the channel. Appropriate choice of a communication channel helps people communicate clearly, improving both their productivity and personal relationships.

sending auxiliary messages, such as a "heads-up" phone call or email, that could increase your main message's chances of success.

While these activities tend to form a linear pattern, the communicator often needs to revisit earlier steps while moving through the different activities. In other words, solving a communication problem can be a **recursive** process. This is particularly true for situations that have many possible solutions or heavily involve the audience in the communication process. A communicator may begin a communication event with a certain view of the situation and then find, upon further analysis or the discovery of additional facts, that this view needs to be revised in order to accommodate all the involved parties and their goals.

If all goes as planned, here is what will happen on the recipient's end:

7. *Receiving the message.* Your chosen channel has delivered your message to each intended recipient, who has perceived and decided to read or listen to your message.

8. *Interpreting the message.* Just as you had to interpret the situation that prompted your communication, your recipient now has to interpret the message you sent. This activity will involve not only extracting information from the message but also guessing your communication purpose, forming judgments about you and those you represent, and picking up on cues about the relationship you want to promote between yourself and the recipient.

9. *Deciding on a response.* Any time you send a message, you hope for a certain response from your recipient, whether it be increased goodwill, increased knowledge, a specific responding action, or a combination of these. If your message has been carefully adapted to the recipient, it has a good chance of achieving the desired response.

10. *Replying to the message.* The recipient's response to your message will often take the form of replying to your message. When this is the case, the receiver is acting as communicator, following the process that you followed to generate your message.

Exhibit 1-5 lists the main questions to consider when developing a communication strategy. Taking this analytical approach will

---

EXHIBIT 1-5   Planning Your Communication Strategy: A Problem-Solving Approach

What is the situation?

- What has happened to make you think you need to communicate?
- What background and prior knowledge can you apply to this situation? How is this situation like or unlike others you have encountered?
- What do you need to find out in order to understand every facet of this situation? Where can you get this information?

What are some possible communication strategies?

- To whom might you communicate? Who might be your primary and secondary audiences? What are their different organizational, professional, and personal contexts? What would each care about or want to know? What, if any, is your prior relationship with them?
- What purpose might you want to achieve with each recipient? What are your organizational, professional, and personal contexts?
- What are some communication strategies that might help you achieve your goals?
- How might the larger business-economic, sociocultural, and historical contexts affect the success of different strategies?

Which is the best course of action?

- Which strategies are impractical, incomplete, or potentially dangerous? Why?
- Which of the remaining strategies looks like the optimum one? Why?
- What will be the best message type, contents, structure, style, and format for your message?
- What channel will you use to deliver it?

What is the best way to design the chosen message?

- Given your goals for each recipient, what information should your message include?
- What logical structure (ordering and grouping of information) should you use?
- What kind of style should you use? How formal or informal should you be? What image of yourself and your audience should you try to project? What kind of relationship with each recipient should your message promote?
- How can you use formatting, graphics, and/or supporting media to make your message easier to comprehend?
- What are your recipients' expectations for the channel you've chosen?

What is the best way to deliver the message?

- Are there any timing considerations related to delivering your message?
- Should you combine the main message with any other messages?
- How can you best ensure that each intended recipient receives and reads or hears your message?

help you think consciously about each stage of the process and give you the best chance of achieving the desired results.

BUSINESS COMMUNICATION: THE BOTTOM LINE

The theme of this chapter might be summed up this way: The goal of business communication is to create a shared understanding of business situations that will enable people to work successfully together.

Timely and clear transfer of information is critical to businesses, now more than ever. But figuring out what kind of information to send, whom to send it to, how to send it, and what form to use requires good decision making. Since every person has his or her own mental "filters"—preconceptions, frames of reference,

and verbal worlds—wording the information so that it will be understood can be a challenge. You and your audience may even attach completely different meanings to the same words (a problem that the communication literature calls "bypassing").

Complicating this picture is the fact that communication is not just about information transfer. The creation and maintenance of positive human relations is also essential to business and thus to business communication. Every act of communication conveys an image of you and of the way you regard those to whom you're speaking or writing. Successful business communicators pay careful attention to the human relations dimension of their messages.

Yes, business communication can be challenging. It can also be extremely rewarding because of the results you achieve and the relationships you build. The advice, examples, and exercises in this book will jump-start you toward success. But it will be your ability to analyze and solve specific communication problems that will take you the rest of the way there.

jump-start your professional success!

- What codes of ethics do major companies and professional organizations use?
- What are 10 qualities of an effective team member?
- How can you become a better problem solver?

Scan the QR code with your smartphone or use your Web browser to find out at www.mhhe.com/RentzM3e. Choose Chapter 1 > Bizcom Tools & Tips. While you're there, you can view a chapter summary, exercises, PPT slides, and more to jump-start your professional success.

www.mhhe.com/RentzM3e

Understanding the **Writing Process and** the Main Forms of **Business Messages**

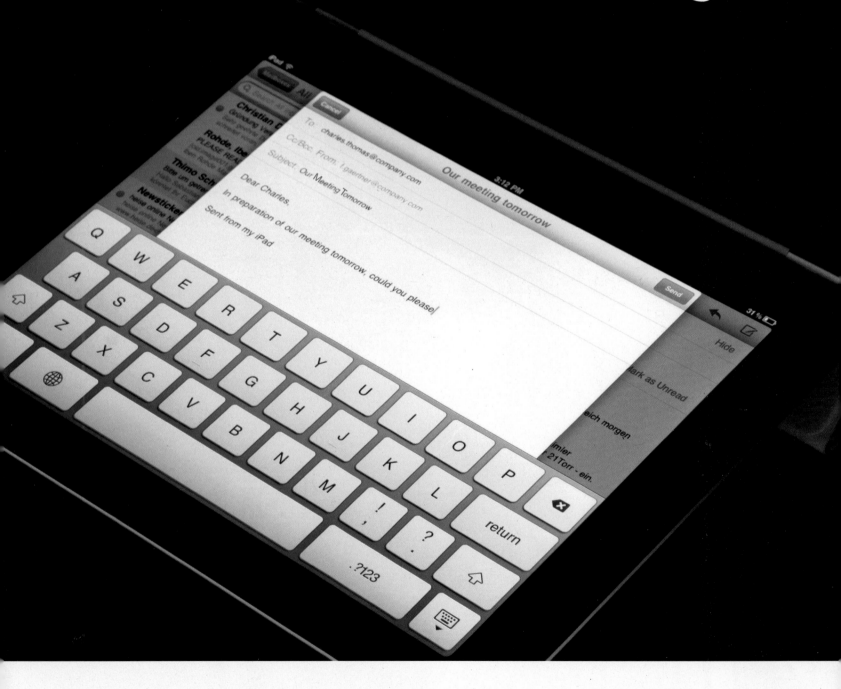

chapter two

Much of this book focuses on writing in business. Is skillful oral communication important? Absolutely. How about visual communication? It's critical. Then why the extra emphasis on writing?

Experienced businesspeople tend to place writing skills ahead of other communication skills when asked what they seek in job applicants. And they seek strong writing skills in particular when considering whom to promote. For example, in one study, a majority of the 305 executives surveyed commented that fewer than half their job applicants were well-versed enough in "global knowledge, self-direction, and writing skills" to be able to advance in their companies.[1] As people move up, they do more knowledge work, and this work often requires expertise in written forms of communication. ■

LEARNING OBJECTIVES

LO 2-1 Describe the writing process and effective writing strategies.

LO 2-2 Describe the development and current usage of the business letter.

LO 2-3 Describe the purpose and form of memorandums (memos).

LO 2-4 Describe the purpose and form of email.

LO 2-5 Understand the nature and business uses of text messaging and instant messaging.

LO 2-6 Understand the nature and business uses of social media.

LO 2-7 Understand the inverted pyramid structure for organizing and writing Web documents.

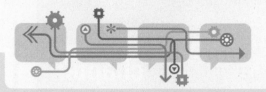
The Nature of Business Messages

Introduce yourself to this chapter by shifting to the role of Julie Evans, a recent college graduate in her first job as an accounts manager at a small company that manufactures windows. You are amazed (and sometimes overwhelmed) by the types of messages you send each day. Every day you process dozens of internal email messages. Occasionally you write and receive memorandums. Then there are the more formal communications you exchange with people outside the company—both email and letters. You also write messages for social media sites and daily rely on text messaging and instant messaging for quick communication. With so many audiences and so many ways to send messages, you often wonder if you're making good choices. This chapter will help you understand the writing process and the main types of business messages so that you are sure to meet various audiences' needs.

THE IMPORTANCE OF SKILLFUL WRITING

Writing is in some ways more difficult to do well than other kinds of communication. Writers essentially have no safety net; they can't rely on their facial expressions, body language, or tone of voice to make up for wording that isn't quite what they mean. The symbols on the page or screen must do the whole communication job. Plus, the symbols used in writing—the alphabet, words, punctuation, and so forth—share no characteristics with the object or concept they represent (unless you count words that sound like the sounds they name, such as "buzz"). Representing something with a photograph is relatively easy. Representing that same thing in words is much harder. Capturing a complex reality by putting one word after another requires ingenuity, discipline, and the ability to anticipate how readers will likely react as they read.

The first major section of this chapter will help you achieve this impressive but commonplace feat in the workplace by showing you how to break the writing process into parts and skillfully manage each part. The remainder of the chapter will discuss the main forms of business messages, which bring with them certain features and conventions of use. These discussions provide the foundation for subsequent chapters on writing different kinds of business messages.

LO 2-1 Describe the writing process and effective writing strategies.

THE PROCESS OF WRITING

Writing researchers have been studying the composing process since the 1970s. They have found, not surprisingly, that each person's way of developing a piece of writing for a given situation is unique. On the other hand, they have also drawn several conclusions about the nature of the writing process and about strategies that can help it. Familiarizing yourself with these findings will help make you a more deliberate, effective writer.

As Exhibit 2-1 shows, preparing any piece of writing involves three stages: **planning**, **drafting**, and **revising**. These stages can be defined roughly as figuring out what you want to say, saying it, and then saying it better. Each of these stages can be divided into various specific activities, which the rest of this section describes. However, as the arrows in the figure suggest, business writers should not think of the three stages as strictly chronological or separate. In practice, the stages are interrelated. Like the steps for solving business communication problems described in Chapter 1, they are **recursive**. For example, a writer in the planning stage may start writing pieces of the draft. Or he or she may find when drafting that gathering more information is necessary. Or he or she may decide that it's necessary to revise a piece of the document carefully before continuing with the drafting. An undue emphasis on keeping the stages separate and chronological will hinder the success of your messages. Allow yourself to blend these stages as necessary.

A good rule of thumb for student writers is to spend roughly a third of their writing time in each of the three stages. A common mistake that writers make is to spend too much time on drafting and too little on the other two stages—planning and revising. Preparing to write and improving what you have written are as critical to success as the drafting stage, and careful attention to all three stages can actually make your writing process more efficient. Once you have become an experienced business writer, you will be able to write many routine messages without as much planning and revising. Even so, some planning and revising will still be essential to getting the best results with your messages.

Planning the Message

Chapter 1 presents a problem-solving approach to business communication. As Exhibit 1-5 (page 18) indicates, you need to develop a definition of the problem that you are trying to solve. Once you have defined your problem, you can plan

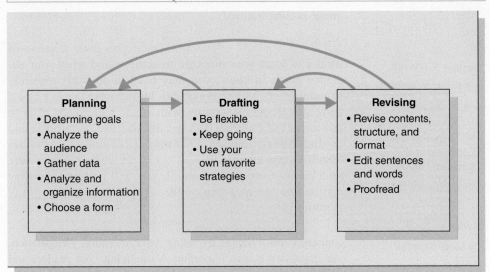

Planning	Drafting	Revising
• Determine goals • Analyze the audience • Gather data • Analyze and organize information • Choose a form	• Be flexible • Keep going • Use your own favorite strategies	• Revise contents, structure, and format • Edit sentences and words • Proofread

(e.g., communicate your message, promote your professional image, build goodwill)—though sometimes, clarifying your writing goals will help you generate business solutions.

analyzing your audience

Once you know your purpose—what you want your message to do—you need to **analyze** the audience who will read your message. Who will be affected by what you write? What organizational, professional, and personal issues or qualities will affect the audience's response to your message? What organizational, professional, and personal issues or qualities do you have that affect how you will write your message? What is your relationship with your reader? Are you writing to your superior? Your colleagues? Your subordinates? Clients? Answers to these questions and others (see Exhibit 2-3) will influence your channel of communication, tone, style, content, organization, and format.

In the hotel manager scenario we discussed, for instance, how might your approach in an announcement to guests who are currently at the hotel differ from your approach in a response to a guest's complaint letter a week after the incident? Though you should take time to analyze your audience early in the planning process, you should continue to think of your audience as you proceed through the rest of the planning stage and through the drafting and revising stages, too. Always be thinking about what kind of information will matter most to your audience and

your message by answering several questions regarding your context and audience. As you plan written documents in particular, you can make the planning process more manageable by thinking about it in five smaller steps: determining goals; analyzing your audience; gathering information; analyzing and organizing the information; and choosing the form, channel, and format the document will take.

Planning a good message takes time. The reason is that you have a lot to consider when writing to an audience you may not know all that well. The investment of your time pays dividends when you are able not only to achieve the goal of your message but also to enhance your professional image by writing a coherent, concise, and thorough document.

determining goals Because business writing is largely performed in response to a certain situation, one of your main planning tasks is to figure out what you want to do about that situation. Bear in mind that in business communication, "what to do" means not only what you want your communication to achieve but also any action related to the larger business problem. Let's say, for example, that you manage a hotel where the air conditioning has stopped functioning. You will need to decide what, if anything, to communicate to your guests about this problem. But this decision is related to other decisions. How and when will you get the air conditioning problem solved? In the meantime, will you simply apologize? Make arrangements for each guest to have a free continental breakfast or complimentary beverages? Rent fans for meeting rooms and any guest rooms occupied by people with health problems? As Exhibit 2-2 shows, solving the business problem and solving the communication problem are closely related. You will need to bring your **business goals** (e.g., increase profits, attract new clients) to bear on your **writing goals**

▼ **EXHIBIT 2-2**   The Interrelated Nature of Business Goals and Communication Goals

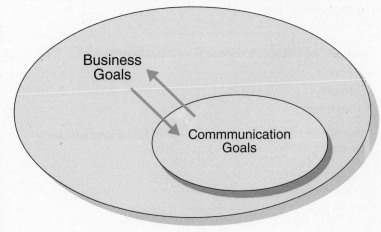

communication matters

Do I Need to Write It?

When you have a substantial message to convey—that is, one not suitable for a quick text or instant message—is it better to write it or speak it? You'll probably want to write it if one or more of the following applies:

• You want a written record of the communication.

• You want the communication to be perceived as somewhat formal.

• You think you can explain better in writing, and you don't want the recipient to interrupt you until you're done.

• Your reader will want to be able to go back over what you said.

• You have to reach a lot of people at once with the same message.

• The situation isn't so sensitive that it requires a richer, more personal communication channel.

adapt your message accordingly. If you fail to meet your audience's needs, your message fails as well, and your professional image is compromised.

gathering information Once you have a sense of what you want your message to achieve and what your audience needs to know, you may need to do some research. In many cases this research can be informal—finding past correspondence; consulting with other employees or with outside advisors; or reviewing sales records, warranties, and product descriptions. In other cases you will do formal research such as conducting surveys or reviewing the literature on a certain subject. In general, you will collect any information that can help you decide what to do and what to say in your message.

Gathering information by using your memory, imagination, and creativity is also important. Visualizing your readers and bearing their interests in mind is an excellent planning technique. Making a list of pertinent facts is helpful. Brainstorming (generating possible solutions without censoring them) will allow you to develop creative solutions. Drawing a diagram of

---

▼**EXHIBIT 2-3**   Audience Analysis Checklist

What is my relationship *to* my audience?
- ☐ Subordinate
- ☐ Colleague
- ☐ Superior
- ☐ Client/Customer
- ☐ Other: _____

What is my relationship *with* my audience?
- ☐ Friendly and informal. I know my audience well. We communicate often and have a social business relationship.
- ☐ Friendly and formal. We've met and have a cordial, business-like relationship.
- ☐ Neutral or no relationship. I don't know my audience personally.
- ☐ Unfriendly or hostile.
- ☐ Other: _____

What will my audience's reaction to my message be?
- ☐ Positive
- ☐ Negative
- ☐ Neutral

What factors in my company culture or other background information should I take into account?
- ☐ _____
- ☐ _____
- ☐ _____

What factors in my audience's culture or background should I take into account?
- ☐ _____
- ☐ _____
- ☐ _____

What does my audience need to know?
- ☐ _____
- ☐ _____
- ☐ _____

What does my audience already know?
- ☐ _____
- ☐ _____
- ☐ _____

What do I want my audience to think, feel, do, know, or believe as a result of my message?
- ☐ _____
- ☐ _____
- ☐ _____

_____ would be the best channel for

delivering the message because _____ .

your ideas can also enable you to collect your thoughts. Use any strategy that shows promise of contributing to a solution.

analyzing and organizing the information

Once you have a number of ideas, you can start to analyze them. If your data are numerical, you will do the calculations that enable you to see patterns and meaning in the numbers. You will put other kinds of data together as well to see what course of action they might indicate, weighing what the parties involved stand to gain or lose from each possible solution.

As you ponder what to do and say with your message, you will, of course, keep your readers in mind. What kind of information will most matter to them? In the scenario described above, will the hotel guests want information about what caused the air conditioning problem or about when it will be fixed and what they can do to stay comfortable in the meantime? As always, your intended readers are your best guide to what information to include.

They are also your guide for organizing the information. Whatever order will draw the most positive reaction from your readers is the best order to use. If you have information that your readers will consider routine, neutral, or positive, put it

content. Business writers do not launch into writing a document without some sense of what kind of document it will be. The medium itself helps them know what to say and how to say it. On the job, choosing the type of document to be written is an important part of planning.

Specific decisions about a document's format or visual design can be made at any point in the writing process, but usually the planning stage involves preliminary decisions along these lines. How can you make the information easily readable and accessible to your audience? Will you be dividing up the contents with headings? How about with a bulleted or numbered list? How long or short will the paragraphs be? Will there be any visual elements such as a logo or picture or diagram? Anticipating the format can help you plan an inviting and readable message.

Formatting devices have a large impact on readers' reactions, making decisions about formatting an integral part of the business writer's writing process, even in the planning stage.

For example, Exhibit 2-4 (page 27) shows the starting text of a memo (sent by email) from a university registrar to the faculty with the subject line " 'X' and 'WX' Grades Effective for Autumn '14 Grading." How inviting do you find the format,

> ## Regardless of the situation, all readers appreciate a logical pattern for the information.

first. This plan, called using the **direct order**, is discussed in Chapter 5. On the other hand, if you think your information could run the risk of evoking a negative response, you will use an **indirect order**, using your message's opening to prepare the reader to receive the news as positively as possible. As you will see in Chapter 6, such a message usually requires a more skillful use of organization and word choice than one written in direct order. Regardless of the situation, all readers appreciate a logical pattern for the information.

choosing a form, channel, and format

Writers in school typically produce writing of two types: essays or research papers. But on the job you have a wide range of established forms of communication **(genres)** to choose from. Which one you use has a huge impact on your planning. For instance, if you want to advertise your company's services, how will you do it? Write potential customers a letter? Email them? Send a brochure? Create a Web site? Post a message on your company's social media sites? Use some combination of these? Each form has its own formatting and stylistic conventions and even conventions about

and how easy is it to extract the information about the two new grades?

Drafting

Writing experts' main advice about drafting boils down to these words: "Be flexible." Writers often hinder themselves by thinking that they have to write a finished document on the first attempt with the parts in their correct order and with perfect results. Writing is such a cognitively difficult task that it is better to concentrate only on one thing at a time. The following suggestions can help you draft your messages as painlessly and effectively as possible.

avoid perfectionism when drafting

Trying to make your first draft perfect causes two problems. First, spending too much energy perfecting the early parts can make you forget important pieces and purposes of the later parts. Second, premature perfectionism can make drafting frustrating and slow and thus keep you from wanting to revise your message when you're done. You will be much more inclined to review your message and improve it if you have not agonized over your first draft.

communication matters

Instant Messaging Etiquette in the Workplace

Instant messaging, as its name implies, provides quick and convenient communication. Like any communication channel, its effectiveness depends on a writer's ability to analyze an audience and write a message that helps accomplish his or her business goal.

Amy Levin-Epstein offers seven tips for using instant messaging in the workplace.

- Be mindful of communication preferences. As you would with any communication channel, you should use instant messaging only with audiences who prefer it as a means of communication.

- Be very, very concise. The beauty of instant messaging is that you can convey a lot of meaning in very few words; however, being concise takes work. You should invest the time so that your messages are concise but do not require a lot of exchanges for your reader to understand your message.

- Tell your reader if you are available—or not. Instant messaging lets you tell people whether you're away from your desk or available. Don't let people think you're available if you're not, as this will frustrate them. You also don't want to miss an important message because you forgot to let people know you were available.

- Avoid the smiley face. Emoticons are generally not professional in office correspondence.

- Never say anything you would not read aloud. As with other forms of electronic messages, instant messages can be shared with, seen by, or sent to audiences beyond the one you intended.

- Wait patiently for a response. If a response is not immediate, don't send follow-up messages such as "Hello?" These messages may be seen as anger or sarcasm and not be well received by your audience.

- Stop an angry IM session short. If an instant messaging exchange becomes angry or confrontational, suggest that the conversation be continued face to face or over the phone where body language and other nonverbal cues can help resolve the issue.

Source: Amy Levin-Epstein, "7 Etiquette Tips for IM'ing in the Office," *Money Watch*, CBS Interactive, 23 Oct. 2012, Web, 24 May 2013. Copyright 2013 CBS Moneywatch, CBS Interactive, Inc. All rights reserved.

keep going When turning your planning into a draft, don't let minor problems with wording or grammar distract you from your main goal—to generate your first version of the document. Have an understanding with yourself that you will draft relatively quickly to get the ideas down on paper or onto the screen and then go back and carefully revise. Expressing your points in a somewhat coherent, complete, and orderly fashion is hard enough. Allow yourself to save close reexamination and evaluation of what you've written for the revision stage.

use any other strategies that will keep you working productively The idea with drafting is to keep moving forward at a reasonably steady pace with as little stalling as possible. Do anything you can think of that will make your drafting relatively free and easy. For example, write at your most productive time of day, write in chunks, start with a favorite part, talk aloud or write to yourself to clarify your thoughts, take breaks, let the project sit for a while, create a setting conducive to writing—even promise yourself a little reward for getting a certain amount accomplished. Your goal is to get the first orderly expression of your planned contents written just well enough so that you can go back and work with it.

Revising

Getting your draft ready for your reader requires going back over it carefully—again and again. Do you say what you mean? Could someone misunderstand or take offense at what you have written? Is your organization best for the situation? Is each word the right one for your goals? Are there better, more concise ways of structuring your sentences? Can you move the reader more smoothly from point to point? Does each element of format enhance readability and highlight the structure of the contents? When revising, you turn into your own critic. You challenge what you have written and look for better alternatives. Careful attention to each level will result in a polished, effective message.

> *When revising, you turn into your own critic. You challenge what you have written and look for better alternatives.*

Any given message has so many facets that using what professional writers call "levels of edit" may be helpful. The levels this term refers to are *revising, editing,* and *proofreading.*

When **revising**, you look at top-level concerns: whether or not you included all necessary information, if the pattern of organization is logical and as effective as possible, if the overall meaning of the message comes through, and if the formatting is appropriate and helpful.

You then move to the **editing** level, focusing on your style. You examine your sentences to see if they pace the information in such a way that the reader can easily follow it, if they emphasize the right things, and if they combine pieces of information coherently. You also look at your word choices to see if they best serve your purpose.

Finally, you **proofread**, looking at mechanical and grammatical elements—spelling, typography, punctuation, and any grammar problems that tend to give you trouble. Editing

---

▼**EXHIBIT 2-4**   What a Difference Formatting Makes!

Here is the starting text of a memo (sent by email) from a university registrar to the faculty regarding two new grades about to go into effect. How inviting do you find the format of the following message, and how easy is it to extract the information about the two new grades?

At its October 20, 2014, meeting, the Faculty Senate, having received a favorable recommendation from the Academic Affairs Committee, voted to approve the creation and Autumn Quarter implementation of two new grades: "X" and "WX." Instructors will record an "X" on the final grade roster for students who never attended any classes and did not submit any assigned work. The "X" will appear on the transcript and will carry zero (0.00) quality points, thus computed into the GPA like the grades of "F" and "UW." Instructors will record a "WX" for those students who officially withdrew from the class (as denoted on the grade roster by either EW or W) but who never attended any classes and did not submit any assigned work. The "WX" may be entered to overwrite a "W" appearing on the grade roster. An assignment of "WX" has no impact on the student's GPA. A "W" will appear on the student's online grade report and on the transcript. The "WX" recognizes the student's official withdrawal from the class and only records the fact of nonparticipation. The need to record nonparticipation is defined in "Rationale" below. With the introduction of the "X" and "WX" grades to denote nonparticipation, by definition all other grades can only be awarded to students who had participated in the class in some way. Instructors will record a "UW" (unofficial withdrawal) only for students who cease to attend a class following some participation. Previously, instructors utilized the "UW" both for those students who had never attended classes and for those who had attended and participated initially but had ceased to attend at some point during the term. In cases of official withdrawal, instructors have three options available at the time of grading: "W," "WX," and "F." If the student has officially withdrawn from the class, a "W" (withdrawal) or "EW" (electronic withdrawal) will appear on the grade roster. If the student participated in the class and the withdrawal was in accordance with the instructor's withdrawal policy as communicated by the syllabus, the instructor may retain the student's "W" grade by making no alteration to the grade roster....

Now look at the first part of the actual message that was sent out. What formatting decisions on the part of the writer made this document much more readable?

At its October 20, 2014, meeting, the Faculty Senate, having received a favorable recommendation from the Academic Affairs Committee, voted to approve the creation and Autumn Quarter implementation of two new grades: "X" and "WX."

Definition of "X" and "WX" Grades, Effective Autumn Quarter 2014

- "X" (nonattendance):
- Instructors will record an "X" on the final grade roster for students who never attended any classes and did not submit any assigned work.

    The "X" will appear on the transcript and will carry zero (0.00) quality points, thus computed into the GPA like the grades of "F" and "UW."

- "WX" (official withdrawal, nonattending):

    Instructors will record a "WX" for those students who officially withdrew from the class (as denoted on the grade roster by either EW or W) but who never attended any classes and did not submit any assigned work.

    The "WX" may be entered to overwrite a "W" appearing on the grade roster. An assignment of "WX" has no impact on the student's GPA. A "W" will appear on the student's online grade report and on the transcript. The "WX" recognizes the student's official withdrawal from the class and only records the fact of nonparticipation. The need to record nonparticipation is defined in "Rationale" below.

Participation and Nonparticipation Grades

With the introduction of the "X" and "WX" grades to denote nonparticipation, by definition all other grades can only be awarded to students who had participated in the class in some way.

Instructors will record a "UW" (unofficial withdrawal) only for students who cease to attend a class following some participation. Previously, instructors utilized the "UW" both for those students who had never attended classes and for those who had attended and participated initially but had ceased to attend at some point during the term.

Official Withdrawals

In cases of official withdrawal, instructors have three options available at the time of grading: "W," "WX," and "F."

1. *If the student has officially withdrawn from the class,* a "W" (withdrawal) or "EW" (electronic withdrawal) will appear on the grade roster. If the student participated in the class and the withdrawal was in accordance with the instructor's withdrawal policy as communicated by the syllabus, the instructor may retain the student's "W" grade by making no alteration to the grade roster....

Source: Reprinted with permission of Dr. Douglas K. Burgess, University Registrar, University of Cincinnati.

functions in your word-processing program can help you with this task.

One last word about revision: Get feedback from others. As you may well know, it is difficult to find weaknesses or errors in your own work. Seek assistance from willing colleagues, and if they give you criticism, receive it with an open mind. It is better to hear this feedback from them than from your intended readers when costly mistakes may have already been made.

The remaining sections of this chapter describe specific purposes and traits of different message types. Bonus Chapter A provides in-depth advice about their physical design. No matter what you're writing, taking time to make careful formatting decisions during your writing process will significantly enhance your chances of achieving your communication goals.

Successful writers often seek others' perspectives on important documents.

LO 2-2 Describe the development and current usage of the business letter.

LETTERS

Letters are the oldest form of business messages. The ancient Chinese wrote letters, as did the early Egyptians, Romans, and Greeks. In fact, American businesspeople used letters as early as 1698 to correspond about sales, collections, and other business matters.[2]

From these early days letters have continued to be used in business. Although their use and purpose have evolved as other business communication genres have developed, they are still the best choice for many communication tasks.

Letters Defined

The general purpose of a letter is to represent the writer and his or her topic rather formally to the recipient. For this reason, **letters** are used primarily for corresponding with people outside your organization. When you write to internal readers, they are often familiar to you—and even if they are not, you all share the connection of being in the same company. Your messages to such audiences tend to use less formal media. But when you write to customers, to suppliers, to citizens and community leaders, and to other external audiences, you will often want to present a professional, polished image of your company by choosing the letter format, complete with

an attractive company letterhead and the elements of courtesy built into this traditional format. Your readers will expect this gesture of respect. Once you have established friendly relations with them, you may well conduct your business through emails, phone calls, instant or text messaging, and social media. But especially when corresponding with an external party whom you do not know well, a letter is often the most appropriate form to use.

Letter Form

The format of the business letter is probably already familiar to you. Although some variations in format are generally acceptable, typically these information items are included: date, inside address, salutation (Dear Ms. Smith), body, and complimentary close (Sincerely yours). Other items sometimes needed are attention line, subject line, return address (when letterhead is not used), and enclosure information. Exhibit 2-5 presents one option for formatting a letter. More options are presented in Bonus Chapter A.

▼**EXHIBIT 2-5**   Illustration of a Letter in Full Block Format (Mixed Punctuation)

Doing it right . . . the first time

Ralston's Plumbing and Heating
2424 Medville Road
Urbana, OH 45702
(515) 555-5555
Fax: (515) 555-5544

February 28, 2014

Ms. Diane Taylor
747 Gateway Avenue
Urbana, OH 45702

Dear Ms. Taylor:

Thank you for allowing one of our certified technicians to serve you recently.

Enclosed is a coupon for $25 toward your next purchase or service call from Ralston. It's just our way of saying that we appreciate your business.

Sincerely yours,

Jack Ralston

Jack Ralston
Owner and President

Enclosure

Letter Formality

As formal as letters can be, they are not nearly as formal as they used to be. Business messages have grown more conversational. This is true of letters as well as of other forms of correspondence.

For instance, in the past, if writers did not know the reader's name, wrote to a mass audience, or wrote to someone whose gender could not be determined by the reader's name (e.g., Pat Smith), they might have used a salutation such as "To Whom It May Concern," "Dear Sir/Madame," or "Dear Ladies and Gentlemen." These expressions are now considered stiff and old fashioned. More modern options include "Dear Human Resources" or "Dear Pat Smith." Writers can also omit the salutation, perhaps adding a subject line (a brief phrase stating the writer's main point) instead. Some business writers also consider the use of the terms "Dear," "Sincerely Yours," and even "Sincerely" outdated or excessively formal. These writers will omit "Dear," replace "Sincerely yours" with "Sincerely," omit the complimentary close, or use "Best regards" or some other cordial phrase. Your audience and company culture will determine what is appropriate for you.

Regardless of its formality, the letter should always be regarded as an exchange between real people as well as a strategic means for accomplishing business goals.

LO 2-3 Describe the purpose and form of memorandums (memos).

MEMORANDUMS (MEMOS)

Memorandums Defined

Another business genre is the **memorandum** (or **memo**). It is a hard copy (printed on paper) documents used to communicate inside a business. Though in rare cases they may be used to communicate with those outside the business, they are usually exchanged internally by employees as they conduct their work. Originally, memos were used only in hard copy, but their function of communicating within a business has been largely replaced by email. Even so, they still are a part of many companies' communications. They are especially useful for communicating with employees who do not use computers in their work.

Memos can be used for a wide range of communication tasks. For example, as Chapter 9 points out, some memos communicate factual, problem-related information and can be classified as reports. As with the letter, the purpose and use of the memo have evolved as other business communication genres have emerged, but the memo is still an important means for communicating in many organizations.

Businesses with multiple locations send many of their internal messages by email as well as instant and text messaging.

Memorandum Form

Memorandums can be distinguished from other messages primarily by their form. Some companies have stationery printed especially for memos, while many use standard or customized templates in word processors. Sometimes the word *memorandum* appears at the top. But some companies prefer other titles, such as *Interoffice Memo* or *Interoffice Communication*. Below this main heading come the specific headings common to all memos: *Date, To, From, Subject* (though not necessarily in this order). This simple arrangement is displayed in Exhibit 2-6. As the figure indicates, hard-copy memos are initialed by the writer rather than signed.

Large organizations, especially those with a number of locations and departments, often include additional information on their memorandum stationery. *Department, Plant, Location, Territory, Store Number*, and *Copies to* are examples (see Exhibit 2-7). Since in some companies memos are often addressed to more than one reader, the heading *To* may be followed by enough space to list a number of names.

Lenaghan Financial

Memo

To: Matthew Lenaghan, President

From: Payton Kubicek, Public Relations *PK*

CC: Katheleen Lenaghan, Chair

Date: June 1, 2014

Re: May Meeting of Plant Safety Committee

As we agreed at the March 30 meeting of the Plant Safety
Committee, we will meet again on May 12. I am requesting agenda items
and meeting suggestions from each (etc.) . . .

Penny-Wise Stores, Inc.

MEMORANDUM

To: **Date:**

                                                 From:

Store:

At:

Territory:

Copies to:

Subject: Form for In-house Letters (Memos)

This is an illustration of our memorandum stationery. It should be used
for written communications within the organization.

Notice that the memorandum uses no form of salutation. Neither does it
have any form of complimentary close. The writer does not need to sign
the message. He or she needs only to initial after the typed name in the
heading.

Notice also that the message is single-spaced with double-spacing
between paragraphs.

Memorandum Formality

Because memos usually are messages sent and received by people who work with and know one another, they tend to use casual or informal language. Even so, some memos use highly formal language. As in any business communication, you will use the level of formality appropriate to your audience and writing goals.

LO 2-4   Describe the purpose and form of email.

EMAIL

Although businesspeople routinely communicate via social media, text messaging, and instant messaging, email remains the most widely used means of written communication in the workplace. In fact, even among those businesspeople who use tablets, email is the most frequently used app.[3]

Email Defined

According to one estimate, 92 percent of Americans use **email**—a number that has remained constant even with the advent and popular use of text messaging, instant messaging, and social media.[4] It's easy to see why email remains popular. Email addresses are readily available, and anyone can send a message to any email address (or multiple addresses simultaneously), regardless of who provides the email account. The speed at which readers receive a message can also make email more attractive than a letter or a memo. Consequently, businesses continue to use email as a low-cost, quick, and efficient means of communicating with both internal and external audiences either formally or informally. Furthermore, email provides HTML and other formatting options that text messaging, instant messaging, and social media may not, and it does not limit the writer to any number of characters or amount of text. In addition, emails can be archived and filed for easy access to a written record of correspondence.

Email, however, also presents communication challenges. Sometimes people use email to avoid having difficult face-to-face or phone conversations, which is not a good way to accomplish communication goals or cultivate the audience's goodwill. Emails are also easily forwarded and therefore can never be considered confidential. Additionally, many businesspeople deal with spam—unsolicited messages or mass emails that are not relevant to their work. Moreover, some writers may assume that an informal email message is not held to the same standards of professionalism, clarity, or correctness as a more formal message might be. Finally, when not used properly, email can be costly. According to one source, "In 2010 organizations lost about $1,250 per user a year in productivity due to spam and up to $4,100 per year due to emails [that] were written poorly."[5] It is important, then, that business writers ensure their emails communicate clearly, cultivate goodwill, and promote a professional image.

Email Form

When you look at an email, you likely notice that its form contains elements of both memos and letters. For example, emails generally contain a *Date, To, From, Subject* heading structure similar to that of a memo. They may also contain salutations and complimentary closes similar to those found in letters.

Although the various email systems differ somewhat, email format includes the following:

- **To:** This is where you include the email address of the recipients. Be sure the address is accurate.

- **Cc:** If someone other than the primary recipient is to receive a *courtesy copy,* his or her address goes here. Before people used computers, *cc:* was called a *carbon copy* to reflect the practice of making copies of letters or memos with carbon paper.

- **Bcc:** This line stands for *blind courtesy copy.* This line is also for email addresses of recipients. However, each recipient's message will not show this information; that is, he or she will not know who else is receiving a copy of the message.

- **Subject:** This line describes the message as precisely as the situation permits. The reader should get from it a clear idea of what the message is about. Always include a subject line to get your reader's attention and indicate the topic of the

at all is more common. A friendly generic greeting such as "Greetings" is appropriate for a group of people with whom you communicate. As we discussed in the section on letters, you'll want to avoid outdated expressions such as "To Whom It May Concern." Like a letter, an email message often ends with a complimentary close followed by a signature block containing the writer's name, job

> It is important, then, that business writers ensure their emails communicate clearly, cultivate goodwill, and promote a professional image.

message. In the absence of a subject line, a reader may think your message is junk mail or unimportant and delete it.

- **Attachments:** In this area you can enter a file that you desire to send along with the message. Attach only files the reader needs so that you do not take up unneeded space in his or her inbox.

- **The message:** The information you are sending goes here. Typically, email messages begin with the recipient's name. If the writer and reader are acquainted, you can use the reader's first name. If you would normally address the reader by using a title (Ms., Dr., Mr.), address him or her this way in an initial email. You can change the salutation in subsequent messages if the person indicates that informality is desired. The salutations commonly used in letters (Dear Mr. Dayle, Dear Jane) are sometimes used, but something less formal ("Hi, Ron") or no salutation

title, company, and contact information. Some writers also use the signature block as an opportunity to promote a sale, product, or service. Exhibit 2-8 shows a standard email format.

Email Formality

A discussion of email formality is complicated by the fact that email messages are extremely diverse. They run the range from highly informal to formal. The informal messages often resemble face-to-face oral communication; some even sound like chitchat that occurs between acquaintances and friends. Others, as we have noted, have the increased formality of reports.

A helpful approach is to view email language in terms of three general classifications: **casual, informal,** and **formal.**[6] Your audience should determine which type of language you choose, regardless of your personal style or preference.

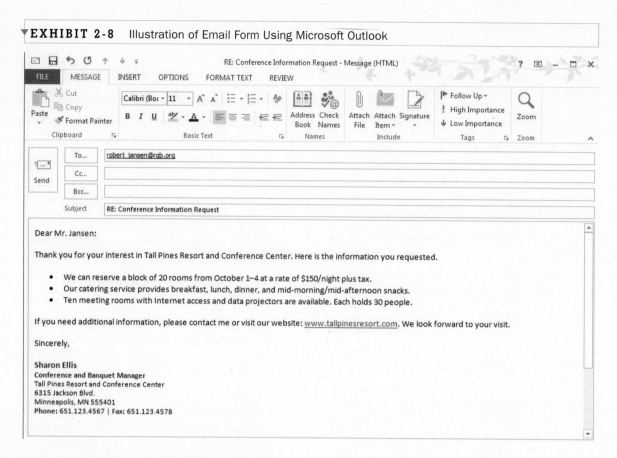

casual By casual language we mean the language we use in talking with close friends in everyday situations. It includes slang, colloquialisms (informal expressions), contractions, and personal pronouns. Its sentences are short—sometimes incomplete—and it may use mechanical emphasis devices and initialisms (e.g., LOL, BTW) freely. Casual language is best limited to your communications with close friends. Following is an example of casual language:

Hey, Cindy,

Props for me! Just back from reps meeting. We totally nailed it ... plan due ASAP. Meet, my office, 10 AM, Wed?

Brandon

Use casual language only when you know your readers well—when you know they expect and prefer casual communication. You should also avoid slang, initialisms, emphasis devices, or other casual elements if you are not certain that they will communicate clearly.

informal Informal language retains some of the qualities of casual writing. It makes some use of personal pronouns and contractions. It occasionally may use colloquialisms

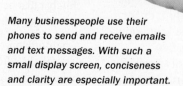

Many businesspeople use their phones to send and receive emails and text messages. With such a small display screen, conciseness and clarity are especially important.

but more selectively than in casual writing. It has the effect of conversation, but it is polished conversation—not chit-chat. Its sentences are short, but they are well structured and organized. They have varied patterns that produce an interesting style. In general, it is the writing that you will find in most of the illustrations in Chapters 5–7, and it is the language that this book uses. You should use it in most of your business email messages, especially when writing to people you know only on a business basis. An example of an email message in informal language is the following:

Cindy:

The management team has approved our marketing plan. They were very complimentary. As you predicted, they want a special plan for the large accounts. They want it as soon as possible, so let's get together to work on it. Can we meet Wednesday, 10 A.M., my office?

Brandon

formal A formal style of writing maintains a greater distance between writer and reader than an informal style. It avoids personal references and contractions, and its sentences are well structured and organized. Formal style is well illustrated in the examples of the more formal reports in Chapter 9 and Bonus Chapter D. It is appropriate to use in email messages resembling formal reports, in messages to people of higher status, and to people not known to the writer.

As with any business message, formal or informal, your emails should achieve your communication goal, promote goodwill, and present a professional image. To do this, follow the advice in Chapter 4 for writing clear, courteous messages. You will also want to follow the guides in Bonus Chapter B to ensure your messages are expressed correctly.

NEWER MEDIA IN BUSINESS WRITING

Sometimes writers in today's fast-paced, global business world need to communicate more immediately and quickly than a letter, a memo, or an email will allow. Technology provides business writers with many more channels for immediate, quick communication including text messaging, instant messaging, and social networking. Exhibit 2-9 illustrates one company's many uses of social media.

However, as with more traditional business writing media such as letters, memos, and emails, the use of these more immediate channels should be driven by audience needs and expectations as well as the writer's goals and purposes.

LO 2-5 Understand the nature and business uses of text messaging and instant messaging.

Text Messaging

Text messaging, also called short message service (SMS), is, as its name suggests, used for sending short messages, generally from a mobile phone. Because the purpose of a text message is to convey a quick message, text messages are much shorter than messages conveyed by more traditional forms. Also, mobile phone service providers may limit the number of characters in a text message.

The need for brevity has led to the use of many abbreviations. So many of these abbreviations have developed that one might say a new language has developed. In fact, a dictionary of over 1,100 text messaging abbreviations has been compiled at Webopedia, an online computer technology encyclopedia (www.webopedia.com/quick_ref/textmessageabbreviations .asp). Some examples are the following:

b4 (before)	NP (no problem)
gr8 (great)	FBM (fine by me)
CU (see you)	TC (take care)
u (you)	HRY (how are you)
BTW (by the way)	TYT (take your time)

from the tech desk

Using Good Email Etiquette Helps Writers Achieve Their Goals

Using proper email etiquette is as easy as applying a bit of empathy to your messages: Send only what you would want to receive. The following additional etiquette questions will help you consider more specific issues when using email.

• Is your message really needed by the recipient(s)?

• Is your message for routine rather than sensitive messages?

• Are you sure your message is not spam (an annoying message) or a chain letter?

• Have you carefully checked that your message is going where you want it to go?

• Has your wording avoided defamatory or libelous language?

• Have you complied with copyright laws and cited sources accurately?

• Have you avoided humor and sarcasm that your reader may not understand as intended?

• Have you proofread your message carefully?

• Is this a message you would not mind having distributed widely?

• Does your signature avoid offensive quotes or illustrations, especially those that are religious, political, or sexual?

• Is your recipient willing or able to accept attached files?

• Are attached files a size that your recipient's system can handle?

• Are the files you are attaching virus free?

EXHIBIT 2-9   Illustration of One Organization's Uses of Social Media

In addition to abbreviations, writers use typed symbols to convey emotions (emoticons), which can also be found at Webopedia:

:-) standard smiley	:-! foot in mouth
;) winking smile	:-(sad or frown
:-0 yell	(((H))) hugs

Whether and when these abbreviations and emoticons are used depends on the writer's relationship with the audience.

Good business writers will compose text messages that not only convey the writer's message but also allow for brief responses from the receiver. Let's say, for example, that you've learned that an important visiting customer is a vegetarian and you have reservations for lunch at Ruth's Chris Steakhouse. You need to let your boss know—before the lunch meeting. However, the boss is leading an important meeting in which a phone call would be disruptive and inappropriate, so you decide to send a text message.

Your immediate thought might be to send the following: *Marina Smith is a vegetarian. Where should we take her for lunch today? Zach.*

Although your message does convey the major fact and is only 77 characters counting spaces, it forces the recipient to enter a long response—the name of another place. It might also result in more message exchanges about availability and time.

A better version might be this: *Marina Smith is a vegetarian. Shall we go to 1) Fish House, 2) Souplantation, 3) Mandarin House? All are available at noon. Zach*

This version conveys the major fact in 130 characters and allows the recipient to respond simply with 1, 2, or 3. As the writer, you took the initiative to anticipate your reader's needs, identify appropriate alternatives, and then gather information—steps that are as important with text messaging as they are with other messages. If your text messages are clear, complete, and concise and have a professional and pleasant tone, you will find them a valuable tool for business use.

Instant Messaging

Instant messaging, commonly referred to as IM-ing or online chatting, is much like telephone conversation in that parties communicate in real time (instantly). It differs primarily in that it is text-based (typed) rather than voice-based communication, though voice-based instant messaging is possible. Many writers will use the same abbreviations and emoticons in instant messages that they use in text messages. Here again, the use of these devices depends on your audience and purpose. Exhibit 2-10 shows Microsoft's Lync instant messaging tool.

Because instant messages are similar to phone conversations, you should write **instant messages** much as you would talk in conversation with another person. If the person is a friend, your language should reflect this friendship. If the person is the president of your company, a business associate, or fellow worker, the relationship should guide you. The message bits presented in instant messaging are determined largely by the flow of the conversation. Responses often are impromptu. Even so, in business situations you should consciously direct the flow toward your objective and keep your language and content professional.

EXHIBIT 2-10 Example of Microsoft's Lync Instant Messaging Tool

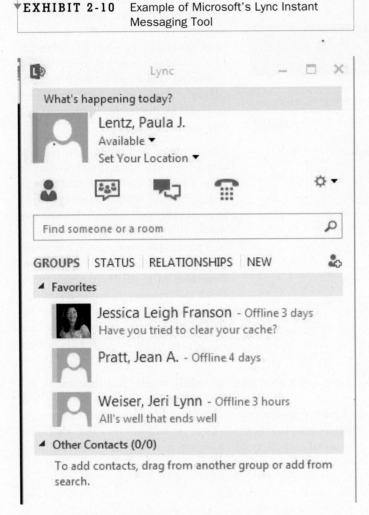

Social Media

You are probably familiar with such **social media** sites as Facebook, Twitter, or LinkedIn. Perhaps you have a blog (a "Web log") where you keep an online diary or journal that you share publicly. Although you may use these sites to connect with friends, family, or classmates, many business writers also use them to connect with clients, customers, colleagues, and supervisors, as they answer questions, promote products, network with other professionals, or interact briefly with co-workers. Business professionals, then, are using social networking sites for purposes that are likely very different from your purpose in using them (see the corporate blog in Exhibit 2-11).

Generally, the messages on social networking sites are brief; one site, Twitter, restricts messages to 140 characters. As with text messaging, messages must not only be brief but concise and clear. If you have only so much space for your message,

you need to make sure your reader immediately knows your point and has enough detail to act on your message. Therefore, messages on social media sites should begin with your main point (what you need your reader to do, think, feel, or believe as a result of reading your message) and then follow with details in order of importance.

In addition, because the messages on these sites are public, you never want to use language or a tone or writing style that you would be embarrassed to have your boss see, that may have legal implications, or that might get you fired. In fact, if you currently have a page on a social networking site where family and friends are your audience, you will want to remove any pictures or language that you wouldn't want a prospective employer, current employer, co-worker, customer, or client to see. No matter how private you believe your page to be, you can never know what your friends and family are sharing with other people. Mashable.com cites a study by Reppler, a social media monitoring service, that found "more than 90 percent of recruiters and hiring managers have visited a potential candidate's profile on a social network." Interestingly, the study found that employers are just as likely to reject a candidate (69 percent) as they are to hire a candidate (68 percent) based on what is on someone's social networking site.[7]

Regardless of the type of business messages you send, remember that on the job, companies often monitor employees' computer activity. They can detect excessive use, inappropriate or unethical behavior, disclosure of proprietary information, use of sexually explicit language, and attachments with viruses. Companies' monitoring systems also have features that protect the company from legal liabilities. As a business professional, you must know your company's computer use policy and avoid writing anything that would reflect poorly on you or your company or put you or your company at risk.

PRINT VERSUS ONLINE DOCUMENTS

As we have discussed, business professionals use a variety of print and electronic genres for business writing. The basic principles of business writing apply to both print and electronic text; that is, your text must be *reader-centered, accessible, complete, concise, and accurate*. Though electronic documents often provide the opportunity for more creative and interactive features than print documents, these features are useless if your reader does not have the technology to access them. Furthermore, no matter how great your blog or your Web site looks, if your message is unclear or incomplete, your blog or site serves no purpose. Business writers also have to realize that many audiences may be viewing a

says that when people look for information in electronic documents, particularly on the Web, they do so not necessarily to read what an author has to say about an issue but to accomplish a specific task (e.g., locate a statistic, fill out a form). Online text, then, needs to facilitate the reader's ability to find and use information.

Furthermore, online text can produce comprehensive data more concisely than a print document. Because technology allows writers to embed links to relevant or related information rather than include that information in a paragraph or on a page, electronic documents can incorporate a lot of information in a relatively small space. Print documents, though, could become quite long and unwieldy if an author were to try to include every fact, statistic, or resource related to the topic at hand.

company's Web site on a mobile device with a small viewing area. In fact, Janice Redish, an expert on Web writing, says, "Understanding your site visitors and their needs is critical to deciding what to write, how much to write, the vocabulary to use, and how to organize the content on your Web site." While this advice also applies when writing print documents, writing for online delivery presents special considerations.[8] The following are the main ones to keep in mind.

Lastly, print documents generally require that thoughts be expressed in complete sentences, with occasional bulleted lists added for clarity and visual appeal. In electronic documents, writers tend to rely much more on bulleted lists and other terse forms of text. Depending on the medium, they may use fragments and frequent abbreviations.

Comparing Print and Online Text

Jakob Nielsen, noted usability expert, has found that Web readers read an average of 20 percent of the words on a page.[9] He says that print text can be distinguished from Web text in that print text tends to be linear, while Web text is nonlinear. That is, when people read print documents, they often start at the beginning and continue reading until they reach the end. By contrast, online readers scan for relevant information and may be diverted by links or other features of the display in their search. In addition, he

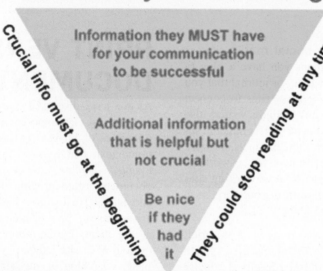

Organizing Content

Redish notes that most people visit Web pages because they need to *do* something. Consequently, they do not read so much as they scan for information, and they want what they need quickly, which means that writers should "think 'information,' not 'document.'"[10] As with print documents, online information must be organized well. Redish advocates organizing Web pages in the **inverted pyramid style**, where the main point is presented first, followed by supporting information and then by any historical or background information (see Exhibit 2-12).

▼EXHIBIT 2-13   Example of an Online Article Written in the Inverted Pyramid Structure

Home > Internet

News

SAP to pay $4.3B for cloud B2B vendor Ariba

By Chris Kanaracus
June 4, 2012 06:00 AM ET   ◯ Add a comment   **T**   + Briefcase   What's this?

Computerworld - SAP has agreed to purchase cloud-based e-commerce vendor Ariba for $4.3 billion.

The deal has been unanimously approved by Ariba's board and is expected to close by Sept. 30.

The addition of Ariba's business-to-business commerce platform will expand SAP's cloud software lineup, which got an earlier boost from the ERP vendor's recent $3.4 billion acquisition of SuccessFactors, a provider of human resources services.

SAP's cloud computing strategy is aimed at enabling the company to compete against traditional rivals such as Oracle, as well as pure cloud vendors like Workday and NetSuite.

Ariba reported $444 million in 2011 revenue and said its trading network is involved with more than $319 billion in "commerce transactions, collaborations, and intelligence among more than 730,000 companies," according to a statement.

SAP said it will consolidate all "cloud-related supplier assets" under the auspices of Ariba, which will operate as an independent subsidiary.

The company plans to keep Ariba's platform open, allowing it to accept data from "any source," according to SAP co-CEO Bill McDermott. SAP also plans to use the platform it bought last year from Crossgate, which has some of the same features as Ariba's.

Source: Used with permission of Computerworld Online. Copyright © 2013. All rights reserved.

Other kinds of electronic documents can benefit from this advice as well. If readers are merely scanning for information, they may not scroll for information, which means that the main point must stand out. Similar to information in printed business documents, information in electronic documents should be chunked in short paragraphs and contain headings and lists that emphasize a logical structure and presentation of the information. Exhibit 2-13 provides an example of writing in the inverted pyramid style.

Presenting the Content

Your choice of design elements (font, color, and graphics) depends on your audience. Redish notes that one of the important differences between print and electronic documents is that the resolution (sharpness of the letters and images) is lower on a screen than in print. Thus, while you can use a variety of **serif** and **sans serif fonts** in a variety of sizes in print documents, you will want to choose sans serif fonts (those without tails at the ends of the letters) in at least a 12-point size for most online documents so that the serifs do not obscure the text. In addition, readers usually see printed documents on 8½ × 11 paper, but online readers may see electronic documents in windows of varying sizes. For this reason, Redish recommends a line length of 50 to 70 characters, or 8 to 10 words. Short lines are also more quickly and easily read, though you do not want lines so short that your text does not capture the main point.

Likewise, text-formatting conventions differ in print and electronic environments. In both environments, writers can use emphasis devices such as bullets, headings, bold text, or italics. However, because underlining also represents links in electronic environments, writers should favor italics or bold text in that context.

In addition, as with printed text, writers should use colors that are visually appealing and appropriate for the message and audience. For most readers, this will be a dark text on a light background.

Making Your Web Writing Accessible

Many businesses have seen the wisdom of ensuring that their Web sites are accessible to people with disabilities; in fact, accessibility is generally required by law. Many features of the Web that we take for granted may present difficulties for those with disabilities. For instance, how do people with hearing impairments access audio content on a Web site? How do people with visual disabilities access text? How does a person with a motor impairment use a mouse? Incorporating text along with audio files gives people who have hearing impairments access to a site, while incorporating text with visuals enables the screen readers of those with visual impairments to "read" the visual for the user. People with motor disabilities can be helped with voice-activated features or key commands rather than mouse-controlled navigation.

In today's business world, professionals have many options for communication. By following the advice in this chapter, you will be sure to choose the right channels for the many types of written communication your job requires.

write business messages like a pro!

Learn more about the writing process and the various types of business documents.

- What are best practices for texting at work?
- How can you cultivate relationships with audiences for business success?
- What are the basic rules for internal communication in a business?

Scan the QR code with your smartphone or use your Web browser to find out at www.mhhe.com/RentzM3e. Choose Chapter 2 > Bizcom Tools & Tips. While you're there, you can view a chapter summary, exercises, PPT slides, and more to become an expert at writing effective business messages.

www.mhhe.com/RentzM3e

chapter three

Communicating Effectively
with Visuals

Visuals are pervasive in today's media-rich society in both our work and leisure activities. We read and write documents and reports that include images; we hear and give presentations where the audience expects graphs and other visuals. In fact, each day millions of people are clicking on Web sites such as Flickr.com and YouTube.com where they view and post photos and videos. Additionally, other specialty sites, such as SlideShare.net, are becoming increasingly popular. Visuals are even being used as search expressions with smartphone apps, such as Google Goggles, for finding information. This growing use of images for conveying information confirms that being visually literate is extremely important in order to communicate effectively today.

One recent experimental study of business executives examined the impact of using visuals in group work. Its results clearly showed that groups who were using visuals in their work had higher productivity, higher quality of outcomes, and greater knowledge gains for the individuals in the group.[1] ∎

LEARNING OBJECTIVES

LO 3-1 Plan which parts of your document or presentation should be communicated or supported by visuals.

LO 3-2 Explain the factors that are important in the effective presentation of visuals: size, layout, type, rules and borders, color and cross-hatching, clip art, background, numbering, titles, title placement, and footnotes and acknowledgments.

LO 3-3 Construct textual visuals such as tables, pull quotes, flowcharts, and process charts.

LO 3-4 Construct and use visuals such as bar charts, pie charts, line charts, scatter diagrams, and maps.

LO 3-5 Avoid common errors and ethical problems when constructing and using visuals.

LO 3-6 Place and interpret visuals effectively.

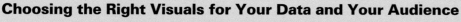
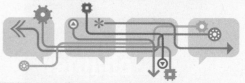
Choosing the Right Visuals for Your Data and Your Audience

In your job as the sales manager for Green Living Industries, you are frequently called upon to gather sales data and market demographics. You share these data with your employees as well as your colleagues and corporate headquarters to support decisions that help achieve business goals.

In fact, you have just finished gathering data for a major report you will give orally and in writing to Green Living's upper management. These data include information on the makeup of your target markets, sales performance by division and geographic area, industry trends, and customer perceptions of your products. Although your primary audience is Green Living's upper management, you also know that these data will be used to inform decisions at various district offices and sales divisions. Essentially, you have a lot of information and several audiences who will use it. You need to think critically regarding how you will present this information so that your audiences quickly, clearly, and correctly understand what you are communicating.

Visual communication is the intentional use of illustrations (e.g., charts, pictures, maps, diagrams) and formatting (e.g., typography, color) to deliver a message. It may seem like you're simply putting numbers in a table, demonstrating a trend in a chart, or getting an audience's attention, but every visual you choose tells your story and leaves the audience with an impression of you, your data, and your ideas. In other words, visuals are never neutral. Rather, they are as essential to communicating meaning and presenting a professional image as the words written in a document or spoken in a presentation.

In addition, as a consumer of visual information, you will want to look critically at the visuals you encounter to ask what message is being conveyed and whether it is being conveyed clearly, accurately, and ethically.

LO 3-1 Plan which parts of your document or presentation should be communicated or supported by visuals.

PLANNING THE VISUALS

You should plan the visuals for a document soon after you organize your findings. Your planning for visuals should be guided largely by your business and communication goals. Visuals can clarify complex or difficult information, emphasize facts, add coherence, summarize data, set the tone of your discussion, inform or persuade an audience, and provide interest. Of course, well-constructed visuals also enhance your document's appearance.

In planning visuals, you should review the information that your document will contain, looking for any possibility of improving communication of the material through the use of visuals. Specifically, you should look for complex information that visual presentation can make clear, for information too detailed to be covered in words, and for information that deserves special emphasis.

Of course, you want to plan with your reader in mind and choose visuals appropriate to both the content and context where they are presented. You should construct visuals to help the intended reader understand the information more quickly, easily, and completely.

As you plan the visuals, remember that some visuals can stand alone, but others will supplement the writing or speaking—not take its place. Visuals in documents or oral presentations should support your words by communicating difficult concepts, emphasizing the important points, and presenting details.

LO 3-2 Explain the factors that are important in the effective presentation of visuals: size, layout, type, rules and borders, color and cross-hatching, clip art, background, numbering, titles, title placement, and footnotes and acknowledgments.

DETERMINING THE GENERAL MECHANICS OF CONSTRUCTION

In constructing visuals, you will make decisions regarding the various conventions for presenting them to business audiences. The most common conventions are summarized in the following paragraphs.

Size

One of the first steps you must take to construct a visual is determining its size. The size of a visual is determined by its contents and importance. If a visual is simple (with only two or three quantities), a quarter page might be enough. But if a visual must display complex or detailed information, a full page might be justified.

With extremely complex, involved information, you may need to use more than a full page. When you do, make certain that this large page is inserted and folded so that the readers can open it easily. You may also consider including large or complex visuals in an appendix or attachment.

Orientation

You should determine the **orientation** of the visual by considering its size and contents. Sometimes a tall orientation (portrait) is the answer; sometimes the answer is a wide orientation (landscape). Simply consider the logical possibilities and select the one that is most easily read.

Type

The type used in visuals throughout a report should generally be consistent in terms of **style and font**. Style refers to the look of the type such as bold or italics; font refers to the look of the letters, such as those with or without feet (*serif* or *sans serif*, respectively). Be aware that even the design of the font you choose will convey a message, a message that should work with the text content and design.

Type size is another variable to watch. The size you choose should look appropriate in the context in which it is used. Your top priority when choosing type style, font, and size should always be readability.

Rules and Borders

You should use **rules (lines) and borders** when they improve the readability of the visual. Rules help distinguish one section or visual from another, while borders help separate visuals from the text. Keep in mind that rules can add clutter, so be sure to use them only when they will enhance the audience's understanding of your visual. And when using borders, be sure to place borders around visuals that occupy less than a full page. You also can place borders around full-page visuals, but such borders serve mostly a decorative function. Except in cases in which visuals simply will not fit into the normal page layout, you should not extend the borders of visuals beyond the normal page margins.

Color and Cross-Hatching

Color and cross-hatching, appropriately used, help readers see comparisons and distinctions (see Exhibit 3-1). In fact, research has found that color in visuals improves the comprehension, retention, and extraction of information. Also, color and cross-hatching can add to the attractiveness of the report. Because color is especially effective for this purpose, you should use it whenever practical and appropriate.

Clip Art

Today you can get good-looking clip art easily—so easily, in fact, that

communication matters

Communicating with Color

When constructing visuals, it is important to choose colors that are appropriate for the audience, context, and message. Colors have culture-specific meanings, so when choosing colors for a U.S. audience, you will want to be aware of the following associations.

Warm colors (e.g., red, orange, yellow) communicate passion, happiness, enthusiasm, energy, importance, or warning. Cool colors (e.g., green, blue, purple) provide a sense of calm, balance, harmony, and professionalism. Neutrals (e.g., black, brown, gray, beige, white) are conservative colors that convey power, sophistication, dependability, reliability, and formality.

However, shades of these colors are also important. Olive green, for instance, is associated with nature, while dark green indicates affluence. Likewise, bright blues can be seen as energizing, while navy is conservative and businesslike, and light purple is associated with romance while dark purple is associated with wealth and royalty. However, colors can also have negative connotations. For example, black is often associated with death, beige can be boring, and green can be associated with jealousy or envy.

As you can see, choosing the right color is important for setting an appropriate tone and conveying your message. Be sure you incorporate color decisions when you plan visuals for your document or presentation.

Source: Cameron Chapman, "Color Theory for Designers, Part 1: The Meaning of Color," *Smashing Magazine*, Smashing Media, 28 Jan. 2010, Web, 30 May 2013.

some writers often overuse it. Although clip art can add interest and bring the reader into a visual effectively, it also can overpower and distract the reader. The general rule is to keep in mind the purpose your clip art is serving: to help the reader understand the content. It should be appropriate in both its nature and size; and it should be appropriate in its representation of gender, race, and age. Also, if the clip art is copyrighted, you may need permission to use it.

Background

Background colors, photos, and art for your visuals should be chosen carefully. The color should provide

▼**EXHIBIT 3-1** Color versus Cross-Hatched Pie

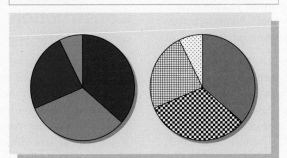

communication matters

Applying Color to Visuals

As we discussed in the Communication Matters feature on page 43, colors help set your tone and communicate your message. While color selection itself is important, so, too, is applying color effectively to your visuals. Stephen Few provides the following rules for ensuring that color enhances your audience's understanding and your communication.

Rule #1: If you want different objects of the same color in a table or graph to look the same, make sure that the background—the color that surrounds them—is consistent.

Rule #2: If you want objects in a table or graph to be easily seen, use a background color that contrasts sufficiently with the object.

Rule #3: Use color only when needed to serve a particular communication goal. Color should be used to enhance data; it should not be used for decoration. In many cases, less color may be better.

Rule #4: Use different colors only when they correspond to differences of meaning in the data. If audiences see similar units of data represented in different colors, they will assume the different colors have meaning, when really, the color was just for decoration.

Rule #5: Use soft, natural colors to display most information and bright and/or dark colors to highlight information that requires greater attention.

Rule #6: When using color to [a sequence of values, such as lowest to highest], stick with a single [color] (or a small set of closely related [colors]) and vary [shades] from pale colors for low values to increasingly darker and brighter colors for high values.

Rule #7: Non-data components of tables and graphs should be displayed just visibly enough to perform their role, but no more so, for excessive salience could cause them to distract attention from the data. For example, use thin gray lines for borders.

Rule #8: To guarantee that most people who are colorblind can distinguish groups of data that are color coded, avoid using a combination of red and green in the same display.

Rule #9: Avoid using visual effects in graphs (e.g., shadows, 3-D effects).

Source: Practical Rules for Using Color in Charts," *Visual Business Intelligence Newsletter*, Stephen Few, Perceptual Edge, February 2008. Reprinted with permission.

> All visuals generally have titles; captions may be used if the reader will need them to interpret or better understand the visual.

high contrast with the data and not distract from the main message. As with any photos or art, backgrounds should create a positive, professional impression. Additionally, when visuals are used cross-culturally, you will want to be sure the message your background sends is the one you intended by testing or reviewing it with your audience.

Numbering

Pull quotes, clip art, and other decorative visuals do not need to be numbered. Neither does a lone table or figure in a document. Otherwise, you should number all the visuals. Many schemes of numbering are available to you, depending on the types of visuals you have.

If you have many visuals that fall into two or three categories, you may number each of the categories consecutively. For example, if your document is illustrated by six tables, five charts, and six maps, you may number these visuals Table I, Table II, … Table VI; Chart 1, Chart 2, … Chart 5; and Map 1, Map 2, … Map 6.

However, if your visuals comprise a wide mixture of types, you may number them in two groups: tables and figures. Figures, a miscellaneous grouping, include all types other than tables. To illustrate, consider a report containing three tables, two maps, three charts, one diagram, and one photograph. You would number these visuals Table I, Table II, and Table III and Figure 1, Figure 2, … Figure 7. Whatever your numbering scheme, remember that tables are numbered separately from other types of visuals.

Construction of Titles and Captions

The **title** is the name you give your visual; a **caption** is a brief description of the visual. All visuals generally have titles; captions may be used if the reader will need them to interpret or better understand the visual. Like the headings used in other parts of the report, the title or caption of the visual has the objective of concisely covering the contents. As a check of content coverage, you might use

communication matters

Infographics: Everything Old Is New Again

According to the folks at Visual.ly, the infographic has been around since humans used cave drawings. Even Florence Nightingale used an infographic (the Coxcomb chart) during the Crimean War to categorize diseases and causes of death. So while the infographic is nothing new, what is new is its popularity as a data visualization tool.

Though Visual.ly acknowledges that infographics serve much the same purpose as other visuals, infographics differ from other visuals in that they (1) "have a flow to them" that makes complex data more accessible to the reader, (2) present data so that the reader can engage with the story the data tell, and (3) are visually engaging illustrations of data stories.

If you search the Internet for tools to create infographics, you'll find several at your disposal. You'll also see you can create beautiful data visualizations on widely varying topics—everything from survey results to résumés. The visual to the right illustrates how data regarding the Professionalism in the Workplace study by the Center for Professional Excellence at York College of Pennsylvania were captured clearly and concisely in an infographic. Should you incorporate infographics in your business writing or presentations? Like any type of visual, an infographic is only effective if it can tell a story more clearly than text alone. You'll also want to keep in mind that some infographics can look informal or cartoonish. As in any business communication, you must analyze your audience, communication goals, and context to know if an infographic is appropriate.

Sources: "History of infographics," *Visual.ly*, Visually, Inc., 2011, Web, 12 July 2012; "Professionalism in the Workplace by the Numbers," *The Undercover Recruiter*, The Undercover Recruiter, n.d., Web, 31 May 2013; "What Is an infographic?," *Visual.ly*, Visually, Inc., 2011, Web, 12 July 2012. Reprinted with permission of iProspect.

---

the journalist's **five Ws:** *who, what, where, when,* and *why.* Sometimes you also might use *how.* But because conciseness is also important, it is not always necessary to include all the Ws in the title. For example, the title of a chart comparing the annual sales volume of the Texas and California territories of the Dell Company for the years 2010–2012 might be constructed as follows:

Who:	Dell Company
What:	Annual sales
Where:	Texas and California branches
When:	2010–12
Why:	For comparison

The title might read, "Comparative Annual Sales of Texas and California Territories of the Dell Company, 2010–12." For even more conciseness, you could use a major title and subtitle. The major title might read, "A Texas and California Sales Comparison"; the subtitle might read, "Dell Company 2010–12." Similarly, the caption might read "A Texas and California Sales Comparison: Dell Company 2010–12."

Placement of Titles and Captions

Titles are placed above tables, but they may be placed either above or below other visuals. However, most software programs automatically place titles at the top. Captions, on the other hand, are generally placed below tables and other visuals. When typing titles and captions, use **title case** (the kind of capitalization used for book titles). For long captions, you may opt to use sentence case (capitalization of the first word and any proper nouns).

Table number and title →

Table I—U.S. Internet User Penetration by Race/Ethnicity 2008–2014 (% of population in each group)							
	Actual			*Projected*			
	2008	2009	2010	2011	2012	2013	2014
Non-Hispanic							
White alone	72.0%	74.0%	76.1%	77.5%	79.0%	80.1%	81.2%
Black alone	58.2%	60.5%	63.8%	66.9%	69.6%	71.7%	72.3%
Asian alone	70.0%	71.2%	73.4%	75.5%	77.5%	79.5%	81.0%
Other*	46.4%	50.0%	52.5%	55.0%	58.0%	62.0%	65.5%
Hispanic**	53.5%	56.5%	59.5%	62.9%	65.0%	67.6%	70.0%

Spanner heads →
Column heads →
Row heads →

Footnote → *Includes native Americans, Alaska natives, Hawaiian and Pacific Islanders, and bi- and multiracial individuals.

**Could be of any race.

Source acknowledgment → SOURCE: eMarketer, March 2010.

Source: Lisa E. Phillips "U.S. Internet User Penetration by Race/Ethnicity 2008–2014," *eMarketer Digital Intelligence*, eMarketer, Inc., 29 Sept. 2010, Web, 24 May 2013. Reprinted with permission of eMarketer, Inc.

Footnotes and Acknowledgments

Parts of a visual sometimes require special explanation or elaboration. When this happens, you should use **footnotes**. These footnotes are concise explanations placed below the illustration by means of a superscript (raised) number or symbol (asterisk, dagger, double dagger, and so on) (see Exhibit 3-2). Footnotes are placed immediately below a visual with no caption. If a visual contains a caption, the footnote is placed below the caption.

Usually, a **source acknowledgment** is the bottom-most component of a visual. By *source acknowledgment* we mean a reference to the body or authority that deserves the credit for gathering the data used in the illustration. The entry consists of the word *Source* followed by a colon and the source information (in some cases, simply the source name will suffice). See Exhibit 3-2 for an illustration of source acknowledgment.

If you or your staff collected the data, you may either omit the source note or give the source as "Primary," in which case the note would read like this:

Source: Primary.

LO 3-3   Construct textual visuals such as tables, pull quotes, flowcharts, and process charts.

CONSTRUCTING TEXTUAL VISUALS

Visuals for communicating report information fall into two general categories: those that communicate primarily through textual content (words and numerals) and those that communicate primarily through visual elements (charts and graphs). Included in the textual group are pull quotes and a variety of process charts (e.g., Gantt, flow, organization).

Tables

A **table** is an orderly arrangement of information in rows and columns. As we have noted, tables are textual visuals (not really pictures), but they communicate like visuals, and they have many of the characteristics of visuals.

Aside from the title, footnotes, and source designation previously discussed, a table contains heads, columns, and rows of data, as shown in Exhibit 3-2. Row heads are the titles of the rows of data, and column heads are the titles of the columns.

The construction of text tables is largely influenced by their purpose. Nevertheless, a few rules generally apply:

- If rows are long, the row heads may be repeated at the right.

- The em dash (—) or the abbreviation *n.a.* (or *N.A.* or *NA*), but not the zero, is used to indicate data not available.

- Since footnote numbers in a table full of numbers might be confusing, footnote references to numbers in the table should be keyed with asterisks (*), daggers (†), double daggers (‡), section marks (§), and so on. Small letters of the alphabet can be used when many references are made.

- Totals and subtotals should appear whenever they help readers interpret the table. The totals may be for each column and sometimes for each row. Row totals are usually placed at the right, but when they need emphasis, they may be placed at the left. Likewise, column totals are generally placed at the bottom of the column, but they may be placed at the top when the writer wants to emphasize them.

A ruled line (usually a double one) separates the totals from their components.

- The units used to represent the data must be clear. Unit descriptions (e.g., bushels, acres, pounds, dollars, or percentages) appear above the columns, as part of the headings or subheadings. If the data are in dollars, however, placing the dollar mark ($) before the first entry in each column can be sufficient.

Tabular information need not always be presented in formal tables. In fact, short arrangements of data may be presented more effectively as parts of the text. Such arrangements are generally made as either leaderwork or text tabulations.

Leaderwork is the presentation of tabular material in the text without titles or rules. (*Leaders* are the repeated dots with intervening spaces.) Typically, a colon precedes the tabulation, as in this illustration:

> The August sales of the representatives in the Western Region were as follows:
>
> Charles B. Brown $33,517
>
> Thelma Capp 39,703
>
> Bill E. Knauth 38,198

Text tabulations are simple tables, usually with column heads and sometimes with rules and borders. But they are not numbered, and they have no titles. They are made to read with the text, as in this example:

> In August the sales of the representatives in the Western Region increased sharply from those for the preceding month, as these figures show:

Representative	July Sales	August Sales	Increase
Charles B. Brown	$32,819	$33,517	$ 698
Thelma Capp	37,225	39,703	2,478
Bill E. Knauth	36,838	38,198	1,360

Pull Quotes

The **pull quote** is a textual visual that is often overlooked yet extremely useful for emphasizing key points. It is also useful when the text or content of the report does not lend itself naturally or easily to other visuals. By selecting a key sentence, copying it to a text box, enlarging it, and perhaps even enhancing it with a new font, style, or color, a writer can break up the visual boredom of a full page or screen of text. Software lets users easily wrap text around shapes as well as along curves and irregular lines. Exhibit 3-3 shows an example that is simple yet effective in both drawing the reader's attention to a key point and adding visual interest to a page.

Bulleted Lists

Bulleted lists are listings of points arranged with bullets (•) to set them off. These lists can have a title that covers all the points, or they can appear without titles, as they appear in various places in this book. When you use this arrangement, make the points grammatically parallel. If the points have subparts, use **sub-bullets** for them. Make the sub-bullets different by color, size, shape, or weight. Darts, check marks, squares, or triangles can be used for the secondary bullets.

Flowcharts and Process Charts

Business professionals use a variety of specialized charts in their work. Often these charts are a part of the

---

EXHIBIT 3-3   Illustration of a Pull Quote

COVER STORY

It's a tragic tale — but not completely accurate, according to some tech-employment experts. The situation is more nuanced than what can be captured in a headline, and both workers and employers share responsibility for the gap, they say.

Most portentous, though, is that the gap, whatever its true nature, is rapidly becoming a yawning chasm — one that IT employees will have to cross sooner rather than later. Many hiring experts, IT managers and CIOs believe that the tech employment landscape will be radically different five years from now as more and more companies outsource IT operations to service providers, perhaps offshore, or move traditional IT jobs to other business units.

In the face of such rapid change, it's becoming clear that the one skill every member of the IT workforce needs is career management.

"Everybody is a free agent, navigating the corporate chaos," says Todd Weinman, president of The Weinman Group, an executive search firm headquartered in Oakland, Calif., that specializes in audit and corporate governance. In the IT job market, he says, "the people who are faring a little bit better are constantly cultivating their careers on a variety of fronts."

Tech employees log long hours, meaning they get a lot of hands-on experience, but they're not getting the training and other types of enrichment they need to develop their careers. "In addition to your 50-plus hours a week, you need in-depth coursework to refresh your skills, plus studying to sit for certifications," says Weinman. At many companies, employees used to be able to take time for those types of pursuits during the workday, but not anymore.

"Those who want to stay relevant have to work very hard" — at work and during off-hours, says Weinman, who is a member of the ISACA Leadership Development Committee. ISACA is an IT professional association that, among other things, provides security certifications.

services, concurs. "In today's marketplace, if you have good references and a strong technical skill set and can communicate how you'll provide ROI, four jobs will be waiting for you," he says.

What amazes, and to some degree frustrates, Cullen are those instances when clients choose not to hire a job applicant because they can't check every box on their wish lists. "We're seeing this huge pent-up demand, and the pool of labor isn't growing. And yet, what's perplexing is just how specific hiring managers still are," he says. "They want this skill, that particular work on the network side, certifications, this many years of experience. Companies are not willing to take a risk. Nobody's jumping out the window to hire the average employee."

Weinman blames the Great Recession for starting IT down the path that led to the skills gap, while cautioning that an improved economy won't much ease the crunch for many workers.

"Companies are getting leaner and leaner. Starting in 2008, they downsized and streamlined, and they haven't replaced those positions," he observes. "If you're the hiring director of one of these very lean teams, you want only A+ workers. In the past, someone could get away with being a solid middle-of-the-road employee. Not anymore."

Charles Williams sees the situation from both sides. As manager of data systems at Georgia System Operations, an electric utility in Tucker, Ga., he wants and expects the people who report to him (currently there are seven) to keep their skills up to date. At the same time, he acknowledges that he is challenged to keep his own knowledge fresh when day-to-day duties take priority over opportunities to investigate up-and-coming technologies.

"In a way, it's natural for a manager to develop a technical skills gap. We're not able to sit down and play with things the way our employees might," he says. And that worries him. "I feel like I need

> **Companies are not willing to take a risk. Nobody's jumping out the window to hire the average employee.**
>
> JACK CULLEN, PRESIDENT, MODIS

▼EXHIBIT 3-4   Illustration of an Organization Chart

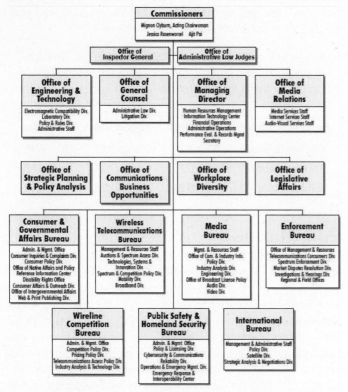

Source: "FCC Organizational Chart," *FCC*, Federal Communications Commission, May 2013, Web, 24 May 2013.

▼EXHIBIT 3-5   Illustration of a Flowchart

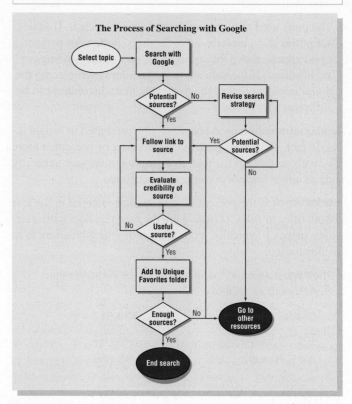

Source: Primary.

▼EXHIBIT 3-6   Illustration of a Gantt Chart

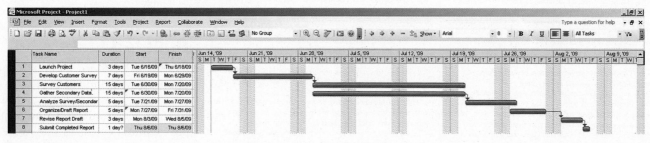

Source: Primary.

information presented in reports. Perhaps the most common of these is the **organization chart** (see Exhibit 3-4). This type of chart shows the hierarchy of levels and positions in an organization. A **flowchart** (see Exhibit 3-5), as the word implies, shows the sequence of activities in a process. Flowcharts use specific designs and symbols to show process paths. A variation of the organization and flowchart is the **decision tree**. This chart helps one follow a path to an appropriate decision. **Gantt charts** are visual presentations that show planning and scheduling activities (see Exhibit 3-6). You can easily construct these charts in a variety of applications.

LO 3-4   Construct and use visuals such as bar charts, pie charts, line charts, scatter diagrams, and maps.

CONSTRUCTING CHARTS, GRAPHS, AND OTHER VISUALS

Visuals built with raw data include bar, pie, and line charts and all their variations and combinations. Illustrations include maps, diagrams, drawings, photos, and cartoons.

Bar and Column Charts

Simple bar and **column charts** compare differences in quantities using differences in the lengths of the bars to represent those quantities. You should use them primarily to show comparisons of qualities at a moment in time.

As shown in Exhibit 3-7, the main parts of the bar chart are the bars and the grid (the field on which the bars are placed). The bars, which may be arranged horizontally or vertically (then called a column chart), should be of equal width. You should identify each bar or column, usually with a caption at the left or bottom. The grid (field) on which the bars are placed is usually needed to show the magnitudes of the bars, and the units (e.g., dollars, pounds, miles) are identified by the scale caption below. It is often a good idea to include the numerical value represented by each bar for easy and precise comprehension, as shown in Exhibits 3-7 and 3-8.

Source: "Earnings Reports by Day This Season," *Think B.I.G,* Bespoke Investment Group, 9 Jan. 2012, Web, 24 May 2013. http://www.bespokeinvest.com/thinkbig/2012/1/9/earnings-reports-by-day-this-season.html

When you need to compare quantities of two or three different values in one chart, you can use a **clustered** (or **multiple**) **bar chart**. Cross-hatching, colors, or other formatting on the bars distinguish the different kinds of information (see Exhibit 3-8). Somewhere within the chart, a **legend** explains what the different bars mean. Because clustered bar charts can become cluttered, you should usually limit comparisons to three to five kinds of information in one of them.

When you need to show plus and minus differences, you can use **bilateral column charts**. The columns of these charts begin at a central point of reference and may go either up or down, as illustrated in Exhibit 3-9. Bar titles appear either within, above, or below the bars, depending on which placement works best. Bilateral column charts are especially good for showing percentage changes, but you may use them for any series that includes plus and minus quantities.

▼ **EXHIBIT 3-8**   Illustration of a Clustered Bar Chart

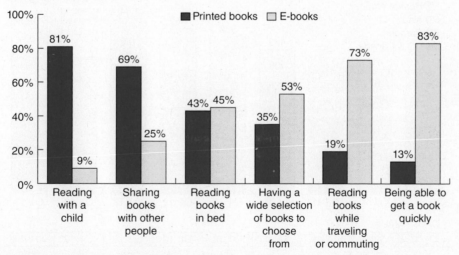

Source: Lee Rainie, et al., "The Rise of e-Reading," *Pew Internet & American Life Project,* Pew Research Center, 4 Apr. 2012, Web, 24 May 2013.

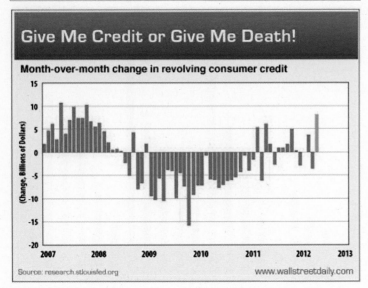

Source: Louis Basenese, "Friday Charts: Beer, Earnings, Recessions and Peak Oil Nonsense," *Wall St. Daily*, Wall Street Daily, LLC, 13 July 2012, Web, 24 May 2013. Reprinted with permission of Wall Street Daily.

If you need to compare subdivisions of columns, you can use a **stacked (subdivided) column chart.** As shown in Exhibit 3-10, such a chart divides each column into its parts. It distinguishes these parts with color, cross-hatching, or other formatting; and it explains these differences in a legend. Subdivided columns may be difficult for your reader to interpret since both the beginning and ending points need to be found.

Then the reader has to subtract to find the size of the column component. Clustered bar charts or pie charts avoid this possibility for error.

Another feature that can lead to reader error in interpreting bar and column chart data is the use of three dimensions when only two variables are being compared. Therefore, unless more than two variables are used, choosing the two-dimensional presentation over the three-dimensional form is usually better. (Exhibit 3-20 illustrates the appropriate use of a three-dimensional visual to compare three dimensions.)

A special form of stacked (subdivided) column chart is used to compare the subdivisions of percentages. In this form, all the bars are equal in length because each represents 100 percent. Only the subdivisions within the bars vary. The objective of this form is to compare differences in how wholes are divided. The component parts may be labeled, as in Exhibit 3-11, or explained in a legend.

Pictographs

A **pictograph** is a bar or column chart that uses bars made of pictures. The pictures are typically drawings of the items being compared. For example, the number of senators in Exhibit 3-12 is represented by the image of a single person for each senator instead of by ordinary bars.

In constructing a pictograph, you should follow the procedures you used in constructing bar and column charts. In addition, you must make all the picture units equal in size. The human eye cannot accurately compare geometric designs that vary in

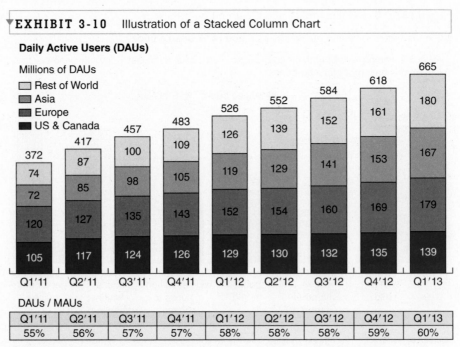

Daily Active Users (DAUs)

DAUs / MAUs	Q1'11	Q2'11	Q3'11	Q4'11	Q1'12	Q2'12	Q3'12	Q4'12	Q1'13
	55%	56%	57%	57%	58%	58%	58%	59%	60%

Source: CNET News, "Facebook Quarterly Earnings Slides," *Scribd.*, Scribd Inc., n.d., Web, 24 May 2013.

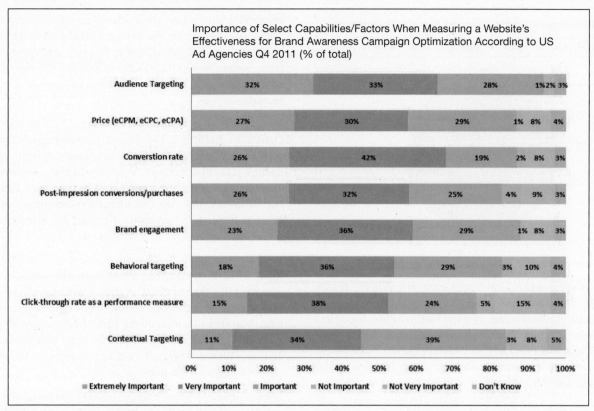

Importance of Select Capabilities/Factors When Measuring a Website's Effectiveness for Brand Awareness Campaign Optimization According to US Ad Agencies Q4 2011 (% of total)

Source: Data from Maxifier, "Optimization Research," *eMarketer,* eMarketer, Inc., 14 Feb. 2012, Web, 24 May 2013. Reprinted with permission of eMarketer, Inc.

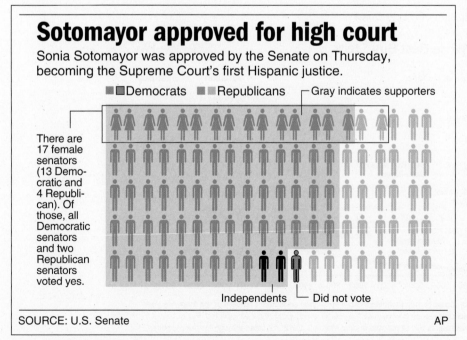

Sotomayor approved for high court

Sonia Sotomayor was approved by the Senate on Thursday, becoming the Supreme Court's first Hispanic justice.

■■Democrats   ■■Republicans   ─Gray indicates supporters

There are 17 female senators (13 Democratic and 4 Republican). Of those, all Democratic senators and two Republican senators voted yes.

Independents   └─ Did not vote

SOURCE: U.S. Senate   AP

Source: D. Morris, "Sotomayor Approved for High Court," *The Washington Post,* The Washington Post Company, 7 Aug. 2009, Web, 24 May 2013. Reprinted with permission of AP Images.

more than one dimension, so show differences by varying the number, not the size, of the picture units. Also, be sure you select pictures or symbols that fit the information to be illustrated. In comparing the cruise lines of the world, for example, you might use ships. In comparing computers used in the world's major countries, you might use computers. The meaning of the drawings you use must be immediately clear to the readers.

Pie Charts

The most frequently used chart in comparing the subdivisions of wholes is the **pie chart** (see Exhibit 3-13). As the name implies, pie charts show the whole of the information being studied as a pie (circle) and the parts of this whole as slices of the pie. The slices may be distinguished by labeling and color or cross-hatching. A single slice can be emphasized by pulling it out from the pie or enlarging it. Because it is hard to judge the values of the slices with the naked eye, it is good to include the percentage values within or near

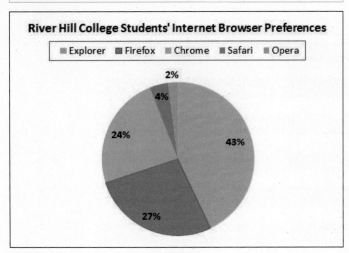

River Hill College Students' Internet Browser Preferences

■ Explorer ■ Firefox ■ Chrome ■ Safari ■ Opera

2%
4%
24%
43%
27%

Source: Primary.

each slice. Also, placing a label near each slice lets the reader more quickly understand the items being compared than using a legend to identify components. A good rule to follow for ordering the pieces of the pie is to begin by slicing the largest piece at the 12 o'clock position and then ordering the pieces largest to smallest; however, you should order the pieces in the way that will make the most visual sense to your audience.

Line Charts

Line charts are useful for showing changes of information over time. For example, changes in prices, sales totals, employment, or production over a period of years can be shown well in a line chart.

In constructing a line chart, draw the information to be illustrated as a continuous line on a grid that is scaled to show time changes from left to right (the X-axis) and quantity changes from bottom to top (the Y-axis). You should clearly mark the scale values and the time periods. They should be in equal increments.

You also may compare two or more series on the same line chart (see Exhibit 3-14). In such a comparison, you should clearly distinguish the lines by color or form (e.g., dots, dashes, dots and dashes). You should also label them on the chart or with a legend somewhere in the chart. But the number of series that you can effectively compare on one line chart is limited. As a practical guide, the maximum number is five.

It is also possible to show parts of a series using an **area** chart. Such a chart, though, can show only one series. You should construct this type of chart, as shown in Exhibit 3-15, with a top line representing the total of the series. Then, starting from the base, you should cumulate the parts, beginning with the largest and ending with the smallest or beginning with the smallest and ending with the largest. You may use cross-hatching or coloring to distinguish the parts.

Line charts that show a range of data for particular times are called *variance* or *hi-lo* charts. Some variance charts show high and low points as well as the mean, median, or mode. When used to chart daily stock prices, they typically include closing price in addition to the high and low. When you use

Amazon Sales Surge While Best Buy Stalls   ◙ReadWriteWeb

Annual* revenue, billions of dollars.   — Amazon   — Best Buy

Circuit City closes.
Best Buy opens first China store.
Best Buy acquires Geek Squad.
BestBuy.com launches.
Kindle launches.
Best Buy reaches 250 stores.
Amazon Web Services launches.
Amazon gets 1-click patent.
Amazon Prime launches.

$60
$50
$40
$30
$20
$10
$0

1995 1996 1997 1998 1999 2000 2001 2002 2003 2004 2005 2006 2007 2008 2009 2010 2011

*This isn't an exact, apples-to-apples comparison: Best Buy's fiscal year ends near February of the following year. For simplicity's sake, BBY's annual revenue is shown as the year in which *most* of it was reported. For example, BBY's "2011" revenue really represents the 12 months ending March 3, 2012. It's close enough for our purposes.

Source: Dan Frommer, "Amazon vs. Best Buy: A Tale of Two Retailers," *ReadWriteWeb*, SAY Media, Inc., 18 Apr. 2012, Web, 24 May 2013. *http://readwrite.com/2012/04/18/amazon_vs_best_buy_a_tale_of_two_retailers#awesm=~obtcV9k1klQipt*

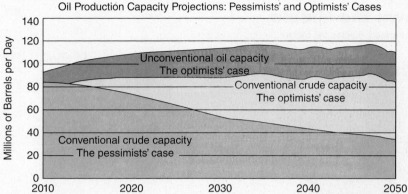

Oil Production Capacity Projections: Pessimists' and Optimists' Cases

Source: Adapted with special permission from "The Argument for and Against Oil Abundance," *Bloomberg BusinessWeek,* 18 January 2010, p. 48.

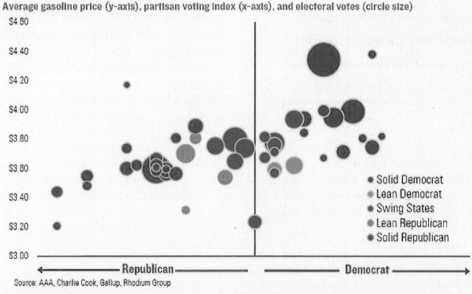

The Bluer the State, the Higher the Price
Average gasoline price (y-axis), partisan voting index (x-axis), and electoral votes (circle size)

Source: AAA, Charlie Cook, Gallup, Rhodium Group

Source: Trevor Houser, "Gasoline Prices and Electoral Politics in the Age of Unconventional Oil," *Notes,* Rhodium Group, LLC, 7 Mar. 2012, Web, 24 May 2013. *http://rhg.com/notes/gasoline-prices-and-electoral-politics-in-the-age-of-unconventional-oil.* Reprinted with permission.

points other than high and low, be sure to make it clear what these points are.

Scatter Diagrams

Scatter diagrams are often considered another variation of the line chart. Although they do use X- and Y-axes to plot paired values, the points stand alone without a line drawn through them. For example, a writer might use a scatter diagram in a report on digital cameras to plot values for price and resolution of several cameras. While clustering the points allows users to validate hunches about cause and effect, they can only be interpreted for correlation—the direction and strength

relationships. The points can reveal positive, negative, or no relationships. Additionally, by examining the tightness of the points, the user can see the strength of the relationship. The closer the points are to a straight line, the stronger the relationship. In Exhibit 3-16, the paired values are gas prices and political party affiliation.

Maps

You also may use maps to communicate quantitative as well as physical (or geographic) information. **Statistical maps** are useful primarily when quantitative information is to be compared by geographic areas. On such maps, the geographic areas

EXHIBIT 3-17   Illustration of a Map (Quantitative)

Employed Down the Middle
Rural unemployment in January divides the nation

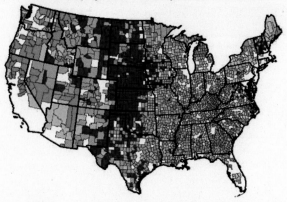

Unemployment in rural and exurban counties in January 2012
National rate was 8.8%; rate in rural counties was 9.1%
(Blank areas above are metro regions.)

Low Unemployment. 6% or below (531 counties)

Below Average. 6.1% to 8.8% (824 counties)

Above Average. 8.9% to 10.9% (613 counties)

High Unemployment. Above 11% (590 counties)

Source: "International, U.S. and Local Economic Indicators for May 2012," Scott Financial Group, 29 Apr. 2012, Web, 24 May 2013. *http://www.samuelscottfg.com/domestic-and-international-economic-indicators-for-may-2012*

are clearly outlined, and formatting techniques are used to show the differences between areas (see Exhibit 3-17). Statistical maps are particularly useful in illustrating and analyzing complex data. **Physical or geographic** maps (see Exhibit 3-18) can show distributions as well as specific locations. Of the numerous formatting techniques available to you, these are the most common:

- Showing different areas with color, shading, or cross-hatching (see Exhibit 3-17). Maps using this technique must have a legend to explain the quantitative meanings of the various colors or cross-hatchings.

- Placing visuals, symbols, or clip art within each geographic area to depict the quantity for that area or geographic location.

- Placing the quantities in numerical form within each geographic area.

Combination Charts

Combination charts often serve readers extremely well by allowing them to see relationships of different kinds of data. The

EXHIBIT 3-18   Illustration of a Map (Physical)

Source: "Google Maps," *Google.com.* Google, 28 May 2013, Web, 28 May 2013.

example in Exhibit 3-19 shows the reader the price of stock over time (the trend), the volume of sales over time (comparisons), and the MACD (an indicator of trends in stock performance). It allows the reader to detect whether the change in volume affects the price of the stock. This kind of information would be difficult to get from raw data alone.

Three-Dimensional Visuals

Earlier we said that **three-dimensional graphs** are generally undesirable. However, we have mostly been referring to the three-dimensional presentation of visuals with two variables. But when you actually have three or more variables, presenting them in three dimensions is an option if doing so will help your readers see the data from multiple perspectives and gain additional information. In fact, Francis Crick, who won a Nobel prize for discovering the structure of DNA, once revealed that he and his collaborators understood the configuration of DNA only when they took a sheet of paper, cut it, and twisted it. Today we have sophisticated statistics, visuals, and data-mining tools to help us see our data from multiple perspectives.

EXHIBIT 3-19   Illustration of a Combination Chart

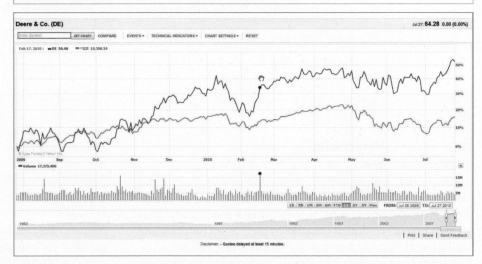

Source: Reprinted with permission from Yahoo! Inc. © 2013 Yahoo Inc. YAHOO! and the YAHOO! Logo are trademarks of Yahoo! Inc.

from the tech desk

Making the Most of Excel

Microsoft recently released Excel 2013. Though earlier versions of Excel were powerful, Excel 2013 presents even more exciting options for data visualization—and the best part is that you don't have to know a lot about Excel to use them. Among other things, Excel 2013 suggests visuals appropriate for your data, provides numerous templates to ensure your data look sharp, and offers a Quick Analysis feature, which, as its name indicates, lets you quickly calculate data and create visuals. A sample of a template and color palette options, and the Quick Analysis feature, appear here.

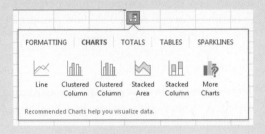

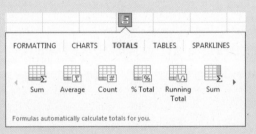

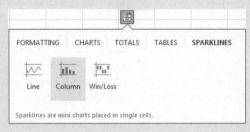

These three-dimensional tools are making their way from science labs into business settings. Thanks to the increased use of data-gathering tools, businesses large and small are collecting and attempting to analyze extremely large amounts of detailed data. They are analyzing not only their own data but also data on their competitors. And advances in hardware, software, and Web-based applications are making it easier to visually represent both quantitative and qualitative data.

Although 3D visuals help writers display the results of their data analysis, they change how readers look at information and may take some time to get used to. These tools enable users both to see data from new perspectives and to interact with them. They allow users to free themselves from two dimensions and give them ways to stretch their insights and see new possibilities.

Exhibit 3-20 shows a three-dimensional visual plot of factors identified as the major ones consumers use when deciding which slate computer to purchase. Five products are plotted on three variables: cost, battery life, and slate weight. This visual could help a company identify its major competitors and help consumers identify those products that are best suited to their needs. The more products or data that are plotted, the more valuable the graph is at helping the reader extract meaning. If these data had been displayed on a two-dimensional graph, the lines would have overlapped too much to be distinguishable, thus requiring you to accompany this visual with a table, enable the reader to rotate the visual, or both in order to make the data clear.

In deciding whether to use a three-dimensional representation or a two-dimensional one, you need to consider your audience, the context, and the goal of your communication. Overall, multidimensional presentation on paper is difficult; multiple representations can be made from separate two-dimensional views, but not always effectively. Moreover, if the 3D visual is being presented online or digitally where the reader can rotate it to see perspectives, it is likely to be much more effective.

Photographs

Cameras are everywhere today, enabling anyone to capture the images he or she needs. And royalty-free photos intended for commercial use are readily available on the Internet, too. Photos can serve useful communication purposes. They can be used to document events as well as show products, processes, or services. Exhibit 3-21 illustrates how a photo creates a message. What does the context created by the split image (half fall/half summer) suggest to you?

Today photos, like data-generated visuals, can easily be manipulated. A writer's job is to use them ethically, including getting permission when needed.

▼**EXHIBIT 3-21**   Illustration of a Photo

▼**EXHIBIT 3-20**   Illustration of a Three-Dimensional Visual

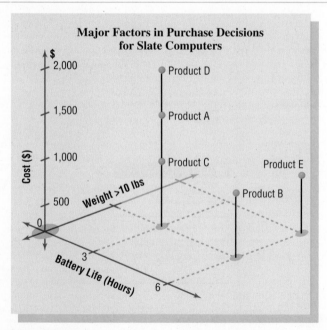

Major Factors in Purchase Decisions for Slate Computers

Source: Primary.

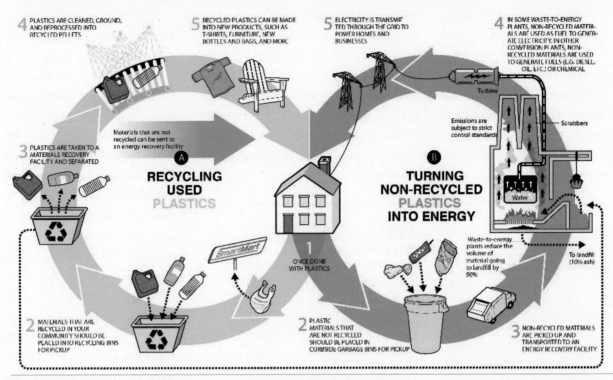

Source: From "Diverting Plastics from Landfills: A Two-Pronged Approach," *Plastics Make It Possible*, American Chemistry Council, 21 Sept. 2012, Web, 24 May 2013. http://plasticsmakeitpossible.com/2011/09/diverting-plastics-from-landfills-a-two-pronged-approach/

Other Visuals

The types of visuals discussed thus far are the ones most commonly used in business. Other types also may be helpful. **Diagrams** (see Exhibit 3-22) and drawings may help simplify a complicated explanation or description. **Icons** are another useful type of visual. You can create new icons, or you can select one from an existing body of icons with easily recognized meanings, such as ⊘ . Even carefully selected **cartoons** can be used effectively. **Video clips** and **animation** are now used in many electronic documents. For all practical purposes, any visual is acceptable as long as it helps communicate the intended story.

LO 3-5   Avoid common errors and ethical problems when constructing and using visuals.

Visual Integrity

In writing a business document, you are ethically bound to present data and visuals in ways that enable readers to interpret them easily and accurately. By being aware of some of the common errors made in presenting visuals, you learn how to avoid them and how to spot them in other documents. Keep in mind that any errors—deliberate or not—compromise your credibility, casting doubt on the document as well as on other work you

have completed. Therefore, writers need to ensure that visuals accurately and honestly represent the data they contain.

avoiding errors in graphing data   Common graphing errors are errors of scale and errors of format. Another category of error is inaccurate or misleading presentation of context.

Errors of scale occur whenever the dimensions from left to right (X-axis) or bottom to top (Y-axis) are unequal. Three sources of scale errors include no uniform scale size, scale distortion, and violating the zero beginning.

No uniform scale size occurs when intervals in data points on the X- or Y-axis are not consistent. Note in the graphs below how the unequal intervals on the Y-axis in the graph on the right create a different presentation of the data.

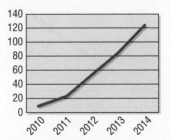

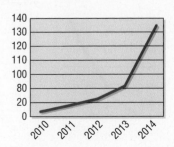

communication matters

Avoiding Chartjunk

In 1983 Edward Tufte, a pioneer and leading expert in data visualization, invented the term *chartjunk* to refer to any elements in a visual that are either unnecessary or irrelevant to the reader's understanding of the data or that impede or distract from a reader's ability to understand the data.

Technology enables business communicators to incorporate many features (e.g., colors, lines, images) into their visuals—some that will enhance a visual's message and some that will not. Charley Kyd from the ExcelUsers blog explains that the distinction between visual elements that enhance a visual versus those that are chartjunk is the extent to which visual elements make the reader work to understand the information.

To illustrate, Kyd presents variations on the same visual about Juicy sales from the April 24, 2012, edition of *The Wall Street Journal*. The one on the left contains chartjunk; the one on the right is Kyd's rendering of the visual minus the chartjunk.

Kyd acknowledges that neither visual makes the data impossible to understand; what makes the elements chartjunk in the visual on the left is that the visual elements "act as noise where your readers need silence." That is, the shape of the bite in the apple appears to alter the shape of the bars in the chart, making the reader look twice to see that the bars, indeed, retain their shape. Is this a lot of work for the reader? No, but why make your reader work at all? When you create visuals, just as when you create text, all elements should enhance your reader's understanding of your message.

Source: Reprinted with permission of Charley Kyd, ExcelUser.com.

Sources: Charley Kyd, "Oh, No! Chart Junk from *The Wall Street Journal*," *ExcelUser Blog: Insight for Business Users of Microsoft Excel*, WordPress 12 May 2012, Web, 12 July 2012; Jessica E. Vascellaro and Ian Sherr, "Apple Rides iPhone Frenzy: Quarterly Profit Nearly Doubles as Tech Giant Taps China, New Markets," *The Wall Street Journal*, Dow Jones & Company, Inc., 24 Apr. 2012, Web, 24 May 2013.

Scale distortion occurs when a visual is stretched excessively horizontally or vertically to change the meaning it conveys to the reader. Expanding a scale can change the appearance of the line. For example, if the values on a chart are plotted one-half unit apart, changes appear to be much more dramatic. Determining the distances that present the most accurate picture is a matter of judgment. Notice the different looks of the visuals at the bottom of the left column when they are stretched vertically and horizontally.

Finally, another type of scale error is the **missing zero beginning** of the series. For accuracy you should begin the scale at zero. But when all the information shown in the chart has high values, it is awkward to show the entire scale from zero to the highest value. For example, if the quantities compared range from 1,320 to 1,350 and the chart shows the entire area from zero to 1,350, the line showing these quantities would be almost straight and very high on the chart. Your solution in this case is not to begin the scale at a high number (say 1,300),

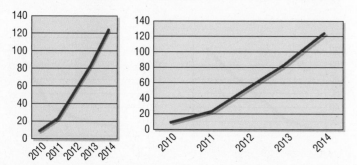

which would distort the information, but to begin at zero and show a scale break. Realize, however, that while this makes the differences easier to see, it does exaggerate the differences. You can see this effect here.

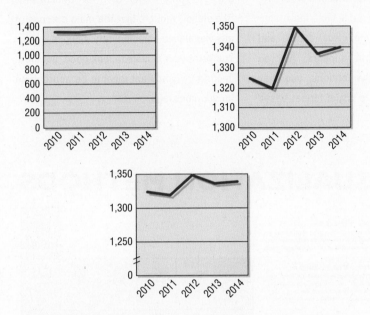

Computer applications enable writers to easily create a wide variety of visuals from small to huge data sets.

Occasionally, though, a writer needs to use his or her judgment when the guides for starting at the zero point are not practical. For example Exhibit 3-23 compares race times of three runners, all of whom have times that cluster in the 14–17 second range. Beginning the time scale at 0 would hinder the reader's interpretation of the information. In this context, presenting the data at a scale that begins at 14 efficiently present the data while maintaining integrity.

Errors of format come in a wide variety. Some of the more common ones are choice of wrong chart type, distracting use of grids and shading, misuse of typeface, and

▼**EXHIBIT 3-23**   Illustration of a Line Graph That Considers Context

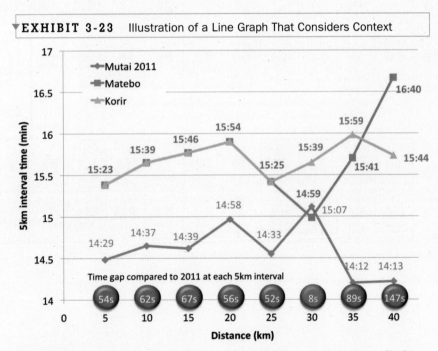

Source: From Ross Tucker and Jonathan Dugas, "Boston Strikes Back: The Boston 2012 Meltdown, *The Science of Sport,* blog, 16 April 2012, *http://www.sportsscientists.com/2012/04/boston-strikes-back-boston-2012.html*

communication matters

The Periodic Table of Visualization Methods

This chapter presents many of the most common options for using visuals to present your text and data. The Periodic Table of Visualization Methods provides an array of many other possibilities to help you choose the best visual as well. To see your options, just visit the Web site www.visual-literacy.org/periodic_table/periodic_table.html, move your mouse over any of the visual types, and view the example that appears. As you view the many creative options, you may be tempted to choose a visual format based on its novelty rather than its functionality. Remember, though, that the main purpose of any visual is to communicate. Choose wisely. You want your visuals to look good, but more importantly you want them to be appropriate for your message.

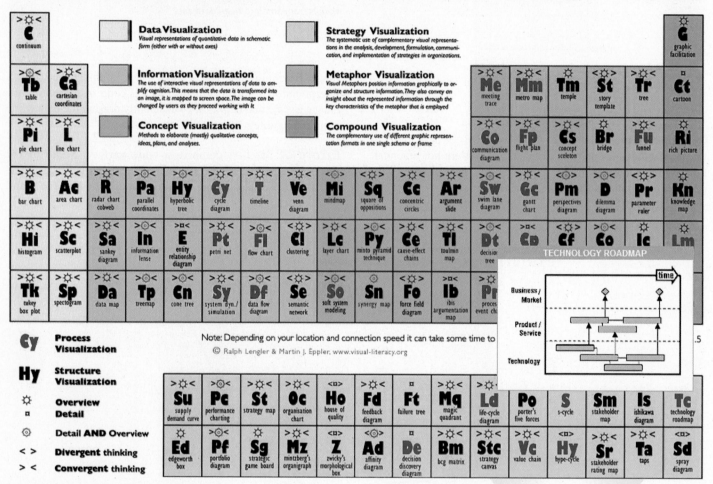

Source: Ralph Lengler and Martin J. Eppler, www.visual-literacy.org. Reprinted with permission of Professor Dr. Martin Eppler.

communication matters

Practicing Visual Ethics

As you have learned in this chapter, visuals can serve several useful purposes for the business writer. However, the writer needs to be accountable in using visuals to present images that in the eye and mind of the reader communicate accurately and completely. To do this, the careful writer pays attention to both the design and content of the visual. These are particularly important, for readers often skim text but read the visuals. Research shows that people remember images much better and longer than text.

Donna Kienzler offers the following guides to help you evaluate the visuals you use:

• Does the visual's design create accurate expectations?

• Does the story told match the data?

• Is the implied message congruent with the actual message?

• Will the impact of the visual on your audience be appropriate?

• Does the visual convey all critical information free of distortion?

• Are the data depicted accurately?

We would also add that ethical use of visuals requires you to cite the source of any data that you did not produce. In addition, you must get permission to use visuals that you do not create or own—including those found through Web searches.

Source: Adapted from Donna S. Kienzler, "Visual Ethics," *Journal of Business Communication* 34 (1997): 171–87, print.

problems with labels. For example, if a company used pie charts to compare expenses from one year to the next, readers might be tempted to draw conclusions that would be inappropriate because, although the pies would both represent 100 percent of the expenses, the size of the business and the expenses may have grown or shrunk drastically in a year's time. If one piece of the pie is colored or shaded in such a way as to make it stand out from the others, it could mislead readers. And, of course, small type or unlabeled, inconsistently labeled, or inappropriately labeled visuals confuse readers.

Another ethical challenge is accounting for the context. For example, as we have discussed, the number and size of visuals should be proportionate to the importance of the topic and appropriate for the emphasis a topic deserves.

avoiding other ethical problems There are other ethical issues to consider. Writers need to be careful when choosing the information to represent and the visual elements to represent it. One area writers need to watch is appropriate selection of the contents. Are people or things over- or underrepresented? Are the numbers of men and women appropriate for the context? Are their ages appropriate? Is ethnicity represented appropriately? Have colors been used appropriately and not to evoke or manipulate emotions? Writers need to carefully select and design visuals to maintain integrity.

LO 3-6 Place and interpret visuals effectively.

PLACING AND INTERPRETING THE VISUALS

For the best communication effect, you should place each visual near the place where it is discussed. Exactly where on the page you should place it, however, should be determined by its size. If the visual is small, you should place it within the text that discusses it. If it is a full page, you should place it on the page following the first reference to the information it covers.

Some writers like to place all visuals at the end of a document, usually in the appendix. This arrangement may save time in preparing the document, but it makes the readers' task more difficult because they have to flip through pages every time they want to see a visual. Therefore, place visuals where they are most helpful to the reader.

That said, sometimes you may have a visual that is necessary for completeness but is not discussed in the document (e.g., a print-out of an online survey or list of questions from an interview cited in a report). Or you may have summary charts or tables that apply to the entire document but to no specific place in it. When such visuals are appropriate, you should place them in an appendix, and you should refer to the appendix at an appropriate point in the document.

Visuals communicate most effectively when the readers view them at the right point in their reading. Thus, you should tell the readers when to look at a visual and what to see. Of the many wordings used for this purpose, these are the most common:

As Figure 4 shows, ...

. . . , indicated in Figure 4, ...

. . . , as a glance at Figure 4 reveals, ...

. . . (see Figure 4) . . . .

If your visual is carrying the primary message, as in a detailed table, you can just refer the reader to the information in the visual, as in "As Table 1 illustrates, our increased sales over the last three years. . . ." No further explanation or discussion may be necessary.

However, sometimes the visual is part of a more detailed discussion or presentation of your data. In these cases, you will start with a summary statement that reveals the big picture. If you were discussing Exhibit 3-17 (page 54), you might say, "As Exhibit 3-17 shows, areas of the country with the lowest unemployment rate are in the High, Central, and Southern Plains states." After presenting the figure, you would call your readers' attention to more specific points in the visual. Then you would give the exception to the general trend, if there is one.

Your readers will appreciate well-chosen, well-designed, and well-explained visuals, and you will achieve powerful communication results.

learn more about visual communication!

- What are some tips and tricks for visualizing data?
- What technologies can help you learn to create visuals?
- Where can you find a tutorial for using data visualization software such as Excel, PowerPoint, or Visio?

Scan the QR code with your smartphone or use your Web browser to find out at www.mhhe.com/RentzM3e. Choose Chapter 3 > Bizcom Tools & Tips. While you're there, you can view a chapter summary, exercises, PPT slides, and more to use visuals effectively in your business messages.

www.mhhe.com/RentzM3e

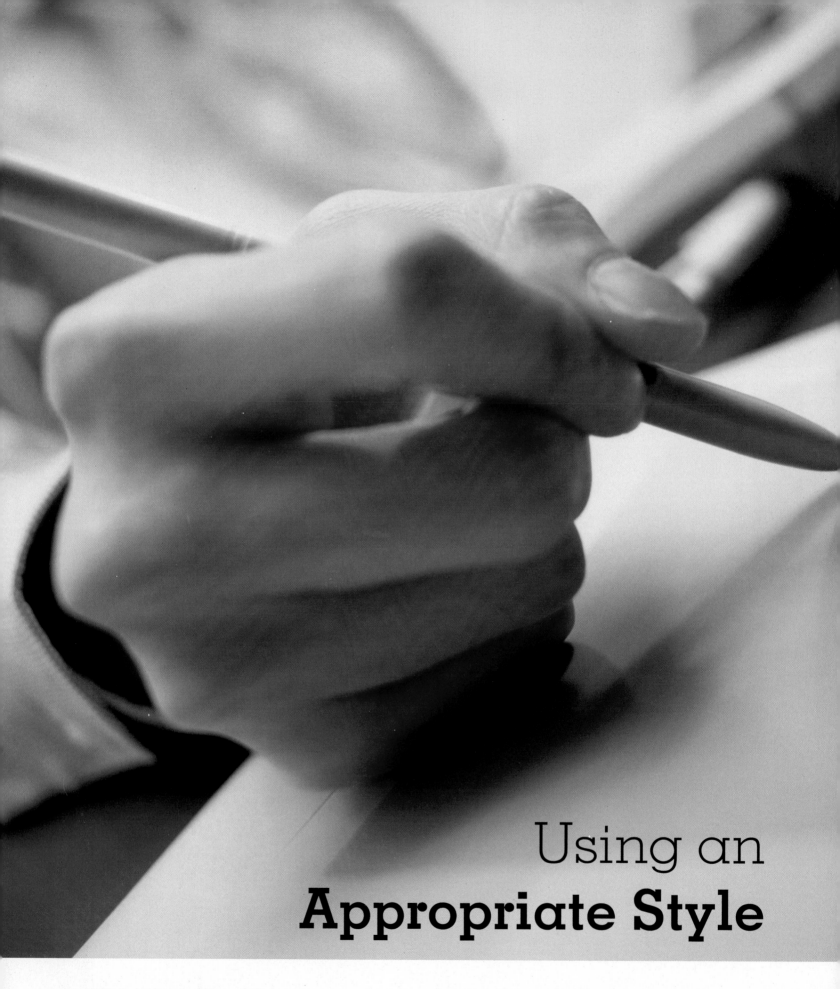

Using an
Appropriate Style

four

O nce you have analyzed your communication task, decided what kind of message you need to write, and planned your verbal and visual contents, you're ready to get down to the challenge of writing—putting one word, sentence, and paragraph after another to communicate what you want to say.

While each document you write will need to respond to the unique features of the situation, keeping in mind certain guidelines can help you make good writing choices. This chapter offers advice on selecting appropriate words, writing clear sentences and paragraphs, and achieving the desired effect with your readers. The goal is documents that communicate clearly, completely, efficiently, and engagingly. ∎

LEARNING OBJECTIVES

LO 4-1 Simplify writing by selecting familiar and short words.

LO 4-2 Use slang and popular clichés with caution.

LO 4-3 Use technical words and acronyms appropriately.

LO 4-4 Use concrete, specific words with the right shades of meaning.

LO 4-5 Avoid misusing similar words and use idioms correctly.

LO 4-6 Use active verbs.

LO 4-7 Use words that do not discriminate.

LO 4-8 Write short, clear sentences by limiting sentence content and economizing on words.

LO 4-9 Design sentences that give the right emphasis to content.

LO 4-10 Employ unity and good logic in writing effective sentences.

LO 4-11 Compose paragraphs that are short and unified, use topic sentences effectively, and communicate coherently.

LO 4-12 Use a conversational style that has the appropriate level of formality and eliminates "rubber stamps."

LO 4-13 Use the you-viewpoint to build goodwill.

LO 4-14 Employ positive language to achieve goodwill and other desired effects.

LO 4-15 Explain and use the elements of courtesy.

LO 4-16 Use the three major techniques for emphasizing the positive and de-emphasizing the negative.

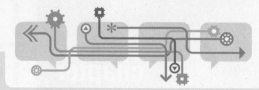
Writing with Clarity and Courtesy

This summer you're making some college money by working as a groundskeeper at a local hotel. Recently, you and the other hotel staff—including the rest of the maintenance crew and the housekeepers—received this written message from the new assistant manager:

> It has come to my attention that certain standards of quality are not being met by personnel in service positions. For successful operations, it is imperative that we adhere to the service guidelines for staff in each functional area, as set forth by corporate in the training materials that were reviewed during your orientation and onboarding. Be advised that there will be two mandatory training sessions on May 1, one at 4:00 p.m. for daytime employees and the other at 3:00 for the late shift, to reinforce your understanding of performance standards. Just as a machine cannot work properly if the pistons are all firing at different times, we cannot achieve our goals if individuals are setting their own criteria for how to serve our guests. I assume that I will see each and every one of you at a training session so that we may move forward with better comprehension of our roles and responsibilities.

You're no management pro, but you're sure this is a faulty message. In addition to being impersonal and insulting, it is difficult to understand, and the key point—that everyone will need to attend a training session—is buried in the middle of the message. How can you avoid writing like this? The advice in this chapter will help.

ADAPTING YOUR STYLE TO YOUR READERS

This chapter provides many tips for writing effectively. But the most important advice is this: Write in a style that is appropriate for the situation. As Chapter 1 explains, readers occupy particular organizational, professional, and personal contexts. They do not all have the same vocabulary, knowledge, or values. And you do not have the same relationship with all of them.

To communicate clearly and with the appropriate tone, you should learn everything possible about those with whom you wish to communicate and consider any prior correspondence with them. Then you should word and organize your message so that it is easy for them to understand it and respond favorably. Tailoring your message to your readers is not only strategically necessary; it is also a sign of consideration for their time and energy. Everyone benefits when your writing is reader focused.

SELECTING APPROPRIATE WORDS

Choosing the best words requires thinking about what you want to achieve and with whom. Do you and your readers know each other, or are you writing to strangers? How well educated are your readers, and what kinds of knowledge can they be presumed to have? How do you want your readers to feel about you, your company, and what you're writing about?

Adaptation requires asking yourself these question and more. The following sections will help you think through your choices.

LO 4-1 Simplify writing by selecting familiar and short words.

Use Familiar Words

To communicate clearly, you must use words that your readers are familiar with. Because words that are familiar to some people may be unfamiliar to others, you will need to decide which ones your readers will understand.

Tailor your style to your intended reader.

EXHIBIT 4-1 A Guide to Plain Language

A great resource for business writers—and a great model of the advice it gives—is the U.S. Securities and Exchange Commission's *A Plain English Handbook: How to Create Clear SEC Disclosure Documents* (available at www.sec.gov/pdf/handbook.pdf).

Here's what the handbook says about using familiar words:

Surround complex ideas with short, common words. For example, use *end* instead of *terminate, explain* rather than *elucidate*, and *use* instead of *utilize*. When a shorter, simpler synonym exists, use it.

In general, using familiar words means using the language that most of us use in everyday conversation. The U.S. government calls this kind of wording **plain language** and defines it as "communication your audience can understand the first time they read or hear it."[1] (Read more about plain language in Exhibit 4-1.) To write clearly, avoid the stiff, more difficult words that do not communicate precisely or quickly. For example, instead of using the less common word *endeavor*, use *try*. Prefer *do* to *perform, begin* to *initiate, find out* to *ascertain, stop* to *discontinue*, and *show* to *demonstrate*.

The suggestion to use familiar words does not rule out using more difficult words in some cases. You should use them whenever their meanings fit your purpose best and your readers understand them clearly. A good suggestion is to use the simplest words that carry the meaning without offending the readers' intelligence.

The following contrasting examples illustrate the communication advantages of familiar words.

Unfamiliar Words	Familiar Words
This machine has a tendency to develop excessive and unpleasant audio symptoms when operating at elevated temperatures.	This machine tends to get noisy when it runs hot.
Purchase of a new fleet is not actionable at this juncture.	Buying new trucks is not practical now.
We must leverage our core competencies to maximize our competitiveness.	Relying on what we do best will make us the most competitive.
Company operations for the preceding accounting period terminated with a deficit.	The company lost money last year.

Prefer Short Words

According to studies of readability, short words generally communicate better than long words. Part of the explanation is that short words tend to be familiar words. But there is another explanation: A heavy use of long words—even long words that are understood—creates an impression of difficulty that hinders communication.

communication matters

The Most Annoying Business Clichés

Blogger and writing expert Mary Cullen surveyed a wide range of clients from various industries to ask them "which overused phrases they would like to see banished." Here were their top replies:

1. At the end of the day
2. 30,000-foot view
3. Give 110%
4. Think outside of the box
5. FYI
6. 800-pound gorilla
7. Throw under the bus
8. My bad
9. Rightsizing
10. Reaching out
11. Low-hanging fruit
12. Paradigm shift
13. Take it offline
14. At this point in time
15. Synergy
16. Action item

Cullen adds one more that particularly bothers her: "Going forward." "Where else would we go?" she asks. "Backward?"

Source: "Top 25 Jargon and Gobbledygook Phrases 2011," *Instructional Solutions*, www.instructionalsolutions.com, 21 Feb. 2011, Web, 20 May 2013. Reprinted with permission.

This point is illustrated by the following examples. Notice how much easier to understand the short-word versions are.

Long Words	Short Words
The *proposed enhancement is under consideration*.	We are *considering your suggestion*.
They *acceded* to *the proposition to undertake a collaborative venture*.	They *agreed* to *work with* us.
Prior to *accelerating productive operation*, the supervisor inspected the machinery.	Before *speeding up* production, the supervisor inspected the machinery.
This *antiquated merchandising strategy* is *ineffectual in contemporary* business *operations*.	This *old sales* strategy *will not work* with *today's* customers.

LO 4-2 Use slang and popular clichés with caution.

Use Slang and Popular Clichés with Caution

At any given time in any society, certain slang expressions and clichés are in vogue. In the United States, for example, you might currently hear "for real" (or, "for *real*?"), "no worries," and "all over it" (in control), while other such expressions—"no way," "get out," and "bodacious"—now sound dated.

> Clichés catch on because they represent popular concepts, but with overuse, they begin to sound like a replacement for thinking.

Business clichés come and go as well. "State of the art," "cutting edge," and "world class" have given way to "moving forward," "thought leaders," and "best practices." More examples are listed in the Communication Matters feature on page 67.

It is true that business clichés can sometimes increase your credibility with other businesspeople and make you sound like "one of them." These expressions can also add color to your language and quickly convey an idea. But they can also work against you. As Harvard professor Marjorie Garber puts it, "Jargon marks the place where thinking has been."[2] Clichés catch on because they represent popular concepts, but with overuse, they begin to sound like a replacement for thinking. Plus, they run the risk of sounding out of date, and they can create problems in cross-cultural communication.

Google "annoying clichés" and you will see that popular expressions can soon become unpopular. Use them sparingly and only in communication with people who will understand and appreciate them.

LO 4-3   Use technical words and acronyms appropriately.

Use Technical Words and Acronyms Appropriately

Every field of business—accounting, information systems, finance, marketing, and management—has its technical language. As you work in your chosen field, you will learn its technical words and acronyms and use them often when communicating with others in your field, as well you should. Frequently, one such word will communicate a concept that would otherwise take dozens of words to describe. Moreover, specialized language can signal to other specialists that you are qualified to communicate on their level.

▼**EXHIBIT 4-2**   Alphabet Soup

Do you know what the following business abbreviations stand for?

| B2B | CRM | EPS | FIFO | SWOT |
| GAAP | ROI | SAAS | TQM | ERP |

Find out by typing the acronym followed by "stands for" in the Google search box or Chrome address bar—and then be sure your readers also know them before you use them without defining them.

However, communication can fail when you use technical terms with people outside your field. Remember that not everyone will have your specialized vocabulary. For example, does everyone know what an *annuity* is? How about *supply chain management*? Or *opportunity cost*? When possible, use your reader's language, and if you must use a technical term, explain it.

Initials (including acronyms) should be used with caution, too. While some initials, such as IBM, are widely recognized, others, such as SEO (search-engine optimization) and CRM (customer relationship management), are not. Exhibit 4-2 lists common business acronyms that may be unfamiliar to your readers. If you have any doubt that your reader will understand the initials, the best practice is to spell out the words the first time you use them and follow them with the initials. You may also need to go one step further and define them.

LO 4-4   Use concrete, specific words with the right shades of meaning.

Use Precise Language

Good business communicators use words that have sharp, clear meanings for their intended readers, as well as the right emotional tone. Choosing such words means being concrete, specific, and sensitive to shades of meaning.

Concrete is the opposite of **abstract**. While abstract words are vague, concrete words stand for things the reader can see, feel, taste, smell, or count.

The most concrete words are those that stand for things that exist in the real world, such as *chair, desk, computer, Bill Gates,* and the *Empire State Building*. Abstract nouns, on the other hand, cover broad, general meanings, as in these examples: *administration, negotiation, wealth, inconsistency, loyalty, compatibility, conservation, discrimination, incompetence,* and *communication*. It is difficult to visualize what these words stand for.

Notice how much clearer the concrete words are in the following examples:

Abstract	Concrete
A significant loss	A 53 percent loss
The leading company	First among 3,212 companies
The majority	62 percent
In the near future	By noon Thursday
Substantial amount	$3,517,000

from the tech desk

Grammar and Style Checkers Help Writers with Word Selection

Today's word processors can help you with grammar and style as well as with spelling. You can even tell some of them what issues to look for. In Word 2010 or 2013, go to File > Options > Proofing and then click the Settings button, indicated in the red box below. That will open an extensive list of options in the Grammar Settings box. Word will look for whatever options you check.

In the example shown here, a writer using Word 2010 checked "Passive sentences" to have Word catch instances of passive voice. Then, when she had written the first part of a document, she clicked Review > Spelling & Grammar to have Word check it. The review identified an instance of passive voice. To see how to correct it, she clicked the "Explain …" button, and an explanation appeared. (In Word 2013, the explanation will automatically appear.) Now she will need to decide if she wants to keep the sentence as is or change it to active voice.

If you use your word processor to help you identify potential problems, just remember that no computer application can tell you if your wording is appropriate; only you can determine that.

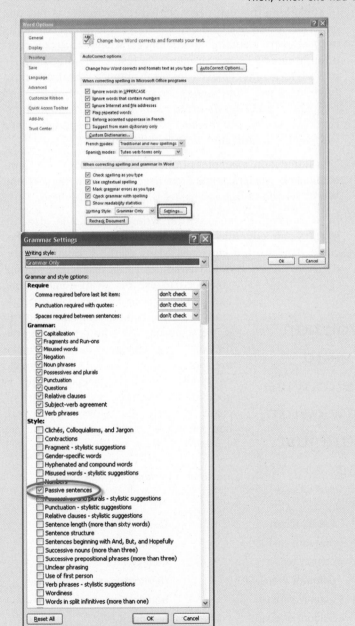

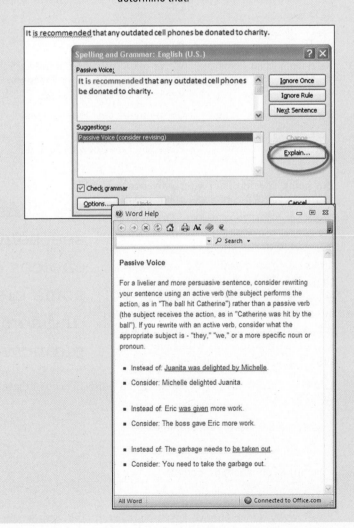

Like different clothing styles, connotations project different moods and meanings.

Closely related to being concrete is being **specific**. Even if you are talking about something intangible, you can still make your wording as precise as possible. These examples show what we mean:

Vague	Specific
We have a great company.	We've been voted one of the *Business Courier's* "Best Places to Work" for the last five years.
Our batteries are better.	Our batteries cost less and last longer.
Please respond soon.	Will you let me know by June 1?

But being specific isn't your only concern. You also need to elicit the right emotional response from your reader. Good writers possess a sensitivity to words' shades of meaning. Some words are forceful and some timid; some are positive and some negative; some are formal and some informal. Any given word can occupy a place on many different scales of tone and meaning. To achieve your communcation goals, you need to choose the words that will achieve the desired response from your intended readers.

Consider the different associations of the words in each of these groups:

- sell, market, advertise, promote
- money, funds, cash, finances
- improve, enhance, fix, correct
- concern, issue, problem, incident
- secretary, administrative assistant, support staff, coordinator

> " As you can see, some word choices are unwise, some are awkward, and some are just plain wrong. "

Though the words in each list share the same **denotation** (basic meaning), they vary widely in their **connotations** (their social and emotional associations). Being attentive to what different words imply will make you a more skillful and effective writer.

LO 4-5   Avoid misusing similar words and use idioms correctly.

Select Words for Appropriate Usage

Certain pairs of words in English can cause trouble for writers. For example, do you know the difference between *fewer* and *less*? *Fewer* is used with items that can be counted (e.g., customers), while *less* is used to refer to an overall quantity of something that can't be counted (e.g., traffic). *Affect* and *effect* are often used interchangeably. But *affect* is usually used as a verb meaning "to influence," whereas *effect* is most often used as a noun that means "a result" of something (in its verb form, *effect* means "to bring about"). Similarly, careful writers use *continual* to mean "repeated regularly and frequently" and *continuous* to mean "repeated without interruption." They write *farther* to express geographic distance and *further* to indicate "more, in addition."

You'll find more examples of often misused words in the Communication Matters feature on page 71 and in the "Wrong Word" section of Bonus Chapter B. Watch out for them.

In your effort to be a skillful writer, you should also use **idioms** correctly. Idioms are word combinations that have become standard in a language. Many of these seem arbitrary, but to avoid unclear or distracting writing, you need to use the word combinations that people expect. For example, there is really no logic behind using the word *up* in the sentence "Look up her name in the directory," but leaving it out would make the sentence nonsensical. Other idiomatic errors just sound bad. "Independent of" is good idiomatic usage; "independent from" is not. Similarly, you "agree to" a proposal, but you "agree with" a person. You are "careful about" a sensitive situation, but you are "careful with" your money. Here are some additional illustrations:

Faulty Idiom	Correct Idiom
authority about	authority on
comply to	comply with
different than	different from
equally as bad	equally bad
seldom or ever	seldom if ever
based off of	based on

communication matters

Words that sound alike (or nearly alike) but are spelled and used differently are called *homophones*. Some English homophones, like "beat" and "beet" or "allowed" and "aloud," are easy for native speakers to keep straight—but many can be tricky. Be sure you know when to use each word below:

Accept	To take in
Except	Other than
Affect	To influence
Effect	A result (or to bring about)
Aid	To help
Aide	An assistant
Cite	To quote or refer to
Site	A specific location

Compliment	Words of praise (or to give words of praise)
Complement	Something that completes (or to complete)
Discreet	Tactful
Discrete	Distinct
Elicit	To evoke
Illicit	Illegal
It's	A contraction meaning "it is"
Its	A possessive pronoun (e.g., "its handle")
Its'	[no such word]
Led	Guided
Lead	A type of metal (or to guide)
Passed	Moved ahead

Past	The time before the present (or beyond, as in "we went past the school")
Personal	Private or pertaining to a particular person
Personnel	Employees or staff
Principle	A rule or standard
Principal	The highest ranking person; main or primary
Their	A possessive pronoun (e.g., "their employees")
There	Beginning of a clause or sentence (e.g., "There are …")
They're	A contraction meaning "they are"
Your	A possessive pronoun (e.g., "your application")
You're	A contraction meaning "you are"

As you can see, some word choices are unwise, some are awkward, and some are just plain wrong. If you are unsure which word you need or would have the best effect, consult a dictionary.

LO 4-6 Use active verbs.

Prefer Active Verbs

Of all parts of speech, verbs do the most to make your writing interesting and lively, and for good reason: They contain the action of the sentence.

But not all verbs add vigor to your writing. Overuse of the verb "to be" and passive voice can sap the energy from your sentences. To see the difference between writing that relies heavily on forms of "to be" and writing that uses active verbs, compare the following two passages (the forms of "to be" and their replacements are *italicized*):

Version A (weak verbs)

There *are* over 300 customers served by our help desk each day. The help desk personnel's main tasks *are* to answer questions, solve problems, and educate the callers about the software. Without their expert work, our customer satisfaction ratings *would be* much lower than they *are*.

Version B (strong verbs)

Our help desk personnel *serve* over 300 customers each day. They *answer* questions, *solve* problems, and *educate* the users about the software. Without their expert work, our customer satisfaction ratings *would drop* significantly.

As these examples show, using active verbs adds impact to your writing, and it usually saves words as well.

In addition to minimizing your use of "to be" verbs, you can make your verbs more lively by using **active voice**. As the Communication Matters feature on page 72 explains, a sentence with a verb that can take a direct object (the recipient of the action) can be written either in a direct (active) pattern or an indirect (passive) pattern. Here are some examples:

Passive	Active
The results were reported in our July 9 letter.	We reported the results in our July 9 letter.
The new process is believed to be superior by the investigators.	The investigators believe that the new process is superior.
The policy was enforced by the committee.	The committee enforced the policy.
The office will be inspected on Tuesday.	Mr. Hall will inspect the office on Tuesday.
It is desired by the director that this problem be brought before the board.	The director wants the secretary to bring this problem before the board.

communication matters

Everything You Wanted to Know about Active and Passive Voice

Students are often confused by the terms *active voice* and *passive voice*. Here's the lowdown:

Broadly speaking, there are two main categories of verbs in English: those that can take direct objects and those that can't. To illustrate, the verb *repair* can take a direct object (that is, you can repair something), while the verb *happen* cannot (you can't happen anything).

Sentences with verbs that can take direct objects are the ones that can be written in either active or passive voice. When you write in active voice, the sentence is in "who + does/did what + to what/whom" order, as in this example:

An authorized technician repaired the new laser printer.

[who]   [did what]   [to what]

When you write the same idea in passive voice, the direct object moves to the start of the sentence and bumps the real subject to a phrase at the end of it (or out of it altogether). With this move, you now have

The new laser printer was repaired

[what]   [had something done to it]

by an authorized technician.

[by whom]

Or even just

The new laser printer was repaired. [real subject removed]

As you can see, inverting the word order this way makes the sentence less energetic, more roundabout, and sometimes less informative.

You can find instances of passive voice in your own writing by looking for two- and three-word verbs that consist of

- a form of the verb to be (for example, *is, was, has been, will be*) *and*
- a verb in past-tense form (for example, *installed, reduced, chosen, sent*).

When you find such verbs—*was installed, has been reduced, will be chosen*—see if your meaning would be clearer and sharper if you wrote in the active voice instead.

Give your writing impact by using strong verbs.

As you can see, the active versions are clearer and usually shorter.

The suggestion to prefer active voice does not mean that passive voice is incorrect or that you should never use it. Sometimes passive voice is preferable.

For example, when the doer of the action is unimportant to the message, passive voice properly de-emphasizes the doer.

Advertising is often criticized for its effect on price.

The copier has now been repaired.

Passive voice may enable you to avoid accusing your reader of an action:

The damage was caused by exposing the material to sunlight.

The choice of color was not specified in your order.

Passive voice also may be preferable when the performer is unknown, as in this example:

During the past year, the equipment has been sabotaged seven times.

In general, though, you should write your sentences in the active, "who does what" order.

Avoid Camouflaged Verbs

Another construction that should be avoided is the **camouflaged verb**. When a verb is camouflaged, the verb describing the action in a sentence takes the form of a noun. Then other words have to be added. For example, suppose you want to write a sentence in which *eliminate* is the action to be expressed. If you change *eliminate* into its noun form, *elimination*, you must add more words to have a sentence. Your sentence might then be "The staff *effected an elimination of* the surplus." The sentence is wordy and hard to understand. You could have avoided the camouflaged construction with a sentence using the verb *eliminate*: "The staff eliminated the surplus." You'll find more examples in the Communication Matters feature on page 73.

Verbs give your writing interest. Don't bury them in long-winded, roundabout expressions.

LO 4-7 Use words that do not discriminate.

AVOIDING DISCRIMINATORY WRITING

As the workforce has grown more diverse, it has become increasingly important to avoid discriminatory words. By **discriminatory words** we mean words that do not treat all people with equal respect. More specifically, they are words that refer negatively to groups of people because of their gender, race, nationality, sexual orientation, age, physical ability, or some other trait. Such words do not promote good business ethics or good business and thus have no place in business communication.

Many discriminatory words are a part of the vocabularies we have acquired from our environments. We often use them innocently, not realizing how they affect others. We can eliminate discriminatory words from our vocabularies by examining them carefully and placing ourselves in the shoes of those to whom they refer. The following review of the major forms of discriminatory words should help you achieve this goal.

Use Gender-Neutral Words

Our language developed in a society in which it was customary for women to work in the home and for men to be the breadwinners and decision makers. But times have changed, and the language you use in business needs to acknowledge the gender-diverse nature of most workplaces today.

This means avoiding words implying that only one gender can be in charge or perform certain jobs. Such job titles as *fireman, waitress, congressman*, and *chairman* should be replaced with the more neutral labels *firefighter, server, representative,* and *chairperson*. It also means avoiding modifiers that call attention to gender, as in *lady lawyer* or *male nurse*. When tempted to use such masculine-specific words as *manpower* and *man-made,* see if you can find a more neutral expression (*personnel, manufactured*). And be sure to give men and women in the same role or group the same level of respect; don't use "Mr." when referring to a male but the first name only (e.g., "Betty") when referring to a female.

Perhaps the most troublesome sexist words are the masculine pronouns (*he*, *his*, *him*) when they are used to refer to both sexes, as in this example: "The typical State University student eats *his* lunch at the student center." Assuming that State is coeducational, the use of *his* excludes the female students. It used to be acceptable to use *his* to refer to both sexes, but most modern-day businesspeople are offended by the use of the masculine pronoun in this way.

communication matters

Will the Real Verb Please Stand Up?

Be sure your verbs are action words, not camouflaged verbs like those in the left column.

Camouflaged Verb	Strong Verb Form
An *arrangement was made* to meet for breakfast.	We *arranged* to meet for breakfast.
We should *have a discussion about* our technology needs.	We should *discuss* our technology needs.
The new policy *brought about the standardization of* the procedures.	The new policy *standardized* the procedures.
We must *bring about a reconciliation* of our differences.	We must *reconcile* our differences.
The *establishment* of a wellness center *has been accomplished* by the company.	The company *has established* a wellness center.

You can avoid the use of masculine pronouns in such cases in three ways (summarized in Exhibit 4-3). First, you can reword the sentence to eliminate the offending word. Thus, the illustration above could be reworded as follows: "The typical State University student eats lunch at the student center." Here are other examples:

Sexist	Gender-Neutral
If a customer pays promptly, *he* is placed on our preferred list.	A customer who pays promptly is placed on our preferred list.
When an unauthorized employee enters the security area, *he* is subject to dismissal.	An unauthorized employee who enters the security area is subject to dismissal.
A supervisor is not responsible for such losses if *he* is not negligent.	A supervisor who is not negligent is not responsible for such losses.

EXHIBIT 4-3 Getting Around *He, His,* and *Him*

You can avoid sexist use of *he, his,* and *him* by

- eliminating the personal pronoun altogether.
- using the plural personal pronouns (*they, their, them*).
- using a neutral expression, such as *he or she* or *you.*

A second way to avoid sexist use of the masculine pronoun is to make the reference plural. Fortunately, the English language has plural pronouns (*their, them, they*) that refer to both sexes. Making the references plural in the examples given above, we have these nonsexist revisions:

If customers pay promptly, *they* are placed on our preferred list.

When unauthorized employees enter *the* security area, *they* are subject to dismissal.

Supervisors are not responsible for such losses if *they* are not negligent.

A third way to avoid sexist use of *he, his,* or *him* is to substitute any of a number of gender-neutral expressions. The most common are *he or she, he/she, s/he, you, one,* and *person.* Using neutral expressions in the problem sentences, we have these revisions:

If a customer pays promptly, *he or she* is placed on our preferred list.

If *you* are unauthorized to enter the security area, *you* will be subject to dismissal.

A supervisor is not responsible for such losses if *he/she* is not negligent.

But do not mix the second and third solutions, as in this sentence:

If *a customer pays* promptly, *they are* placed on our preferred list.

The sentence avoids sexist language, but it has another problem: The plural pronoun *they* is referring to the singular noun *customer.* You must make the pronoun and its antecedent (what it refers to) either both singular or both plural, or else you will have an agreement error.

Avoid Words That Stereotype by Race, Nationality, or Sexual Orientation

Words that characterize all members of a group based on their race, nationality, or sexual orientation can be especially harmful because they frequently reinforce negative stereotypes about this group. Members of any group vary widely in all characteristics. Thus, it is unfair to suggest that Americans are materialistic, that Jews are miserly, that Italians flout the law, that gays are too fussy about details, and so on.

Also unfair are words suggesting that a minority member has struggled to achieve something that is taken for granted in the majority group. Usually well intended, words of this kind can carry subtle discriminatory messages. For example, a reference to a "neatly dressed Hispanic man" may suggest that he is an exception to the rule—that most Hispanics are not neatly dressed, but here is one who is.

Eliminating unfair references from your communication requires two basic steps. First, you must consciously treat all people equally, and you should refer to a person's group membership only in those rare cases in which it is a vital part of the message to be communicated. Second, you must be sensitive to the effects of your words. Specifically, you should ask yourself how those words would affect you if you were a member of the group to which they refer. You should evaluate your word choices from the viewpoints of others.

Avoid Words That Stereotype by Age

Your avoidance of discriminatory wording should be extended to include age discrimination—against both the old and the young. While those over 65 might be retired from their first jobs, many lead lives that are far from the sedentary roles in which they are sometimes depicted. They also are not

In today's diverse workplaces, mutual respect between genders and across generations is key.

communication matters

How Diverse Is Too Diverse?

Can your employer tell you what to wear, outlaw decorated fingernails, or forbid the display of such body art as tattoos and piercings?

According to EmployeeIssues.com, a website about employee rights, the answer is *yes*—as long as the appearance policies are clearly stated in writing and are applied fairly to all employees.

Just as employers can require the use of uniforms, they can delineate what kinds of personal clothing will be acceptable on the job. For example, they might define "business casual" in a way that explicitly excludes T-shirts, shorts, and flip-flops. And as long as tattoos and body piercings aren't required by your religion, they can be grounds for being disciplined or even fired—as long as the rules have been clearly laid out and communicated.

Looking professional need not mean selling out your cultural or ethnic heritage, argues Kali Evans-Raoul, founder of an image consultancy for minorities. Everyone must "balance self-expression with workplace realities," she asserts. Just as one doesn't wear a uniform at home, one shouldn't expect to bring one's entire personal look to work.

To avoid conflicts over your on-the-job identity, your best bet is to try to choose an employer whose values align with your own. Then find and abide by that company's appearance policy.

Sources: "Dress Code Policy," *EmployeeIssues.com*, EmployeeIssues.com, 2003–2012, Web, 23 May 2013; Dan Woog, "Your Professional Image: Balance Self-Expression with Workplace Expectations," *Monster.com*, Monster.com, 2009, Web, 23 May 2013.

necessarily feeble, forgetful, or slow. While some do not mind being called *senior citizens*, others do. Be sensitive with terms such as *mature* and *elderly* as well; perhaps *retired, experienced*, or *veteran* would be better received. Likewise, when tempted to refer to someone as *young* (*young accountant, accomplished young woman*), be sure that calling attention to the person's age is defensible.

Also be careful when using one of the popular generational labels in your writing. While it makes sense for the popular management literature to use such labels as *Baby Boomer* and *Millennial* as short-hand references to different generations, the same labels can seem discriminatory in business messages. Your co-worker Frank probably does not want to be referred to as the "Baby Boomer in the group," and your manager Courtney probably will not appreciate your saying that she holds the opinions she does because she's a "Generation Xer." Use such labels only when relevant and appropriate. (Read more about the different generations in the workplace in the Communication Matters feature on the next page.)

Avoid Words That Typecast Those with Disabilities

People with disabilities are likely to be sensitive to discriminatory words. Like those in other minority groups, they run the risk of having others exclude them, treat them as strange, or minimize their abilities. But they are the largest minority group in the world; in the U.S., 19 percent of the noninstitutionalized population has a disability.[3] Plus, all of us have different levels of ability in different areas. It is important to keep these facts in mind when choosing your words.

For example, negative descriptions such as *crippled, confined to a wheelchair, wheelchair bound, handicapped*, and *retarded* should be avoided. Instead, use *wheelchair user, developmentally disabled*, or whatever term the people with the disability prefer. In general, that also means saying "those with disabilities" rather than "the disabled" to avoid suggesting that the disability is the only noteworthy trait of people in this group.

Work to develop a nonbiased attitude, and show it through carefully chosen words.

Source: GRANTLAND® Copyright Grantland Enterprises, www.grantland.net.

Understanding the Different Generations in the Workplace

According to *Generations, Inc.,* five generations now comprise the U.S. workforce: the Traditional Generation (born between 1918 and 1945), the Baby Boomers (born right after World War II), Generation X (born between 1960 and 1979), Generation Y or the Millennials (born after 1979), and the Linkster Generation (born after 1995). Different social and historical forces have shaped these generations, with the result that their values and work habits are quite different.

The three main groups—Boomers, Gen Xers, and Gen Yers—have these major traits and preferences:

• The *Boomers* "have the wisdom of experience that can provide historical perspective," have a "tenacity" that can help the organization meet new challenges, and "are team players who can enhance any group in which they participate." But they "need to be engaged," need to feel that they are still valuable, and need to feel free to use their knowledge to make decisions within reasonable limits (they shouldn't be "micromanaged").

• *Gen Xers* "are happy working MTV style," meaning they like activities that require intense bursts of energy and that challenge them to think quickly." They don't like "stupid rules," they do like "individual recognition," and they want to move ahead based on merit, not on schmoozing or seniority. They value work–life balance more than the Boomers did and are thus unwilling to work as hard, but they do value professional development. They want to be free to bring their own style to work but also want clear, fair, quick feedback on their performance.

• *Gen Yers* like to work in supportive environments like the ones that most of them grew up in. Thus, they react better to "coaching" than to being told what to do, and they like frequent reassurance that they are on the right track. Like Gen Xers, they are willing to work hard but not overly hard, they want to understand the value of the work, and they like creativity. They tend to be more fun-loving and less anxious than Gen Xers, though. They are also even more comfortable with communication technologies than Gen Xers are, so much so that they may need to be encouraged to use other forms of communication (e.g., face-to-face and phone conversations).

Source: Meagan Johnson and Larry Johnson, *Generations, Inc.* (New York: AMACOM, 2010), print.

Some Final Words about Words

There's a lot to keep in mind when selecting the most appropriate words. Under time pressure, it can be tempting to take a shortcut and settle—as Mark Twain once put it—for the best word's "second cousin." But remember: Business and business relationships can be won or lost with one word choice. The effort to say what you mean as clearly, readably, and appropriately as you can is effort well spent.

WRITING CLEAR SENTENCES

When you sit down to write a given message, you have many bits of information at hand. How will you turn them into a clear, coherent message?

Your first task will probably be grouping and ordering the information—that is, planning the message's overall organization or structure. But sooner or later, writing a successful message comes down to figuring out how to stitch your contents together in a series of sentences. How much information will you put into each sentence? And in what form and order will that information be?

The advice that follows will help you answer these questions.

LO 4-8 Write short, clear sentences by limiting sentence content and economizing on words.

Limit Sentence Content

Business audiences tend to prefer simple, efficient sentences over long, complex ones. Having too much to do in too little time is a chronic problem in business. No one, whether executive or entry-level employee, wants to read writing that wastes time.

Which of the following passages is easier to understand?

Version A	Version B
Once you have completed the online safety course, which will be available on the company portal until 29 June, your supervisor will automatically be notified, at which point he/she will authorize you to handle these materials, unless your score was below 90, in which case you will need to repeat the course.	You can access the safety course on the company portal any time between now and 29 June. You must score at least a 90 to pass it. Once you have received a passing score, your supervisor will be automatically notified. He/she will then authorize you to handle these materials.

Sometimes writers try to be efficient by using as few sentences as possible, but as these examples show, that can cause trouble. Having more sentences with less in them is often the better strategy.

Preferring short sentences does not mean making all sentences equally short. You will need some moderately long sentences to convey your more complex ideas and add flow to your writing. But take care not to make the long sentences excessively long.

The following examples illustrate the point. The paragraph below is an excerpt from an employee handbook. Obviously the use of one long sentence to convey the information was a poor decision.

> When an employee has changed from one job to another job, the new corresponding coverages will be effective as of the date the change occurs, unless, however, if due to a physical disability or infirmity as a result of advanced age, an employee is changed from one job to another job and such change results in the employee's new job rate coming within a lower hourly job-rate bracket in the table, in which case the employee may, at the discretion of the company, continue the amount of group term life insurance and the amount of accidental death and dismemberment insurance that the employee had prior to such change.

So many words and relationships are in the sentence that they cause confusion. The result is vague communication at best—complete miscommunication at worst.

Now look at the message written in all short sentences. The meanings may be clear, but the choppy effect is distracting and irritating. Imagine reading a long document written in this style.

> An employee may change jobs. The change may result in a lower pay bracket. The new coverage is effective when this happens. The job change should be because of physical disability. It can also be because of infirmity. Old age may be another cause. The company has some discretion in the matter. It can permit continuing the accidental death insurance. It can permit continuing the dismemberment insurance.

The following paragraph takes a course between these two extremes. Generally, it emphasizes short sentences, but it combines content items where appropriate.

> When an employee changes jobs, the new corresponding insurance coverage becomes effective when the change occurs. If the change has occurred because of disability, infirmity, or age and it puts the employee in a lower pay bracket, the company may, at its discretion, permit the previous level of insurance coverage to continue.

The upcoming sections on conciseness, management of emphasis, and sentence unity can help you decide how much content each sentence should carry.

from the tech desk

Readability Statistics Help Writers Evaluate Document Length and Difficulty

Grammar and style checkers give writers the option of viewing readability statistics. These statistics report the number of words, characters, paragraphs, and sentences in a document along with averages of characters per word, words per sentence, and sentences per paragraph.

The report you see here was generated for a scholarly manuscript. It reports an average of 18.5 words per sentence, a bit high for a business document but probably at an acceptable level for a scholarly document's readers. The Flesch-Kincaid score confirms that the reading grade level is 9.4, too high

for business documents but appropriate for a scholarly audience. However, the Flesch Reading Ease score might give the writer cause to review the document for accessibility, even for its targeted audience. The 59.3 score is slightly below the 60–70 range that Microsoft recommends.

To have Word calculate the readability statistics of your documents, select File > Options > Proofing and check "Show readability statistics." Then click Review > Spelling & Grammar. Your statistics will appear at the end of the review.

Readability Statistics

Counts	
Words	1625
Characters	7716
Paragraphs	30
Sentences	85

Averages	
Sentences per Paragraph	3.8
Words per Sentence	18.5
Characters per Word	4.5

Readability	
Passive Sentences	9%
Flesch Reading Ease	59.3
Flesch-Kincaid Grade Level	9.4

OK

communication matters

Avoiding Stringy and See-Saw Sentences

If you try to pack too much information into a sentence, you can wind up with a stringy sentence like this:

> While we welcome all applications, we are particularly interested in candidates who have at least three years' experience, although we will consider those with less experience who have a degree in the field or who have earned a certificate from an industry-certified trainer, and we will also consider fluency in Italian a plus.

A see-saw sentence is one that goes back and forth between two points, like this:

> A blog can add visibility to a business, although it can be labor intensive to maintain, but the time spent on the blog could be worthwhile if it generates a buzz among our potential customers.

In these cases, edit the sentences down to readable size, use helpful transitional phrases (*in addition, on the other hand*) between them, and don't switch directions too often.

Here, for example, are more readable versions of the problem sentences:

> While we welcome all applications, we are particularly interested in candidates who (1) have at least three years' experience or (2) have less experience but have earned a degree or certificate in the field. Fluency in Italian is also a plus.

> A blog can add visibility to a business. True, maintaining a blog takes time, but if the blog generates a buzz among our potential customers, the time will be well spent.

Too many simple sentences create an elementary-sounding style, so combine ideas where appropriate for your adult readers.

Here is an example of a cluttering phrase:

> *In the event that* none of the candidates is acceptable, we will reopen the position.

The phrase *in the event that* is uneconomical. The little word *if* can substitute for it without loss of meaning:

> *If* none of the candidates is acceptable, we will reopen the position.

Similarly, the phrase that begins the following sentence adds unnecessary length:

> *In spite of the fact that* they received help, they failed to exceed the quota.

Although is an economical substitute:

> *Although* they received help, they failed to exceed the quota.

The following partial list of cluttering phrases (with suggested substitutions) should help you avoid them:

Cluttering Phrase	Shorter Substitution
At the present time	Now
For the purpose of	For
For the reason that	Because, since
In the amount of	For
In the meantime	Meanwhile
In the near future	Soon
In the neighborhood of	About
In very few cases	Seldom, rarely
In view of the fact that	Since, because
With regard to, with reference to	About

Economize on Words

A second basic technique for shortening sentences is to use words economically. Anything you write can be expressed in many ways, some shorter than others. In general, the shorter wordings save the reader time and are clearer and more interesting.

To help you recognize instances of uneconomical wording, we cover the most common types below.

cluttering phrases An often-used uneconomical wording is the **cluttering phrase**. This is a phrase that can be replaced by shorter wording without loss of meaning. The little savings achieved in this way add up.

surplus words To write economically, eliminate words that add nothing to sentence meaning. Eliminating these

EXHIBIT 4-4   Can Your Writing Pass the Monotony Test?

If your verbs are strong, your sentences vary in length, and you emphasize the right ideas, your writing will have an interesting and inviting rhythm.

So read your writing aloud. If you find your voice lapsing into a monotone, you probably need to apply one or more of the guidelines presented in this chapter.

surplus words sometimes requires recasting a sentence, but often they can just be left out.

The following is an example of surplus wording from a business report:

> *It will be noted that* the records for the past years show a steady increase in special appropriations.

The beginning words add nothing to the meaning of the sentence. Notice how dropping them makes the sentence stronger—and without loss of meaning:

> The records for the past years show a steady increase in special appropriations.

Here is a second example:

> His performance was good enough to *enable him* to qualify for the promotion.

The words *to enable* add nothing and can be dropped:

> His performance was good enough to qualify him for the promotion.

The following sentences further illustrate the use of surplus words. In each case, the surplus words can be eliminated without changing the meaning.

Contains Surplus Words	Eliminates Surplus Words
There are four rules *that* should be observed.	Four rules should be observed.
In addition to these defects, numerous other defects mar the operating procedure.	Numerous other defects mar the operating procedure.
It is essential that the income be used to retire the debt.	The income *must* be used to retire the debt.
In the period between April and June, we detected the problem.	Between April and June we detected the problem.
He criticized everyone he *came in contact with*.	He criticized everyone he *met*.

communication matters

Is *That* a Surplus Word?

How easy is it to read the following sentence without making a misstep?

> We found the reason for our poor performance was stiff competition from a local supplier.

In such a sentence, adding the word *that* where it is implied would help:

> We found *that* the reason for our poor performance was stiff competition from a local supplier.

On the other hand, sometimes *that* can be omitted, as in this example:

> Check all the items *that* you wish to order.

How do you know whether to include *that*? You'll have to judge by pretending to be the reader. If *that* prevents misreading, keep it. If it seems unnecessary or distracting, leave it out.

unnecessary repetition of words or ideas

Repeating words obviously adds to sentence length. Such repetition sometimes serves a purpose, as when it is used for emphasis or special effect. But all too often it is unnecessary, as this sentence illustrates:

> We have not received your payment covering invoices covering June and July purchases.

It would be better to write the sentence like this:

> We have not received your payment covering invoices for June and July purchases.

Another example is this one:

> He stated that he believes that we are responsible.

The following sentence eliminates one of the *thats:*

> He stated that he believes we are responsible. [See the Communication Matters box above for more advice about *that*.]

Repetitions of ideas through the use of different words that mean the same thing (*free gift, true fact, past history*) also add to sentence length. Known as **redundancies**, such repetitions are rarely needed. Note the redundancy in this sentence:

> The beginning of the speech will open with a welcome.

There Is, There Are ... Do You Really Need Them?

There is/There are sentences are sometimes justified, as in these examples:

> *There is* simply no reason why we cannot achieve our goals this quarter.
>
> When *there are* more than 14 attendees, we use the larger computer lab.

But sometimes *there is* and *there are* just add extra words, as these pairs of sentences illustrate:

> **Wordy:** If *there is* a problem with the copier, the service agreement will cover it.
>
> **Better:** If the copier has a problem, the service agreement will cover it.

> **Wordy:** *There are* three ways in which the Princess Resort is superior to the Breezemont Hotel.
>
> **Better:** The Princess Resort is superior to the Breezemont Hotel in three ways.

> **Wordy:** *There are* varying opinions on the plan depending on whether you are a manager or a staff person.
>
> **Better:** The managers and the staff people have different opinions about the plan.

When tempted to write *there is* or *there are*, see if you can say what you want to say more concisely and directly.

The beginning and *will open* are two ways to say the same thing. The following sentence is better:

> The speech will open with a welcome.

Exhibit 4-5 provides more examples of redundancies and ways to eliminate them.

LO 4-9 Design sentences that give the right emphasis to content.

Manage Emphasis in Sentence Design

Any written business communication contains a number of items of information, not all of which are equally important. Some are very important, such as a conclusion in a report or

---

▼EXHIBIT 4-5 Search Out and Destroy Needless Repetition

Needless Repetition	Repetition Eliminated
Please *endorse your name on the back* of this check.	Please *endorse* this check.
We must *assemble together* at 10:30 AM *in the morning.*	We must assemble at 10:30 AM.
Our new model *is longer in length* than the old one.	Our new model *is longer* than the old one.
If you are not satisfied, *return it back* to us.	If you are not satisfied, *return* it to us.
One should know the *basic fundamentals* of clear writing.	One should know the *fundamentals* of clear writing.
The *consensus of opinion* is that the tax is unfair.	The *consensus* is that the tax is unfair.
At the present time, we are conducting two clinics.	We *are* conducting two clinics.
As a matter of interest, I would like to learn more about your procedure.	I am *interested* in learning more about your procedure.

the objective of the message. Others are relatively unimportant. One of your tasks as a writer is to form your sentences to communicate the importance of each item.

put the main ideas in the main clauses Main (or independent) clauses are called **main clauses** for a reason: they express the main point of the sentence.

Compare these two sentences that might appear in an article for a company newsletter (main clauses are italicized):

> *Mr. Freshley,* who began working for Remington in 1952, *oversees all company purchases.*
>
> *Mr. Freshley,* who oversees all company purchases, *began working for Remington in 1952.*

The sentences contain exactly the same information, but the main clauses make them say different things. The first sentence focuses on Mr. Freshley's importance to the company, while the second sentence emphasizes how long he has been an employee. Which version is preferable? It would depend on what you wanted to emphasize and on what your readers would expect you to emphasize.

use coordination and subordination deliberately When two or more pieces of information need to be included in one sentence, understanding **coordination** and **subordination** will help you decide how to structure the sentence.

When you coordinate ideas, you treat them as equal in importance by joining them with "and," "but," "or," "so," or "yet" (the coordinate conjunctions), or by setting them up as equals in a list. When you subordinate an idea, you treat it as less important than the main idea by putting it into in a modifying clause or phrase. These examples illustrate:

> The company enjoyed record sales last year, *but* it lost money. [The two ideas are both in main clauses and are thus being treated as equally important.]

> *Although the company enjoyed record sales last year*, it lost money. ["Although" turns the first idea into a subordinate clause, making it less important than the second idea.]

> *Despite record sales*, the company lost money last year. [The first idea is now in a phrase, making it less important than the second idea.]

Business leaders are communicating more and more outside the office. They want their incoming messages to communicate clearly and quickly.

use short sentences for emphasis Sentence length affects emphasis. **Short sentences** carry more emphasis than long, involved ones. They call attention to their contents by conveying a single message without the interference of related or supporting information.

For example, notice the impact of the concluding sentence in the following excerpt from a fundraising letter:

> Alumni like you have started a transformation that has increased our academic achievements, increased our enrollment, and created an internationally recognized campus landscape. But we have more to do.

Beginnings and endings of business messages are often good places for short sentences. A short opening enables you to get to the point quickly, and a short closing leaves the reader with an important final thought. Here are some examples:

Openings:

Yes, your insurance is still in effect.

Will you please send me your latest catalog?

Closings:

Thank you again for your generous contribution.

I'm looking forward to your presentation.

LO 4-10   Employ unity and good logic in writing effective sentences.

Give Sentences Unity

Good sentences have **unity**. This means that all the parts of the sentence work together to create one clear point.

Lack of unity in sentences is usually caused by one of two problems: (1) unrelated ideas or (2) excessive detail.

unrelated ideas The ideas in a sentence must have a reason for being together and clearly convey one overall point. If they don't, you should separate them. If you believe they do go together, then make clear how they're related.

The following sentences illustrate:

Unrelated	Improved
The brochure you requested is enclosed, and please let me know if you need any additional information.	The brochure you requested is enclosed. Please let me know if you need additional information.
Our territory is the southern half of the state, and our salespeople cannot cover it thoroughly.	Our territory is the southern half of the state, and it is too large for our salespeople to cover thoroughly.
Using the cost-of-living calculator is simple, but no tool will work well unless it is explained clearly.	Using the cost-of-living calculator is simple, but, like any tool, it will not work well unless it is explained clearly.
We concentrate on energy-saving products, and 70 percent of our business comes from them.	Because we concentrate on energy-saving products, 70 percent of our business comes from them.

excessive detail Putting too much detail into one sentence tends to hide the central thought, and it also makes the sentence too long.

If the detail isn't necessary, remove it. If it is important, keep it, but divide the sentence into two or more sentences, as in the following examples:

Excessive Detail	Improved
Because our New York offices, which were considered plush in the 1990s, are now badly in need of renovation, as is the case with most offices that have not been maintained, I recommend closing them and finding a new location.	Our New York offices, which were furnished in the 1990s, have not been maintained properly. I recommend closing them and finding a new location.
In response to your email inquiry, we searched our records and confirmed your Oct. 1 order, so we are sending you a rush shipment of Plytec insulation immediately.	Thank you for helping us locate your order. We are sending you a rush shipment of Plytec insulation immediately.
In 2009, when I, a small-town girl from a middle-class family, began my studies at Bradley University, which is widely recognized for its business administration program, I set my goal as a career with a large public company.	I selected Bradley because of its widely recognized business administration program. From the beginning, my goal was a career with a large public company.
The fact that I have worked in the chemical industry for over 15 years holding a variety of positions in sales, marketing, and product management has provided me with a solid understanding of how to promote products in today's industry.	I have worked in the chemical industry for over 15 years in sales, marketing, and product management. This experience has given me a solid understanding of how to promote products in the industry.

Word Sentences Logically

At some point, you've probably had a teacher write "awkward" beside one or more of your sentences. Often, the cause of such a problem is illogical wording. The paragraphs that follow will help you avoid some of the most common types of illogical sentences (see Exhibit 4-6). But keep in mind that many awkward sentences defy efforts to label them. The only guards against

▼ **EXHIBIT 4-6** Common Types of Faulty Sentence Logic

- Mixed constructions
- Incomplete constructions
- Dangling or misplaced modifiers
- Faulty parallelism

letting these kinds of sentences slip past you are your own good ear and careful editing.

mixed constructions Sometimes illogical sentences occur when writers mix two different kinds of sentences together. This problem is called a **mixed construction**.

For example, can you describe what's wrong with the following sentence about cutting costs?

> First we found less expensive material, and then a more economical means of production was developed.

If you said that the first half of the sentence used active voice but the second half switched to passive voice, you're right. Shifts of this kind make a sentence hard to follow. Notice how much easier it is to understand this version:

> First we found less expensive material, and then we developed a more economical means of production.

There's a similar problem in the following sentence:

> The consumer should read the nutrition label, but you often don't take the time to do so.

Did you notice that the point of view changed from third person (*consumer*) to second (*you*) in this sentence? The following revision would be much easier to follow:

> Consumers should read nutrition labels, but they often don't take the time to do so.

Sometimes we start writing one kind of sentence and then change it before we get to the end, illogically putting parts of two different sentences together. Here's an example:

> Because our salespeople are inexperienced caused us to miss our quota.

Rewriting the sentence in one of the following ways would eliminate the awkwardness:

> Because our salespeople are inexperienced, we missed our quota.

> Our inexperienced salespeople caused us to miss our quota.

These sentences further illustrate the point:

Mixed Construction	Improved
Some activities that the company participates in are affordable housing, conservation of parks, and litter control. ["Affordable housing" isn't an activity.]	Some causes the company supports are affordable housing, conservation of parks, and litter control.
Job rotation is when you train people by moving them from job to job. [The linking verb "is" has to be followed by a noun or adjective, but here it's followed by an adverb clause].	Job rotation is a training method in which people are moved from job to job.

Mixed Construction	Improved
My education was completed in 2009, and then I began work as a manager for Home Depot. [The sentence switches from passive voice to active voice.]	I completed my education in 2009 and then began work as a manager for Home Depot.
The cost of these desks is cheaper. ["Cheaper" can"t logically refer to cost.]	These desks cost less. (*or* These desks are cheaper.)

incomplete constructions Certain words used early in a sentence signal that the rest of the sentence will provide a certain kind of content. Be careful to fulfill your reader's expectations. Otherwise, you will have written an **incomplete construction**.

For example, the following sentence, while technically a sentence, is incomplete:

> She was so happy with the retirement party we gave her.

She was so happy . . . that what? That she sent everyone a thank-you note? That she made a donation to the library in the company's name? In a sentence like this, either complete the construction or leave "so" out.

Or consider the incomplete opening phrase of this sentence:

> As far as time management, he is a master of multitasking.

You can rectify the problem in one of two ways:

> As far as time management goes [*or* is concerned], he is a master of multitasking.

> As for time management, he is a master of multitasking.

dangling/misplaced modifiers Putting modifiers in the wrong place or giving them nothing to modify in the sentence is another common way that sentence logic can go awry. Consider this sentence:

> Believing the price would drop, the purchasing agents were instructed not to buy.

The sentence seems grammatically correct . . . but it doesn't make sense. It looks as though the purchasing agents believed the price would drop—but if they did, why did someone else have to tell them not to buy? The problem is that the people whom the opening phrase is supposed to modify have been left out, making the opening phrase a **dangling modifier**.

You can correct this problem by putting the right agents after the opening phrase:

> Believing the price would drop, we instructed our purchasing agents not to buy.

What makes this sentence hard to follow?

> We have compiled a list of likely prospects using the information we gathered at the trade show.

Surely the "prospects" aren't really the ones using the information. The sentence would be clearer if the final phrase, a **misplaced modifier**, were more logically placed, as in

> Using the information we gathered at the trade show, we have compiled a list of likely prospects.

faulty parallelism Readers expect the same kinds of content in a sentence to be worded in the same way. **Faulty parallelism** violates this logical expectation.

How might you make the similar items in this sentence more parallel in wording?

> They show their community spirit through yearly donations to the United Way, giving free materials to Habitat for Humanity, and their employees volunteer at local schools.

Here's one way:

> They show their community spirit by donating yearly to the United Way, giving free materials to Habitat for Humanity, and volunteering at local schools.

A sentence has faulty parallelism when an item doesn't match the others in the list.

Can you spot the faulty parallelism in this sentence?

> To create a more appealing Web site, we can gather personal stories, create a new logo, as well as making the layout more readable.

Here's a corrected version:

> To create a more appealing Web site, we can gather personal stories, create a new logo, and make the layout more readable.

Note that if you format your series as a bulleted list, you still need to keep the items parallel. This example has faulty parallelism:

The branding standards include

- The approved logos
- A style guide
- Using the approved color palette
- How to redesign existing materials

This bulleted list has much better parallelism:

The branding standards include

- The approved logos
- A style guide
- The approved color palette
- Instructions for redesigning existing materials

Other rules of grammar besides those mentioned here can help you avoid illogical constructions and write clear sentences. See Bonus Chapter B for more examples and advice.

WRITING CLEAR PARAGRAPHS

Skillful paragraphing is also important to clear communication. Paragraphs show the reader where topics begin and end, thus helping the reader mentally organize the information. Strategic paragraphing also helps you make certain ideas stand out and achieve the desired response to your message (see Exhibit 4-7).

The following advice will help you use paragraphing to your best advantage.

LO 4-11 Compose paragraphs that are short and unified, use topic sentences effectively, and communicate coherently.

Give Paragraphs Unity

Like sentences, paragraphs should have **unity**. When applied to paragraph structure, unity means that a paragraph sticks to a single topic or idea, with everything in the paragraph developing this topic or idea. When you have finished the paragraph,

you should be able to say, "All the points in this paragraph belong together because every part concerns every other part."

A violation of unity is illustrated in the following paragraph from an application letter. Because the goal of the paragraph is to summarize the applicant's coursework, all the sentences should pertain to coursework. By shifting to personal qualities,

Give every paragraph a clear focus.

the third sentence (in italics) violates paragraph unity. Taking this sentence out would correct the problem.

> At the university I studied all the basic accounting courses as well as specialized courses in taxation, international accounting, and computer security. I also took specialized coursework in the behavioral areas, with emphasis on human relations. *Realizing the value of human relations in business, I also actively participated in organizations, such as Sigma Nu (social fraternity), Alpha Kappa Psi (professional fraternity), intramural soccer, and A Cappella.* I selected my elective coursework to round out my general business education. Among my electives were courses in investments, advanced business report writing, financial policy, and management information systems.

Keep Paragraphs Short

As a rule, you should keep your paragraphs short. This suggestion will help with paragraph unity because unified paragraphs tend to be short.

Using short paragraphs also aids comprehension by helping your reader see the structure of your ideas. In addition, such writing is inviting to the eye. People simply prefer to read writing with frequent paragraph breaks.

How long a paragraph should be depends on its purpose and contents. Readability research has suggested an average length of eight lines for longer documents such as reports. Shorter paragraphs are appropriate for messages. And some paragraphs can be very short—as short as one line. One-line paragraphs are an especially appropriate means of emphasizing major points in business messages. As noted earlier in this chapter, a one-line paragraph may be all that is needed for a goodwill closing comment or an attention-grabbing opening.

A good rule to follow is to question the unity of any paragraph longer than eight lines. You will sometimes find that it has more than one topic. When you do, make each topic into a separate paragraph.

Make Good Use of Topic Sentences

One good way of organizing paragraphs is to use topic sentences. The **topic sentence** expresses the main idea of a paragraph, and the remaining sentences build around and support it. In a sense, the topic sentence serves as a headline for the paragraph, and all the other sentences supply the story.

Not every paragraph must have a topic sentence. Some paragraphs introduce ideas, continue the point of the preceding paragraph, or present an assortment of facts that lead to no conclusion. The central thought of such paragraphs is difficult to put into a single sentence. Even so, you should use topic sentences whenever you can. They force you to determine the purpose of each paragraph and help you check for paragraph unity.

Where the topic sentence should be in the paragraph depends on the subject matter and the writer's plan, but you basically have three choices: the beginning, end, or middle.

topic sentence first The most common paragraph arrangement begins with the topic sentence and continues with the supporting material. This arrangement has strong appeal because it enables the reader to see right away how all the sentences in the paragraph will be related.

Let's say, for example, that you're writing a paragraph reporting on economists' replies to a survey question asking their view of business activity for the coming year. The facts to be presented are these: 13 percent of the economists expected an increase; 28 percent expected little or no change; 59 percent expected a downturn; 87 percent of those who expected a downturn thought it would come in the first quarter. The obvious conclusion—and the subject for the topic sentence—is that the majority expected a decline in the first quarter. Following this reasoning, you might develop a paragraph like this:

> *A majority of the economists consulted think that business activity will drop during the first quarter of next year.* Of the 185 economists interviewed, 13 percent looked for continued increases in business activity, and 28 percent anticipated little or no change from the present high level. The remaining 59 percent looked for a recession. Of this group, nearly all (87 percent) believed that the downturn would occur during the first quarter of the year.

topic sentence at the end The second most common paragraph arrangement places the topic sentence at the end. Paragraphs of this kind present the supporting details first,

> ## "the topic sentence serves as a headline for the paragraph, and all the other sentences supply the story."

Beware the Vague or Illogical *This*

When using *this* to add coherence to your writing, be careful to make the reference both clear and logical.

What does *this* refer to in the following example?

> I do not think the donors should be listed on the Web site. *This* may make some viewers feel uncomfortable or pressured to donate.

This almost refers to something clearly, but not quite. We can make the meaning sharp by revising the example in one of these ways:

> I do not think the donors should be listed on the Web site. *Naming the donors* could make some viewers feel uncomfortable or pressured to donate.

> I think we should avoid naming the donors on the Web site. *This practice* could make some viewers feel uncomfortable or pressured to donate.

Here's another example:

> We need exposure to other markets. One of the easiest ways to do *this* is to advertise strategically.

"One of the easiest ways" to do what? There is nothing in the previous sentence that *do this* can refer to. Here's a possible correction:

> We need to gain exposure to other markets. One of the easiest ways to do this is to advertise strategically.

Now *do this* has something it can refer to: "gain exposure."

Any time you use the word *this* to refer to previous content, be sure the reference is clear. If it isn't, reword the content or write "this [something]" (e.g., "this practice").

and from these details they lead readers to the conclusion, as in this example:

> At present, inventories represent 3.8 months' supply, and their dollar value is the highest in history. If considered in relation to increased sales, however, they are not excessive. In fact, they are well within the range generally believed to be safe. *Thus, inventories are not likely to cause a downward swing in the economy.*

topic sentence within the paragraph A third arrangement places the topic sentence somewhere within the paragraph. This arrangement is rarely used, and for good reason: It does not emphasize the topic sentence, which contains

the main point. Still, you can sometimes justify using this arrangement for a special effect, as in this example:

> Numerous materials have been used in manufacturing this part. And many have shown quite satisfactory results. *Material 329, however, is superior to them all.* When built with Material 329, the part is almost twice as strong as when built with the next best material. It is also three ounces lighter. Most important, it is cheaper than any of the other products.

Leave Out Unnecessary Detail

You should include in your paragraphs only the information needed to achieve your purpose.

What you need will be a matter of judgment. You can judge best by putting yourself in your reader's place. Ask yourself questions such as these: How will the information be used? What information will be used? What will not be used? Then make your decisions. If you follow this procedure, you will probably leave out much that you originally intended to use.

The following paragraph from a message to an employee presents excessive information.

> In reviewing the personnel records in our company database, I found that several items in your file were incomplete. The section titled "Work History" has blanks for three items of information. The first is for dates employed. The second is for company name. And the third is for type of work performed. On your record only company name was entered, leaving two items blank. Years employed or your duties were not indicated. This information is important. It is reviewed by your supervisors every time you are considered for promotion or for a pay increase. Therefore, it must be completed. I request that you log into the company portal and update your personnel record at your earliest convenience.

The message says much more than the reader needs to know. The goal is to have the reader update the personnel record, and everything else is of questionable value. This revised message is better:

> A recent review of the personnel records showed that your record is incomplete. Please log into the company portal at your earliest convenience to update it. This information will enable your supervisors to see all your qualifications when considering you for a promotion or a pay increase.

▼**EXHIBIT 4-8**   Enhancing Coherence

After organizing your material logically, the three main ways to give your sentences and paragraphs coherence are

- repetition of key words.
- use of pronouns (to refer back to something earlier).
- use of appropriate transitional words.

Make Paragraphs Coherent

Like well-made sentences, well-made paragraphs move the reader logically and smoothly from point to point. They clearly indicate how the different bits of information are related to each other in terms of logic and the writer's apparent purpose. This quality of enabling readers to proceed easily through your message, without side trips and backward shifts, is called **coherence**.

The best way to give your message coherence is to arrange its information in a logical order—an order appropriate for the situation. So important are such decisions to message writing that we devote whole chapters to different patterns of organization. But logical organization is not enough. Various techniques are needed to tie the information together. These techniques are known as **transitional devices**. Here we will discuss three major ones: repetition of key words and ideas, use of pronouns, and the use of transitional words.

repetition of key words and ideas By repeating key words and ideas from one sentence to the next, you can smoothly connect successive ideas. The following sentences illustrate this transitional device (key words are in italics):

> I am a certified financial planner (CFP) and a member of the Financial Planning Association (FPA). Roughly 70 percent of the *FPA's* 23,800 *members* are *CFPs*. To earn this designation, we had to study for and pass a difficult exam. In addition, about half the *members* have been in business at least 15 years. As an *organization*, we want to establish *planning* as a true profession, one seen in the same light as medicine, the law, and accounting.

use of pronouns Because pronouns refer to words previously used, they make good transitions between ideas. The demonstrative pronouns (*this, that, these, those*) can be especially helpful. The following sentences (with the demonstrative pronouns in italics) illustrate this technique.

> Ever since the introduction of our Model V nine years ago, consumers have suggested only one possible improvement—voice controls. During all *this* time, making *this* improvement has been the objective of Atkins research personnel. Now we proudly report that *these* efforts have been successful.

A word of caution, though: When using *this* or another demonstrative pronoun to refer to an earlier sentence, try to use it with a noun—for example, *this plan*—to make the reference clear (see the Communication Matters box on page 86).

transitional words When you talk in everyday conversation, you connect many of your thoughts with transitional words. But when you write, you may not use them enough. So be alert for places where providing such words will help move your readers through your paragraphs.

Among the commonly used transitional words are *in addition, besides, in spite of, in contrast, however, likewise, thus, therefore, for example*, and *also*. A more extensive list appears in Chapter 8, where we review transitions in report writing. These words bridge thoughts by indicating the nature of the connection between what has been said and what will be said next. *In addition*, for example, tells the reader that what is to be discussed next builds on what has been discussed. *However* clearly signals that a contrasting idea is coming. *Likewise* indicates that what will be said resembles what has just been said.

Notice how the transitional expressions (in italics) in the following paragraph show the relations among the parts and move the reader steadily forward through the ideas:

> Three reasons justify moving from the Crowton site. *First*, the building rock in the Crowton area is questionable. The failure of recent geologic explorations in the area appears to confirm suspicions that the Crowton deposits are nearly exhausted. *Second*, the distances from the Crowton site to major markets make transportation costs unusually high. Obviously, any savings in transportation costs will add to company profits. *Third*, the out-of-date equipment at the Crowton plant makes this an ideal time for relocation. The old equipment at the Crowton plant could be scrapped.

The transition words *first, second*, and *third* bring out the paragraph's pattern of organization and make it easy for the reader to follow along.

Keep in mind that transitional devices can also be used between paragraphs—to tie thoughts together, to keep the focus of the message sharp, and to move the reader smoothly from point to point. Strive for coherence on both the paragraph and the document level.

WRITING FOR A POSITIVE EFFECT

As Chapter 1 explains, clarity is not your only communication goal. The "people" content of your messages often needs as much attention as the informational content—and in some cases, it will be your primary consideration.

The following sections will help you see how to write in a way that elicits positive responses. In other words, it will help you build **goodwill** with your business associates and customers. Businesses cannot survive without goodwill, and you will not last long in business if you do not value it. Courteous, pleasant behavior is part of being a professional. Read on to see how to make your writing meet this professional standard.

> " Like well-made sentences, well-made paragraphs move the reader logically and smoothly from point to point. "

Use a Conversational Style

One technique that helps build goodwill is to write in **conversational language**. Conversational language is warm, natural, and personable. It is also the language that is most easily understood.

In business, a conversational style does not always mean being colloquial. It does mean tailoring your language to your reader and avoiding stiff, impersonal wording.

choosing the right level of formality

Business relationships are much more casual than they used to be, but a certain formality is still expected in many business situations. When to be formal and when to be casual will depend on whom you're writing to, what genre you're using, and what you're saying. If you choose the wrong level of formality for the situation, you run the risk of offending your reader. A too-formal style can sound impersonal and parental, while a too-informal style can make you sound as though you aren't taking the reader seriously.

The more formal style is appropriate when you are

- Communicating with someone you don't know.

- Communicating with someone at a higher level than you.

- Using a relatively formal genre, such as a letter, long report, or external proposal.

- Writing a ceremonial message, such as a commendation or inspirational announcement.

- Writing an extremely serious message, such as a crisis response or official reprimand.

When you are writing in less formal situations, you can bring your formality down a notch. Co-workers and other associates who know each other well and are using an informal medium, such as texting, often joke and use emoticons (e.g., a smiley face) and initialisms (e.g., BTW) in their correspondence. When appropriate, such touches add goodwill.

Adjusting your level of formality can sometimes be as simple as substituting one word or phrase for another, as shown in Exhibit 4-9.

In trying to sound more formal, do not make the mistake of using stilted and unnecessarily difficult words. You can sound conversational while also being respectful and clear, as these contrasting examples illustrate:

Stiff and Dull	Conversational
Enclosed please find the brochure about which you inquired.	Enclosed is the brochure you requested.
In reply to your July 11 letter, please be advised that your adherence to the following instructions will facilitate the processing of your return.	Here are the procedures for returning your purchase and obtaining a refund.
This is to acknowledge receipt of your letter dated 5 May 2013.	We received your May 5 letter and have forwarded it to the claims department for immediate attention.

Chapter 2 offers more advice about choosing the right level of formality for different message types.

cutting out "rubber stamps"

Rubber stamps are expressions used by habit every time a certain type of situation occurs. They are used without thought and are not adapted to the specific situation. As the term indicates, they are used much as you would use a rubber stamp.

Because they are used routinely, rubber stamps communicate the effect of routine treatment, which is not likely to impress readers favorably. Such treatment tells readers that the writer has no special concern for them—that the present case is being handled in the same way as any other. In contrast, words specially selected for this case show the writer's concern for and interest in the readers.

Pages 67–68 discuss the problems that slang and popular clichés can cause. Here, we focus on the type of business clichés that are more routine and less colorful than those—the kind of

▼**EXHIBIT 4-9**   Equivalent Words with Different Levels of Formality

More Formal	Less Formal
Studied, investigated, analyzed	Looked into
Rearranged	Juggled
Ensure	Make sure
Exceptional, superior	Great, terrific
As a result, therefore	So
Confirm	Double check
Consult with	Check with
Correct, appropriate	Right
Thank you	Thanks
We will	We'll
I am	I'm
Let me know	Keep me posted
By June 3	ASAP

wording that makes a message sound like a form letter. One common example is the "thank you for your letter" opening. Its intent may be sincere, but its overuse make it a rubber stamp. Another is the closing sentence "if I can be of any further assistance, do not hesitate to call on me." Other examples of rubber stamps are the following:

I have received your message.

This will acknowledge receipt of ...

This is to inform you that ...

In accordance with your instructions ...

Thank you for your time.

Please let me know if you have any questions.

Thank you in advance for ...

Perhaps you are asking yourself "What's wrong with thanking the reader for his or her time or with offering to answer questions?" The answer is that such rubber stamps are not specific enough. They signal that you have quit thinking about the reader's actual situation. A better ending is one that thanks the reader for something in particular or that offers to answer questions about a particular topic.

You do not need to know all the rubber stamps to stop using them. You only need to write in the language of good conversation, addressing your comments to a real person in a specific situation.

LO 4-13   Use the you-viewpoint to build goodwill.

Use the You-Viewpoint

Writing from the **you-viewpoint** (also called **you-attitude**) is another technique for building goodwill in written messages. As you will see in following chapters, it means focusing on the reader's interests, no matter what type of message you are preparing. It is fundamental to the practice of good business communication.

In the broadest sense, you-viewpoint writing emphasizes the reader's perspective. Yes, it emphasizes *you* and *your* and

from the tech desk

Grammar and Style Checkers Help Writers Identify Clichés, Colloquialisms, and Jargon

While not perfect, grammar and style checkers can help writers identify some clichés, colloquialisms, and jargon that have crept into their writing. Here, the writer told Microsoft Word 2010 to look for such expressions (the Tech Desk box on page 69 explains how). When she ran a review of her document, the checker found one and offered two suggestions for revising it. Clicking the Explain button will bring up the reason behind the suggestion. (In Word 2013, the explanation will automatically appear.)

Although this software can help, writers still need to be able to identify the trite and overused expressions the software misses. Also, writers need to be able to recast the sentences for clarity and sincerity.

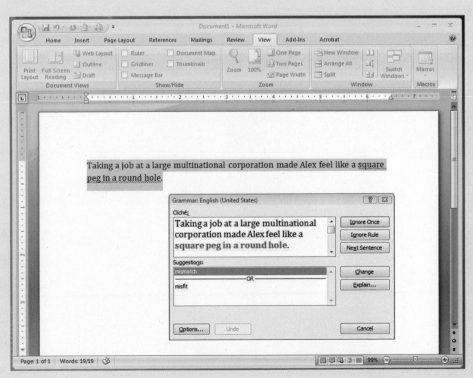

de-emphasizes *we* and *our*, but it is more than a matter of just using second-person pronouns. *You* and *your* can appear prominently in sentences that emphasize the we-viewpoint, as in this example: "If you do not pay by the 15th, you must pay a penalty." Likewise, *we* and *mine* can appear in sentences that emphasize the you-viewpoint, as in this example: "We will do whatever we can to protect your investment."

The point is that the you-viewpoint is an attitude of mind that places the reader at the center of the message. Sometimes it just involves being friendly and treating people the way they like to be treated. Sometimes it involves skillfully managing people's response with carefully chosen words in a carefully designed order. How you apply it will depend on each situation and your own judgment.

The following examples illustrate the value of using the you-viewpoint. Imagine the contrasting effects that the difference in wording would create.

We-Viewpoint	You-Viewpoint
We are pleased to have your new account.	Your new charge account is now open for your convenience.
Our policy prohibits us from permitting outside groups to use our equipment except on a cash-rental basis.	Our policy of cutting operating costs by renting our equipment helps us make efficient use of your tax dollars.

We-Viewpoint	You-Viewpoint
We have been quite tolerant of your past-due account and must now demand payment.	If you are to continue to enjoy the benefits of credit buying, you must clear your account now.
We have received your report of May 1.	Thank you for your report of May 1.
We require that you sign the sales slip before we will charge to your account.	For your protection, you are charged only after you have signed the sales slip.

Some critics of the you-viewpoint point out that it can be insincere and manipulative. It is better, they say, just to "tell it like it is."

Without question, the you-viewpoint can be used insincerely, and it can be used to pursue unethical goals. Our advice is to use the you-viewpoint when it is friendly and sincere and when your goals are ethical. In such cases, using the you-viewpoint is "telling it like it is." If you have your readers' feelings genuinely in mind, writing from the you-viewpoint should come naturally.

LO 4-14 Employ positive language to achieve goodwill and other desired effects.

Accent the Positive

In most situations, it is better to use positive than negative wording in your business messages. (See the supporting research in the Commuication Matters box on page 91.)

In face-to-face communication, words, voice, facial expressions, and gestures combine to create the desired communication effect. In writing, the printed word alone must do the job.

This is not to say that negative words have no place in business writing. Such words are powerful, and you will sometimes want to use them. But positive words tend to put the reader in a cooperative frame of mind, and they emphasize the pleasant aspects of the goal. They also create the goodwill that helps build relationships.

Consider the case of a company executive who had to deny a local civic group's request to use the company's meeting facilities. Even though he had an alternative to offer, he came up with this totally negative response:

> We *regret* to inform you that we *cannot* permit you to use our auditorium for your meeting, as the Sun City Investment Club asked for it first. We can, however, let you use our conference room, but it seats *only* 60.

The positively intended message "We *regret* to inform you" is an unmistakable sign of coming bad news. "*Cannot* permit" is unnecessarily harsh. And notice how the good-news part of the message is weakened by the limiting word *only*.

Had the executive searched for more positive ways of covering the same situation, he or she might have written

> Although the SunCity Investment Club has reserved the auditorium for Saturday, we can offer you our conference room, which seats 60.

Not a single negative word appears in this version. Both approaches achieve the primary objective of denying a request, but their effects on the reader would differ sharply. Clearly the second approach, which focuses on the positive, does the better job of building goodwill.

Here are additional examples (the negative words are in italics):

Negative	Positive
Smoking is *not* permitted anywhere except in the lobby.	Smoking is permitted in the lobby only.
We *cannot* deliver until Friday.	We can deliver the goods on Friday.
We *regret* that we *overlooked* your coverage on this equipment and apologize for the *trouble* and *concern* it must have caused you.	You were quite right in believing that you have coverage on the equipment. We have now credited your account for…
We *regret* to inform you that the guest room is not available on the date you requested.	The guest room is already booked for the evening of August 7. Would August 8 be a possibility? We would be able to accommodate your party at any time on that date.

© 2011 Ted Goff www.tedgoff.com
Reprinted with permission.

"You'll be happy to know that we've redefined the word impossible, and it no longer applies to what you have to do."

communication matters

Parent, Child, or Adult?

In the 1950s, psychologist Eric Berne developed a model of relationships that he called Transactional Analysis. It has proven to be so useful that it is still popular today.

At the core of this model is the idea that, in all our transactions with others (and even within ourselves), people occupy one of three positions: parent, child, or adult.

- A *parent* is patronizing, spoiling, nurturing, blaming, criticizing, and/or punishing.
- A *child* is uninhibited, freely emotional, obedient, whining, irresponsible, and/or selfish.
- An *adult* is reasonable, responsible, considerate, and flexible.

Significantly, the "self" that one projects invites others to occupy the complementary position. Thus, acting "parental" leads others to act "childish" and vice versa, while acting "adult" invites others to be adults.

In both internal and external business messages, strive for "adult–adult" interactions. Your courtesy and professionalism will be likely to elicit the same from your readers.

LO 4-15   Explain and use the elements of courtesy.

Be Courteous

A major contributor to goodwill in business documents is courtesy. By **courtesy** we mean respectful and considerate treatment of others. Courtesy produces friendly relations between people, and the result is a better human climate for solving business problems and doing business.

How to be courteous will depend on the situation. (It also depends on culture, as explained in the Communication Matters feature on page 93.) Including "please," "thank you," "we're sorry," and other standard expressions of politeness do not necessarily make a message courteous. Rather than focusing on stock phrases, consider what will make your reader feel most comfortable, understood, and appreciated. A message with no overtly polite expressions whatsoever can still demonstrate great courtesy by being easy to read, focusing on the reader's interests, and conveying the writer's feelings of goodwill.

Still, courtesy generally is enhanced by using certain techniques. We have already discussed three of them: writing in

conversational language, employing the you-viewpoint, and choosing positive words. More follow.

avoid blaming the reader   Customers, co-workers, bosses, and other businesspeople you work with are going to make mistakes—just as you will. When they do, your first reaction is likely to be disappointment, frustration, or even anger. After all, their mistakes will cost you time, energy, and possibly even money.

But you must avoid the temptation to blame the reader when you are resolving a problem. No one likes being accused of negligence, wrongdoing, or faulty thinking. It is better to objectively explain the facts and then move on to a solution.

The following examples illustrate:

Blaming Language	More Courteous Language
You failed to indicate which fabric you wanted on the chair you ordered.	To complete your order, please check your choice of fabric on the enclosed card.
If you had read the instructions that came with your cookware, *you would have known* not to submerge it in water.	The instructions explain why the cookware should not be submerged in water.
Your claim that we did not properly maintain the copier *is false.*	Listed below are the dates the copier was serviced and the type of service it received.
Your request for coverage is denied because *you did not follow* the correct appeals procedure.	We need additional information to be able to process your appeal. Please supply [the needed information] and resubmit your request to….

Notice two helpful, related strategies in the better language above. One is to avoid using *you* when doing so would blame the reader. In these situations, you will actually have better you-viewpoint if you do not use *you*. The second strategy is to keep the focus on the facts rather than on the people. So instead of writing "*you should not submerge* the cookware in water," write "the cookware *should not be* submerged in water." Notice that the preferred wording in this example uses passive voice. As noted earlier in this chapter, passive voice (omitting the doer of the action) is not only acceptable but desirable in cases where it will keep you from assigning blame.

do more than is expected   One sure way to gain goodwill is to do a little more than you have to do for your reader. Perhaps in an effort to be concise, we sometimes stop short of being helpful. But the result can be brusque, hurried treatment, which is inconsistent with the effort to build goodwill.

The first step toward creating goodwill with your readers is to put yourself in their shoes.

The writer of a message refusing a request to use company equipment, for example, needs only to say "no" to accomplish the primary goal. But this blunt response would destroy goodwill. To maintain positive relations, the writer should explain and justify the refusal and then suggest alternative steps that the reader might take. A wholesaler's brief extra sentence to wish a retailer good luck on a coming promotion is worth the effort. So are an insurance agent's few words of congratulations in a message to a policyholder who has earned some kind of distinction.

Try to solve your readers' problems, and look for other ways to make your readers feel appreciated.

be sincere Courteous treatment is sincere treatment. If your messages are to be effective, people must believe you. You must convince them that you mean what you say and that your courtesy and friendliness are authentic.

The best way of getting sincerity into your writing is actually to be sincere. If you honestly want to be courteous, if you honestly believe that you-viewpoint treatment leads to harmonious relations, and if you honestly think that tactful treatment spares your reader's feelings, you are likely to apply these techniques sincerely, and this sincerity will show in your writing.

from the tech desk

Courtesy in the Age of Mobile Devices

In the U.S., the name "Emily Post" has been synonymous with good manners for decades. With the publication of *Etiquette in Society, in Business, in Politics, and at Home* in 1922, Post became the undisputed authority on the topic, and she continued to be so the rest of her life. Her children, grandchildren, and great-grandchildren have carried on her tradition. *Emily Post's Etiquette* is now in its 18th edition, and it now includes a chapter on personal communication devices. Here are some excerpts from the chapter:

Cell Phones

• Without exception, turn your device off in a house of worship, restaurant, or theater; during a meeting or presentation; or anytime its use is likely to disturb others.

• If you must be alerted to a call, put your device on silent ring or vibrate, and check your caller ID or voice mail later. (Put it in your pocket; a vibrating phone, skittering across a tabletop, is just as disruptive as a ring.)

• Wherever you are, if you must make or take a call, move to a private space and speak as quietly as you can.

• Keep calls as short as possible; the longer the call, the greater the irritation to those who have no choice but to listen.

• On airplanes, it's a courtesy to everyone on board to quickly wrap up your call when the flight crew instructs passengers to turn off all electronic devices before takeoff. When cell phone use is permitted after landing, keep your calls short, limiting them to information about your arrival. Save any longer calls for a private spot in the terminal.

• Think about what your ring tone says about you. Is your frat boy hip-hop tone the right ring for your new job as a trainee at an accounting firm?

Text Messaging

• Text messaging is a strictly casual communication. You shouldn't use text messaging when informing someone of sad news, business matters, or urgent meetings unless it's to set up a phone call on the subject.

• Be aware of where you are. The backlight will disturb others if you text in a theater or house of worship.

• Keep your message brief. If it's going to be more than a couple of lines, make a call and have a conversation.

• Don't be a pest. Bombarding someone with texts is annoying and assumes they have nothing better to do than read your messages.

• Be very careful when choosing a recipient from your phone book; a slip of the thumb could send a text intended for a friend to your boss.

• Whenever you have a chance, respond to text messages, either by texting back or with a phone call.

• Don't text anything confidential, private, or potentially embarrassing. You never know when your message might get sent to the wrong person or forwarded.

Source: Based on Peggy Post, Anna Post, Lizzie Post, and Daniel Post Senning, *Emily Post's Etiquette: Manners for a New World,* 18th ed. (New York: HarperCollins, 2001), 240–248, print.

Be careful not to let exaggeration undermine your goodwill efforts, as in these two examples:

> If you will help these children, Ms. Collins, you will become a hero in their eyes.

> We are extremely pleased to be able to help you and want you to know that your satisfaction means more than anything to us.

Such flattery will damage your reader's trust in you.

Many exaggerated statements involve the use of superlatives, such as *greatest, most amazing, finest, healthiest,* and *strongest.* Other strong words may have similar effects—for example, *extraordinary, incredible, sensational, terrific, revolutionary, world-class,* and *perfection.* Avoid such wording except in the rare cases when it is warranted. To be regarded as sincere, you must be believable.

LO 4-16 Use the three major techniques for emphasizing the positive and de-emphasizing the negative.

Manage Emphasis for a Positive Effect

Getting the desired effect in writing often involves giving proper emphasis to the items in the message. The three most useful ways to manage emphasis for a positive effect are to use position, sentence structure, and space.

emphasis by position The beginnings and endings of a writing unit carry more emphasis than the center parts. This rule of emphasis applies whether the unit is the

▼EXHIBIT 4-10 Emphasis by Position

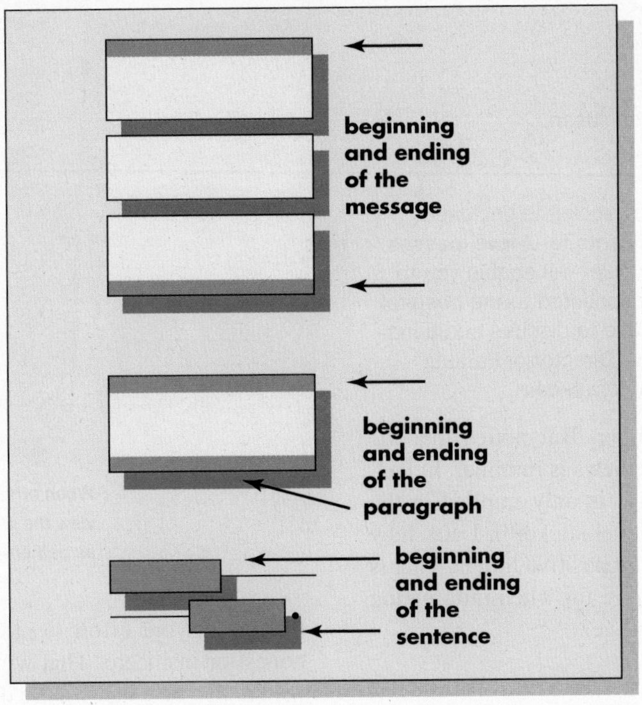

message, a paragraph of the message, or a sentence within the paragraph (see Exhibit 4-10). In light of this fact, you should put your more positive points in beginnings and endings and, if possible, avoid putting negative points in these positions.

If we were to use this technique in a paragraph turning down a suggestion, we might write it like this (the key point is in italics):

> In light of the current budget crunch, we approved those suggestions that would save money while not costing much to implement. While *your plan is not feasible at this time*, we hope you will submit it again next year when we should have more resources for implementing it.

As you can see, putting information in the middle tends to de-emphasize it. Consider position carefully when organizing your positive and negative contents.

sentence structure and emphasis Closely related to the concept of emphasis by position is the concept of managing emphasis through sentence structure. As noted on pages 80–81 in this chapter, short, simple sentences and main clauses call attention to their content. In applying this emphasis technique to your writing, carefully consider the possible arrangements of your information. Place the more positive information in short, simple sentences or main clauses. Put the less important information in subordinate structures such as dependent clauses and modifying phrases. This sentence from the previous example illustrates the technique (the negative point is in a subordinate clause):

> *While your plan is not feasible at this time*, we encourage you to submit it again next year when we are likely to have more resources for implementing it.

Here's another example:

> Your budget will be approved *if you can reduce your planned operating expenses by $2,000.*

space and emphasis In general, the more space you devote to a topic, the more you emphasize it. Therefore, de-emphazing the negative means spending as little space on it as possible.

When we say not to spend much space on negative news, we mean the actual negative point. As Chapter 6 will show, you will often need to preface such news with explanatory, cushioning words in order to prepare your readers to receive it as positively as possible. For this reason, it often takes longer to say "no" than to say "yes," as in these contrasting openings of a message responding to a request:

A Message That Says "Yes"

Your new A-level parking sticker is enclosed.

A Message That Says "No"

Your new University Hospital parking sticker is enclosed. As always, we had many more applicants for A-level passes than we had spaces. Your B-level sticker will enable you to park in the Eden Garage, which is connected to the hospital by a covered skywalk. If you would like to discuss additional options, please contact Ann Barnett, Director of Parking Services, at 555-6666 or ann.barnett@uh.com.

The "no" version certainly took longer. But notice that the space actually devoted to the negative news is minimal. In fact, the negative news isn't even stated; it is only implied in the positive second sentence ("Your B-level sticker"). Look how much of the paragraph focuses on more positive things. That is the allocation of space you should strive for when minimizing the negative and emphasizing the positive.

Use Positive Emphasis Ethically

As with use of the you-viewpoint, emphasis on the positive, when overdone, can lead to fake and manipulative messages. The technique is especially questionable when it causes the reader to overlook an important negative point in the message—the discontinuation of a service, for example, or information about an unsafe product.

When ethical and appropriate, view the glass as half full, not as half empty.

Do not let your effort to please the reader lead you to be dishonest or insincere. That would not only be morally wrong; it would also be a bad way to do business.

On the other hand, the topics we discuss in our communication—whether data, events, people, or situations—can be rightly perceived in multiple ways. In your quest to achieve your communication purpose, think before you let negative feelings make their way into your messages. You will often be able to depict the glass as half full rather than as half empty, and you will probably find that your own perspective has improved in the process.

© Andres Rodriguez/Alamy

make your style work for you!

- How can you find the word with the right connotation?
- What are five common business writing mistakes?
- Where can you find an online exercise to help you avoid faulty parallelism? Redundancy? Wordiness?
- What are seven ways to cultivate good business relationships?
- What are some negative words to avoid in business writing?

Scan the QR code with your smartphone or use your Web browser to find out at www.mhhe.com/RentzM3e. Choose Chapter 4 > Bizcom Tools & Tips. While you're there, you can view a chapter summary, exercises, PPT slides, and more to help you make your style work for you.

www.mhhe.com/RentzM3e

Writing Good-News
and Neutral Messages

chapter five

Most business messages use a direct organizational plan. That is, the message leads with its most important point and then moves to additional or supporting information. If you recall what Chapter 1 observed about the nature of business, you will understand why. Communication is central to organized human activity. Especially in business, people need to know what to do, why, and how. They undertake any job understanding that they have a certain function to perform, and they need information to be able to perform it well. When external audiences interact with companies, they also expect and need certain kinds of information presented as quickly as possible. It is fair to say that direct messages are the lifeblood of virtually any business activity. ■

LEARNING OBJECTIVES

LO 5-1 Properly assess the reader's likely reaction to your message.

LO 5-2 Describe the general plan for direct-order messages.

LO 5-3 Write clear, well-structured routine inquiries.

LO 5-4 Write direct, orderly, and favorable answers to inquiries.

LO 5-5 Write order acknowledgments and other thank-you messages that build goodwill.

LO 5-6 Write direct claims in situations where an adjustment will likely be granted.

LO 5-7 Compose adjustment grants that regain any lost confidence.

LO 5-8 Write clear and effective internal-operational communications.

PRELIMINARY ASSESSMENT

As discussed in Chapter 2, writing any messages other than those for the most mechanical, routine circumstances requires careful thinking about the situation, your readers, and your goals. When determining your message's basic plan, a good

Covering the Remaining Part of the Objective

Whatever else must be covered to complete your objective makes up the bulk of the remainder of the message. If you cover all of your objective in the beginning (as in an inquiry in which a single question is asked), nothing else is needed. If you have to ask or answer additional questions or provide information, do so in the body of your message. Cover your information systematically—perhaps listing the details or arranging them by paragraphs. If these parts have their own explanations or commentary, include them.

> " If you are seeking information, start by asking for it. If you are giving information, start giving it. Whatever your key point is, lead with it. "

beginning is to assess your reader's probable reaction to what you have to say. If the reaction is likely to be positive or even neutral, you will likely use the **direct order**—that is, you will get to the objective right away without delay. If your reader's reaction is likely to be negative, you may need to use the indirect plan, discussed in Chapter 6. The general plan for the direct approach in positive and neutral situations follows.

LO 5-2   Describe the general plan for direct-order messages.

Ending with Goodwill

End the message with some appropriate friendly comment as you would end a face-to-face communication with the reader.

Include a closing that is relevant to the topic of your message. General closings such as "Thank you" or "If you need further information, please don't hesitate to ask" are polite, but they are clichés. You will build more goodwill with a closing that is tailored to your message—for example: "If you will answer these questions about Ms. Hill right away, we can fill the accounting position before our busy tax season."

THE GENERAL DIRECT PLAN

Beginning with the Objective

Begin with your **objective**. If you are seeking information, start by asking for it. If you are giving information, start giving it. Whatever your key point is, lead with it.

In some cases, you might need to open with a brief orienting phrase, clause, or even sentence. Especially if your reader is not expecting to hear from you or is not familiar with you or your company, you may need to preface your main point with a few words of background. But keep any prefatory remarks brief and get to the real message quickly. Then stop the first paragraph. Let the rest of the message fill in the details.

Showing goodwill in writing is as important as showing goodwill in interpersonal situations.

Searching for New Regional Headquarters

Introduce yourself to routine inquiries by assuming you are the assistant to the vice president for administration of White Label Industries (WLI). WLI is the manufacturer and distributor of an assortment of high-quality products.

You and your boss were recently chatting about WLI's plans to relocate its regional headquarters. Your boss tells you that she and other top management have chosen the city but have not been able to find the perfect office space. She says that they have not been happy with what realtors have found for them or with what they have found in their own searches of classified ads and realty agencies' Web sites. When you suggest that they expand their search to something a little less traditional, such as craigslist, your boss says, "Great idea! I don't think any of us have used craigslist, though. Could you find some locations and show them to us at our Friday meeting?"

You're a bit intimidated by the prospect, but you know that this is a great chance to demonstrate your professional skills. You visit craigslist and find what you believe would be the perfect office headquarters. You know you could just show the executives the ad at the meeting, but having read the ad and having analyzed your audience, you know the executives will need more information. To present your best professional image at Friday's meeting, you need to write a routine inquiry seeking additional details about the office space.

Be aware, though, that phrases such as "as soon as possible" or "at your convenience" may have very different meanings for you and your reader. If you need your response by a specific date or time, give your reader that information as well as a reason for the deadline so that your reader understands the importance of a timely response. You may say, for example, "Your answers to these questions by July 1 will help Ms. Hill and us as we meet our deadline for filling the accounting position."

Now let us see how you can adapt this general plan to fit the more common direct-message situations.

LO 5-3 Write clear, well-structured routine inquiries.

ROUTINE INQUIRIES

Choosing from Two Types of Beginnings

The opening of the **routine inquiry** should focus on the main objective. Routine inquiries usually open in one of two ways: (1) with a direct question or request or (2) with a brief statement to orient the reader, followed by the request or question.

If you begin with a direct question or request, you can ask one broad question that sets up other questions you'll ask in the body of the message. For example, if your objective is to get more information about the office space described in the Workplace Scenario above, you might begin with a general question:

> Could you please send me additional information about the Riverdale office space you advertised on craigslist on May 24?

The body of your message would then present a list of the specific information you are seeking.

On the other hand, if you have only one piece of information you are seeking, you could begin with your specific question:

> Could you please send me the dimensions of the first- and second-floor corner offices of the Riverdale office space you advertised on craigslist on May 24?

Answering inquiries that do not include adequate explanation can be frustrating.

communication matters

Choosing the Right Font

Of all the issues a writer considers when writing an effective business message, the type of font to use may be at the bottom of the list (if it is on the list at all). However, choosing the right font can make your documents look as professional as they sound.

- *What are my choices?* The main choice is either a serif or sans serif font. Letters in serif fonts such as Book Antiqua (shown in the box) have "tails" (serifs). Letters in sans serif fonts such as Verdana (also shown in the box) do not. What do serifs do? Serifs connect letters, which makes the space between words more distinguishable and the text therefore more readable—at least in printed documents. In electronic documents, the serifs may actually hinder readability depending on the font size and monitor resolution. Sans serif fonts, however, allow for more white space, which makes letters and words stand out. A possible choice, then, is to use a sans serif font for headings and a serif font for body text.

- *How many fonts can I use?* Limit yourself to not more than two fonts. It's fine to use only one font. However, if you use more than two, you will have a document that looks cluttered and visually confusing. If you do choose two fonts, be sure that one is a serif font and one is a sans serif font. This way the fonts complement rather than compete with each other. Remember that excessive formatting of your fonts (bold, italics, underlining) will also undermine the professional look of your document.

- *How big should my fonts be?* This depends on the font. Start with the body text at 9–12 points. Make your headings two points larger than your body text. Whatever size you choose, be sure the text is readable and looks professional. Fonts that are too small are hard to read. Fonts that are too big look amateurish and visually attack the reader.

- *What style should I choose?* That will depend on what kind of document you're writing and to whom. Look at the sample fonts below. Which would be more appropriate in a print ad for party supplies? In an annual report to investors? In an invitation to a formal event? As you can see, each typeface has its own personality. Choose yours carefully to match your situation.

This font is 12-point Verdana.

This font is 12-point Script MT Bold.

This font is 12-point Book Antiqua.

THIS FONT IS 12-POINT GOUDY STOUT.

> If you do not explain enough or if you misjudge the reader's knowledge, you make the reader's task difficult.

You might then offer some explanation of what you're looking for, or you might conclude your message.

If you think your reader would first need some background information or an orienting statement, provide one. This information helps reduce any startling effect that a direct opening question might have and can help soften the tone if the direct opening question sounds demanding or blunt.

> The 3,200-square-foot Riverdale office space you advertised on May 24 on craigslist seems like a great fit for our regional headquarters. To help us decide on a new office space, could you please answer a few questions about the Riverdale offices?

Regardless of how you begin, be sure your reader has a clear sense of your message's purpose.

Informing and Explaining Adequately

To help your reader answer your questions, you may need to include explanation or information. If you do not explain enough or if you misjudge the reader's knowledge, you make the reader's task difficult. For example, answers to your questions about office space for WLI may depend on characteristics or specific needs of the company. Without knowing how WLI will use the

Shortcut Tools Help Writers Improve Productivity and Quality

Shortcuts help writers save time and improve quality. One of the easiest to use is the AutoCorrect tool in Word (shown here). This tool will automatically replace a word you type with another word you have entered to replace that particular word. The default setting is generally set to correct common misspellings and typos. However, it also can be used to expand acronyms or phrases used repeatedly.

If you worked frequently with the Association for Business Communication, you might set up the AutoCorrect tool to replace the acronym ABC with the full name. Not only will this shortcut enable you to save time, but it also will improve the quality of your work by inserting a correctly spelled and typed replacement every time.

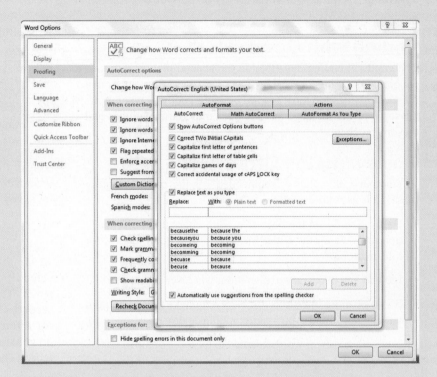

space, even the best realtor or property manager may not know how to answer your questions or perhaps direct you to other office space that better meets your needs.

Where and how you include the necessary explanatory information depend on the nature of your message. Usually, a good place for general explanatory material is before or after the direct request in the opening paragraph. In messages that ask more than one question, include any necessary explanatory material with the questions. Such messages may alternate questions and explanations in the body of the message.

Structuring the Questions

After you ask your initial question and provide any relevant background information, your message will take one of two directions. If your inquiry involves only one question, you have achieved your objective, and you may move to a goodwill ending to finish your message. If you have to ask several questions, develop an organized, logical list in the body of your message.

First, if you have two or more questions, make them stand out. Combining two or more questions in a sentence de-emphasizes each and invites the reader to overlook some. You can call attention to your questions in a number of ways. First, you can make each question a separate sentence with a bullet (for example, ●, ○, ■) to call attention to it.

Second, you can give each question a separate paragraph whenever your explanation and other comments about each question justify a paragraph.

Third, you can order or rank your questions with numbers. By using words (*first, second, third*, etc.), numerals (*1, 2, 3*, etc.), or letters (*a, b, c*, etc.), you make the questions stand out. Also, you provide the reader with a convenient checklist for answering.

Fourth, you can structure your questions in **true question form**. Sentences that merely hint at a need for information do not attract much attention. The statements "It would be nice if you would tell me . . ." and "I would like to know . . ." are really not questions. They do not ask—they merely suggest. The questions that stand out are those written in question form: "Will you please tell me . . .?" "How much would one be able to save . . .?" "How many contract problems have you had . . .?"

Avoid questions that can be answered with a simple *yes* or *no* unless you really want a simple *yes* or *no* answer. For example, the question "Is the chair available in blue?" may not be what you really want to know. Better wording might be "In what colors is the chair available?" Often, combining a yes/no question with its explanation yields a better, more concise question. To illustrate, the wording "Would your software let us deliver our training modules in any format? We need to deliver them in HTML5" could be improved by asking "Would your software let us deliver our training modules in HTML5?"

Ending with Goodwill

The goodwill ending described in the general plan is appropriate here, just as it is in most business messages. Remember that the closing does the most toward creating goodwill when it fits the topic of the message and includes important deadlines and reasons for them.

Reviewing the Order

In summary, the plan recommended for the routine inquiry message is as follows:

- Focus directly on the objective, with either a specific question that sets up the entire message or a general request for information.

- Include any necessary explanation, wherever it best fits.

- If two or more questions are involved, make them stand out with bullets, numbering, paragraphing, and/or question form.

- End with goodwill words adapted to the topic of the message.

Contrasting Examples of a Routine Inquiry

At the bottom of this page and page 105 are two routine inquiry messages that illustrate bad and good approaches to requesting information about office space for a new WLI regional headquarters (recall the Workplace Scenario) on page 101. The first example follows the indirect pattern. The second is direct and more appropriate for this neutral message. You can also study the Case Illustrations on pages 106 and 107. The margin comments help you see how these sample inquiries follow the advice in this chapter.

As you read the first example, note that it is marked by a "🚦" icon in the side panel. We use this icon throughout the book wherever we show bad examples. The good examples will be indicated by a "🚦" icon.

the indirect message The less effective message begins slowly and gives obvious information. Even if the writer thinks that this information needs to be communicated, it does not deserve the emphasis of the opening sentence. The writer gets to the point of the message in the second paragraph. There are no questions here—just hints for information. The items of information the writer wants do not stand out but are listed in rapid succession in one sentence. The close is selfish and stiff.

the direct and effective message The second example (presented below) begins directly by asking for information. The explanation is brief but complete. The questions

This message's indirect and vague beginning slows reading.

Mr. Piper:

We saw the advertisement for 3,200 square feet of Riverdale office space that you posted a couple of weeks ago on craigslist. As we are interested, we would like additional information.

Specifically, we would like to know the interior layout, annual cost, availability of transportation, length of lease agreement, escalation provisions, and any other information you think pertinent.

If the information you give us is favorable, we will inspect the property. Please send your reply as soon as possible.

Sincerely,

> *When a response involves answering a single question,*
> *you begin by answering that question.*

are formatted to stand out; thus, they help make answering easy. The message closes with a courteous and appropriate request for quick action.

LO 5-4 Write direct, orderly, and favorable answers to inquiries.

FAVORABLE RESPONSES

When your answer to inquiries is positive, your primary goal will be to tell your readers what they want to know. Because your message will be a **favorable response**, directness is appropriate.

Identifying the Message Being Answered

Because this message is a response to another message, you should identify the message you are answering. Such identification helps the reader recall or find the message being answered. If you are writing an email response, the original message is appended to your message. Of course, in an email message, your subject line will identify the message you are

answering, but in hard copy messages, you may also use a subject line (Subject: Your April 2 Inquiry about Chem-Treat). Or you can refer to the message incidentally in the text ("as requested in your April 2 inquiry"). Preferably you should identify the message you are responding to early in your message.

Beginning with the Answer

Directness here means giving the readers what they want at the beginning. Thus you begin by answering. When a response involves answering a single question, you begin by answering that question. When it involves answering two or more questions, one good plan is to begin by answering one of them—preferably the most important. In the Chem-Treat case, this opening would get the response off to a fast start:

> Yes, WLI's Chem-Treat acrylic latex paint is among the safest on the market.

This direct and orderly message is better.

Dear Mr. Piper:

Will you please answer the following questions about the 3,200-square-foot Riverdale office suite advertised May 24 on craigslist? This space may be suitable for the new regional headquarters we are opening in your city in August.

- Is the layout of these offices suitable for a work force of two administrators, a receptionist, and seven office employees? (If possible, please send us a diagram of the space.)

- What are the dimensions of the corner offices on the first and second floors?

- What is the annual rental charge?

- Are housekeeping, maintenance, and utilities included?

- What type of flooring and walls does the office space have?

- Does the location provide easy access to mass transportation and the airport?

- What is the length of the lease agreement?

- What escalation provisions are included in the lease agreement?

We look forward to learning more about your property. We hope to secure a space that meets our needs by June 30.

Sincerely,

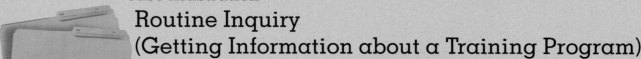

case illustration
Routine Inquiry
(Getting Information about a Training Program)

This email message is from a company training director to the director of a management-training program. The company training director has received literature on the program but needs additional information. The message seeks this information.

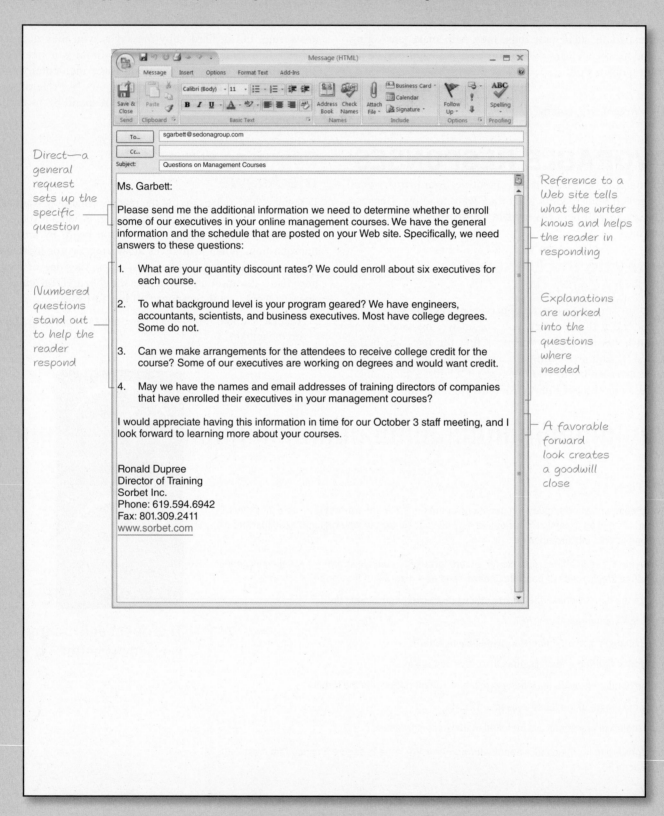

Direct—a general request sets up the specific question

Numbered questions stand out to help the reader respond

Reference to a Web site tells what the writer knows and helps the reader in responding

Explanations are worked into the questions where needed

A favorable forward look creates a goodwill close

To...   sgarbett@sedonagroup.com

Cc...

Subject: Questions on Management Courses

Ms. Garbett:

Please send me the additional information we need to determine whether to enroll some of our executives in your online management courses. We have the general information and the schedule that are posted on your Web site. Specifically, we need answers to these questions:

1.   What are your quantity discount rates? We could enroll about six executives for each course.

2.   To what background level is your program geared? We have engineers, accountants, scientists, and business executives. Most have college degrees. Some do not.

3.   Can we make arrangements for the attendees to receive college credit for the course? Some of our executives are working on degrees and would want credit.

4.   May we have the names and email addresses of training directors of companies that have enrolled their executives in your management courses?

I would appreciate having this information in time for our October 3 staff meeting, and I look forward to learning more about your courses.

Ronald Dupree
Director of Training
Sorbet Inc.
Phone: 619.594.6942
Fax: 801.309.2411
www.sorbet.com

case illustration
Routine Inquiry
(An Inquiry about Hotel Accommodations)

This letter to a hotel inquires about accommodations for a company's annual meeting. In selecting a hotel, the company's managers need answers to specific questions. The message covers these questions.

Visit us: www.womensmedia.com

July 17, 2014

Ms. Connie Briggs, Manager
Drake Hotel
140 East Walton Place
Chicago, IL 60611

Dear Ms. Briggs:

Direct—a courteous general request sets up the specific questions

We have selected Chicago for our 2015 meeting on August 16, 17, and 18 and are interested in holding it at the Drake. I have found helpful information on your website but would like answers to the following questions:

Opening provides brief background information and sets up the specific questions

Can you accommodate a group of about 600 employees on these dates?

Would you be able to set aside a block of 450 rooms? We could guarantee 400.

Do you offer discounted room rates for meetings and conferences? If so, what are the rates and what are the eligibility requirements?

Specific explanations are provided where needed

What are your charges for conference rooms? We will need eight for each of the three days, and each should have a minimum capacity of 60. On the 18th, for the half-hour business meeting, we will need a large ballroom with a capacity of at least 500.

Questions stand out in separate paragraphs

Also, will you please send me your menu selections and prices for group dinners? On the 17th we plan to hold our presidential dinner. About 500 can be expected for this event.

As meeting plans must be announced by September, may we have your response by August 1? We look forward to the possibility of being with you in 2015.

Individually tailored close builds goodwill

Sincerely,

Patti Wolff

Patti Wolff
Site Selection Committee Chair

workplace scenario

Answering a Potential Customer's Questions

Continue in your role as assistant to the vice president for operations of White Label Industries (WLI). This time, your task is to respond to a customer's message.

In your email inbox this morning, you have an inquiry from a veterinarian, a Dr. Motley, who wants to know more about WLI's Chem-Treat paint. In response to an advertisement, this prospective customer asks a number of specific questions about Chem-Treat. Foremost, she wants to know whether the paint is safe to use around the animals (and their owners) who visit her clinic. Do you have supporting evidence? Do you guarantee the results? Does the paint resist dirt and stains? How much does a gallon cost? Will one coat do the job?

You can answer all but one of the questions positively. Of course, you will report this one negative point (that two coats are needed to do most jobs), but you will take care to de-emphasize it. The response will be primarily a good-news message. Because the reader is a potential customer, you will work to create the best goodwill effect.

An alternative is to begin by stating that you are giving the reader what he or she wants—that you are complying with the request. This example illustrates this type of beginning:

> Here are the answers to your questions about Chem-Treat.

Logically Arranging the Answers

If you are answering just one question, you have little to do after handling that question in the opening. You answer it as completely as the situation requires, and you present whatever explanation or other information is needed. Then you are ready to close the message.

If, on the other hand, you are answering two or more questions, the body of your message becomes a series of answers. You should order them logically, perhaps answering the questions in the order your reader used in asking them. You may even number your answers, especially if your reader numbered the questions. Or you may decide to arrange your answers by paragraphs so that each stands out clearly.

Skillfully Handling the Negatives

When your response will include some bad news along with the good news, you will need to handle the bad news with care. Unless you are careful, it is likely to receive more emphasis than it deserves.

> " You should place the good news in positions of high emphasis—at paragraph beginnings and endings and at the beginning and ending of the message as a whole. "

In giving proper emphasis to the good- and bad-news parts, you should use the techniques discussed in Chapter 4, especially positioning. That is, you should place the good news in positions of high emphasis—at paragraph beginnings and endings and at the beginning and ending of the message as a whole. You should place the bad news in secondary positions. In addition, you should use space emphasis to your advantage. This means giving less space to bad-news parts and more space to good-news parts. You also should select words and build sentences that communicate the effect you want. Generally, this means using positive words and avoiding negative words and putting bad news in modifying phrases or clauses rather than in main clauses. Your overall goal should be to present the information in your response so that your readers feel good about you and your company.

Considering Extras

To create goodwill, as well as future business, you should consider including extras with your answers. These are the things you say and do that are not actually required. Examples are a comment or question showing an interest in the reader's situation, some additional information that may prove valuable, or a suggestion for use of the information supplied. In fact, extras can be anything that presents more than the routine response. A business executive answering a college professor's request for information on company operations could supplement the requested information with suggestions of other sources. A technical writer could explain highly technical information in simpler language. In responding to the Chem-Treat problem described in the Workplace Scenario on page 108, you could provide additional information (e.g., how much surface area a gallon covers) would be helpful. Such extras encourage readers to build a business relationship with you.

Closing Cordially

As in the other types of direct messages, your ending should be cordial, friendly words that fit the case. For example, you might close the Chem-Treat message with these words:

> If I can help you further in deciding whether Chem-Treat will meet your needs, please let me know.

Reviewing the Plan

To write a favorable response message, you should use the following plan:

- Identify, either incidentally or in the subject line, the message being answered.

- Begin with the answer or state that you are complying with the request.

- Continue to respond in a way that is logical and orderly.

- De-emphasize any negative information.

- Consider including extras.

- End with a friendly comment adapted to your reader.

Contrasting Examples of a Favorable Response

The contrasting email messages on pages 112 and 113 illustrate two strategies for answering routine inquiries. The first message violates much of the advice in this and earlier chapters. The second meets the requirements of a good business message. It meets the reader's needs and supports the writer's business goals.

an indirect and hurried response
The not-so-good message on page 112 begins indirectly with an obvious statement referring to receipt of the inquiry. Though well intended, the second sentence continues to delay the answers. The second paragraph begins to respond to the reader's request, but it emphasizes the most negative answer by position and by wording. This answer is followed by hurried and routine answers to the other questions asked. Only the barest information is presented. There is no goodwill close.

an effective direct response
The better message on page 113 begins directly with the most favorable answer. Then it presents the other answers, giving each the emphasis and positive language it deserves. It subordinates the one

negative answer by position, use of space, and structure. More pleasant information follows the negative answer. The close is goodwill talk with some subtle selling strategy included.

LO 5-5 Write order acknowledgments and other thank-you messages that build goodwill.

ORDER ACKNOWLEDGMENTS AND OTHER THANK-YOU MESSAGES

In the course of your professional career, you will find yourself in situations where business and social etiquette require thank-you messages. Such messages may be long or short, formal or informal. They may be also combined with other purposes such as confirming an order. In this section we focus on one specific

Routine Response
(Favorable Response to a Professor's Request)

This email message responds to a professor's request for production records that will be used in a research project. The writer is giving the information wanted but must restrict its use. Notice that the message emphasizes the positive points.

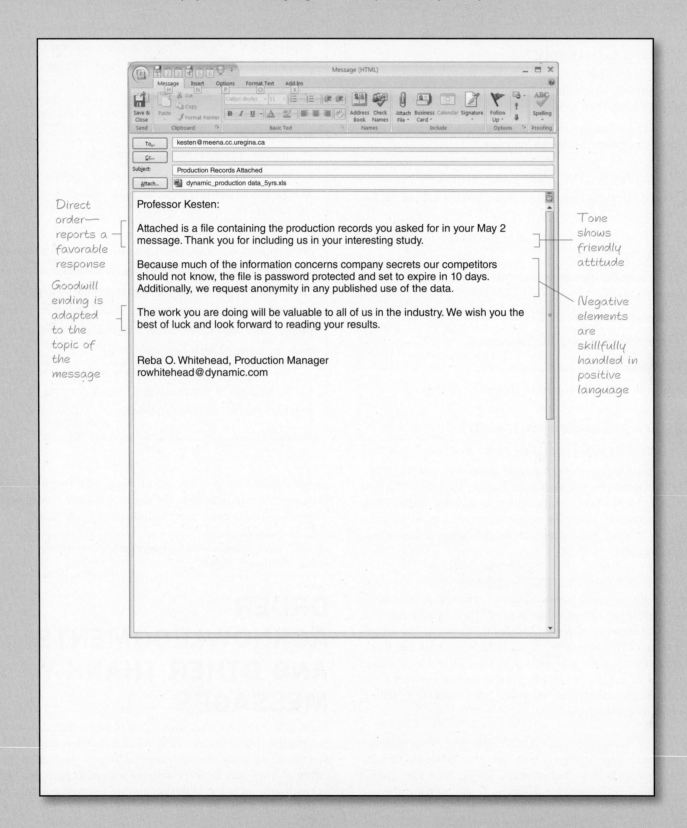

Direct order—reports a favorable response

Goodwill ending is adapted to the topic of the message

To... kesten@meena.cc.uregina.ca

Cc...

Subject: Production Records Attached

Attach... dynamic_production data_5yrs.xls

Professor Kesten:

Attached is a file containing the production records you asked for in your May 2 message. Thank you for including us in your interesting study.

Because much of the information concerns company secrets our competitors should not know, the file is password protected and set to expire in 10 days. Additionally, we request anonymity in any published use of the data.

The work you are doing will be valuable to all of us in the industry. We wish you the best of luck and look forward to reading your results.

Reba O. Whitehead, Production Manager
rowhitehead@dynamic.com

Tone shows friendly attitude

Negative elements are skillfully handled in positive language

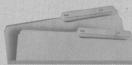

case illustration
Routine Response
(Answering a Request for Detailed Information)

Answering an inquiry about a company's experience with executive suites, this letter numbers the answers as the questions were numbered in the inquiry. The opening appropriately sets up the numbered answers with a statement that indicates a favorable response.

Merck & Co., Inc.
One Merck Drive
P.O. Box 100, WS1A-46
Whitehouse Station NJ 08889

⬤ MERCK

August 7, 2014

Ms. Ida Casey, Sales Manager
Liberty Insurance Company
1165 Second Ave.
Des Moines, IA 50318-9631

Dear Ms. Casey:

Direct opening makes the purpose clear

Here is the information about our use of temporary executive suites that you requested in our August 3 phone conversation.

List format aids comprehension

1. Our executives have mixed feelings about the effectiveness of the suites. At the beginning, the majority opinion was negative, but it appears now that most believe the suites meet our needs.

2. The suites option definitely has saved us money. Rental costs in the suburbs are much lower than downtown costs; annual savings are estimated at nearly 30 percent.

3. We began using executive suites at the request of several sales representatives who had read about other companies using them. We pilot tested the program in one territory for a year using volunteers before we implemented it companywide.

4. We are quite willing to share with you the list of facilities we plan to use again. Additionally, I am enclosing a copy of our corporate policy, which describes our guidelines for using executive suites.

Answers are complete yet concise

Friendly close is adapted to the one case

If after reviewing this information you have any other questions, please write me again. If you want to contact our sales representatives for firsthand information, please do so. I wish you the best of luck in using these suites in your operations.

This extra builds goodwill

Sincerely,

David M. Earp

David M. Earp
Office Manager

Enclosure

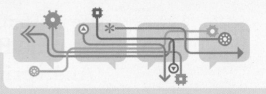
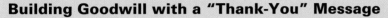

Building Goodwill with a "Thank-You" Message

The next work you take from your inbox is an order for paints and painting supplies. It is from Mr. Tony Lee of Central City Paint Company, a new customer whom White Label Industries (WLI) has been trying to attract for months. You usually acknowledge orders with routine messages, but this case is different. You feel the need to welcome this new customer and to cultivate future business with him.

After checking your current inventory and making certain that the goods will be on the way to Mr. Lee today, you are ready to write him a special acknowledgment and thank him for his business.

kind of thank-you message—the **order acknowledgment**—as well as more general thank-you messages for other business occasions.

Order Acknowledgments

Acknowledgments are sent to let people who order goods know the status of their orders. Most acknowledgments are routine. They simply tell when the goods are being shipped. Many companies use form or computer-generated messages for such situations. Some use printed, standard notes with check-off or write-in blanks. But individually written acknowledgments are sometimes justified, especially with new accounts or large orders.

Skillfully composed acknowledgments can do more than acknowledge orders, though this task remains their primary goal. These messages can also build goodwill through their warm, personal, human tone. They can make the reader feel good about doing business with a company that cares and want to continue doing business with that company. To maintain this goodwill for repeat customers, you will want to revise your form acknowledgments regularly.

Directness and Goodwill Building in Order Acknowledgments

Like the preceding direct message types, the acknowledgment message appropriately begins with its good news—that the goods are being shipped—and it ends on a goodwill note. Except when some of the goods ordered must be delayed, the remainder of the message is devoted to goodwill building. This goodwill building can begin in the opening by emphasizing receipt of the goods rather than merely the shipment of the goods:

> The Protect-O paints and supplies you ordered April 4 should reach you by Wednesday. They are leaving our Walden warehouse today by Arrow Freight.

This response to an inquiry is indirect and ineffective.

Subject: Your Inquiry of April 3

Dr. Motley,

I have received your April 3 message, in which you inquire about our Chem-Treat paint. I want you to know that we appreciate your interest and will welcome your business.

In response to your question about how many coats are needed to cover new surfaces, I regret to report that two are usually required. The paint has been well tested in our laboratories and is safe to use as directed.

Ray Lindner
Customer Service Representative

from the tech de[sk]

Tables Help Writers Organize Data for Easy Reading

Setting up tables within a document is an easy task thanks to tools in today's word-processing programs. This feature allows writers to create tables as well as import spreadsheet and database files. In both instances, information can be arranged in columns and rows.

Headings can be formatted, and formulas can be entered to generate data in the cells. The table you see here could be one the writer created for use in a favorable response to an inquiry about possible locations for a meeting in Chicago.

Organizing information with tables make[s] it easier for the reader to get the information quickly and concisely. A careful writer will include column and row labels as needed, helping the reader extract information both quickly and accurately.

Hotel Name	Address	Convention Room Rate for Standard Rooms	Guest Rating
Chicago Marriott Downtown	540 North Michigan Avenue, Chicago, IL 60611-3869	$409	4.2
Drake Hotel	140 East Walton Street, Chicago, IL 60611-1545	$309	4.3
Palmer House Hilton	17 East Monroe Street, Chicago, IL 60603-5605	$252	4.4

Subject: Your April 3 Inquiry about Chem-Treat

Dr. Motley:

Yes, Chem-Treat's low-odor, low-VOC acrylic latex paint is among the safest, most environmentally friendly paints on the market. Several hospitals and clinics have used Chem-Treat successfully and have reported no reactions to it.

Chem-Treat's latex formula also makes it ideal for high-traffic areas. Cleaning usually requires nothing more than a little soap and water.

One gallon of Chem-Treat is usually enough for one-coat coverage of 500 square feet of previously painted surface. For the best results on new surfaces, you will want to apply two coats. For such surfaces, you should figure about 200 square feet per gallon for a long-lasting coating.

We appreciate your interest in Chem-Treat, Dr. Motley. You can view Chem-Treat's safety ratings and customer reviews on our Web site: www.wli.com/chemtreat/features.

Ray Lindner
Customer Service Representative

This direct response does a better job of answering the reader's questions directly and concisely.

of thanks for the order,
...ved. Anything else you
...ader is appropriate—
..., or opportunities for
...usiness relations is a

...ments

...licated by your
...way. You could be
...u not give you all the infor-
...o send them. In either case, a delay
... cases, delays are routine and expected and
... a serious problem. In these situations, you can use
...direct approach. However, you will still want to minimize any negative news so that your routine message does not become a negative-news message. You can do this by using positive language that focuses on what *can* or *will* happen rather than what didn't or won't happen.

In the case of a vague order, for example, you should request the information you need without appearing to accuse the reader of giving insufficient information. To illustrate, you risk offending the reader by writing "You failed to specify the color of phones you want." But you gain goodwill by writing "So that we can send you the phones you want, please check your choice of colors on the space below." This sentence handles the matter positively and makes the action easy to take. It also shows a courteous attitude.

Similarly, you can handle back-order information tactfully by emphasizing the positive part of the message. For example, instead of writing "We can't ship the ink jet cartridges until the 9th," you can write "We will rush the ink jet cartridges to you as soon as our stock is replenished by a shipment due May 9." If the back-order period is longer than the customer expects or longer than the 30 days allowed by law, you may choose to give your customer an alternative, such as a substitute product or service. Giving the customer a choice builds goodwill.

In some cases delays will lead to major disappointments, which means you will have to write a bad-news message. A more complete discussion of how to handle such negative news is provided in Chapter 6.

Strategies for Other Thank-You Messages

One of the first thank-you messages you write will be the one for a job interview, which is discussed in Chapter 11. Once you are employed, you may send thank-you messages after a meeting or when someone does a favor for you or gives you a gift, when you want to acknowledge others' efforts that have somehow benefited you, when you want to thank customers for their business, or perhaps when someone has donated time or money to your organization or a cause it supports. The possibilities for situations when you might send thank-you notes are many, and sending a message of sincere thanks is a great way to promote goodwill and build your and your company's professional image.

This order acknowledgment delays the important news.

Dear Mr. Lee:

Your April 4 order for $1,743.30 worth of Protect-O paints and supplies has been received. We are pleased to have this nice order and hope that it marks the beginning of a long relationship.

As you instructed, we will bill you for this amount. We are shipping the goods today by Blue Darter Motor Freight.

We look forward to your future orders.

Sincerely,

> ## " Sending a message of sincere thanks is a great way to promote goodwill and build your and your company's professional image. "

Thank-you messages are often brief, and because they are positive messages, they are written directly. You can begin with a specific statement of thanks:

> Thank you for attending the American Cancer Society fundraiser lunch for Relay for Life last week and for donating money to the cause.

Follow with a personalized comment relevant to the reader:

> With your support, the 2014 Relay for Life will be our most successful yet . . . [details follow].

Conclude with a forward-looking statement:

> I look forward to joining you on June 12 for this worthy cause.

Your tone should be informal and friendly. If you are on a first-name basis with the reader, you may omit a salutation or use the reader's first name, but if your relationship with the reader is a formal one, do not use the reader's first name to create a contrived sense of closeness.

Whether you hand write the thank-you, send an email, or use company stationery depends on the audience. If you have poor handwriting or believe your handwriting does not convey a professional image, you may choose to type your message. Though you should always check your own spelling, grammar, and punctuation before sending any message, doing so is especially important in handwritten notes when you have no computer software to alert you to possible errors.

Summarizing the Structure of Order Acknowledgments and Other Thank-You Messages

To write an order acknowledgment or thank-you message,

- Use the direct order: Begin by thanking the reader for something specific (e.g., an order).
- Continue with your thanks or with further information.
- Use positive, tactful language to address vague or delayed orders.
- If appropriate, achieve a secondary goal (e.g., reselling or confirming a mutual understanding).
- Close with a goodwill-building comment, adapted to the topic of the message.

Contrasting Examples of an Order Acknowledgment

The two messages on pages 114 and 115 show bad and good ways to acknowledge Mr. Lee's order. As you would expect, the good version follows the plan described in the preceding paragraphs.

Dear Mr. Lee:

Your selection of Protect-O paints and supplies was shipped today by Blue Darter Freight and should reach you by Wednesday. As you requested, we are sending you an invoice for $1,743.30, including sales tax.

Welcome to the Protect-O circle of dealers. Our representative, Ms. Cindy Wooley, will call from time to time to offer whatever assistance she can. She is a highly competent technical adviser on paint and painting.

Here in the home plant we also will do what we can to help you profit from Protect-O products. We'll do our best to give you the most efficient service. And we'll continue to develop the best possible paints—like our new Chem-Treat line. As you will see from the enclosed brochure, Chem-Treat is a real breakthrough in mildew protection.

Thank you for your order, Mr. Lee. We are determined to serve you well in the years ahead.

Sincerely,

This direct response thanks the reader immediately and builds goodwill.

communication matters

A Workplace without Email? One Company's Strategy

French tech company Atos discovered that its 80,000 employees were spending 15–20 hours per week on email but finding only 15 percent of the email actually necessary.

The company's solution to making communication more productive is to phase out email as an internal communication tool, with the goal of eliminating it altogether by 2014. Instead, employees will use social media, instant messaging, and collaboration tools such as Microsoft's Live Meeting for their internal communication. In addition to eliminating unnecessary communication in the office, the company says moving to communication technologies such as instant messaging means that people will better balance their work and personal lives, as they cannot check for messages outside the office as easily as they could if they were using email.

Responses to this strategy have ranged from enthusiasm and support for the idea to skepticism, the latter arguing that people will just substitute time spent on instant messaging and social media for the time they would have spent on email. Many responses note (rightly so) that email, like instant messaging and social media, is just the technology—ultimately the users control how productively it is used.

Source: "IT Firm Phasing Out Email to Boost Productivity," *CBC News Technology* and *Science,* CBS News, 16 Dec. 2011, Web, 15 May 2013.

> [Many times the easiest and quickest way for you to address these claims is simply to call the company directly to settle the matter.]

slow route to a favorable message The bad example begins indirectly, emphasizing receipt of the order. Although intended to produce goodwill, the second sentence further delays what the reader wants most to hear. Moreover, the letter is written from the writer's point of view (note the we-emphasis).

fast-moving presentation of the good news The better message begins directly, telling Mr. Lee that he is getting what he wants. The remainder of the message is a customer welcome and subtle selling. Notice the good use of reader emphasis and positive language. The message closes with a note of appreciation and a friendly, forward look.

LO 5-6 Write direct claims in situations where an adjustment will likely be granted.

DIRECT CLAIMS

Occasionally things go wrong between a business and its customers (e.g., merchandise is lost or broken during shipment, customers are inaccurately billed for goods or services). Such situations are not routine for a business; for most businesses, the routine practice is to fulfill their customers' expectations.

Because claim messages are not about routine circumstances and because they involve unhappy news, many are written in the indirect approach discussed in Chapter 6. Nevertheless, there are some instances where directness in writing a claim is appropriate, and for this reason we discuss the direct claim in this chapter.

Using Directness for Claims

Most businesses want to know when something is wrong with their products or services so they can correct the matter and satisfy their customers. Many times the easiest and quickest way for you to address these claims is simply to call the company directly to settle the matter. Sometimes, though, you will want to write a claim in order to have a written record of the request. Or, depending on a company's phone options for accessing customer service, a written claim sent via email or the company Web site may be more efficient than a phone call.

When writing a claim in cases where you anticipate that the reader will grant an adjustment of your claim, you may use the direct approach (e.g., adjusting an incorrect charge to an invoice). Be sure that when you write the claim, you keep your tone objective and professional so that you preserve your reader's goodwill. If you use words such as *complaint* or *disappointment*, you will compromise your chances of receiving an adjustment quickly.

Online Order Acknowledgment
(Order Confirmation with a Second Purpose)

This email message thanks the reader for her order and invites her to participate in this company's online product review. It begins directly with a thank you to the reader. Then, to accomplish its secondary goal (inviting participation in the survey), the message incorporates reader benefits and motivates the reader to action by offering an incentive..

From: Gardeners Supply [mailto:gardeners@e-news.gardeners.com]
Sent: Thursday, January 08, 2014 9:08 AM
To: KATHRYN.RENTZ@UC.EDU
Subject: Tell Us What You Think About Our Products

GARDENER'S SUPPLY COMPANY

New Feature: Customer Reviews

Dear Kathryn,

Thanks the reader and indicates a shared interest

Thank you for your purchase from Gardener's Supply. We hope you are enjoying your items and that this year's garden will be your best ever!

Your satisfaction with our products is important to us, and we want to hear what you have to say about them. We recently added customer reviews to our Web site, which helps us improve our product selection and helps other gardeners find the best products to suit their needs.

Moves to another goal of the message

Adds a reader benefit and incentive

We're hoping you'll take a moment to rate and review some or all of the items you have purchased from us. Other gardeners will appreciate your opinions and advice, and you may also enjoy reading what fellow gardeners have to say!

Each time you submit a product review to our Web site, your name will be entered in a monthly drawing for a $1,000 prize (see information below).

Here are the item(s) you recently purchased. Just click on an item to write a review.

Men's Waterproof Gloves
★ <u>Rate and review it</u>

Glove Set, 3 Pairs
★ <u>Rate and review it</u>

Pictures provide a quick visual confirmation of the order

Links make participation easy

Forward-looking ending builds goodwill

Thank you again for shopping with us.

The Employee-Owners at Gardener's Supply

Copyright @2008 America's Gardening Resource, Inc.

Order Acknowledgment
(Acknowledgment with a Problem)

This email letter concerns an order that cannot be handled exactly as the customer would like. Some items are being sent, but one must be placed on back order and another cannot be shipped because the customer did not give the information needed. The message skillfully handles the negative points.

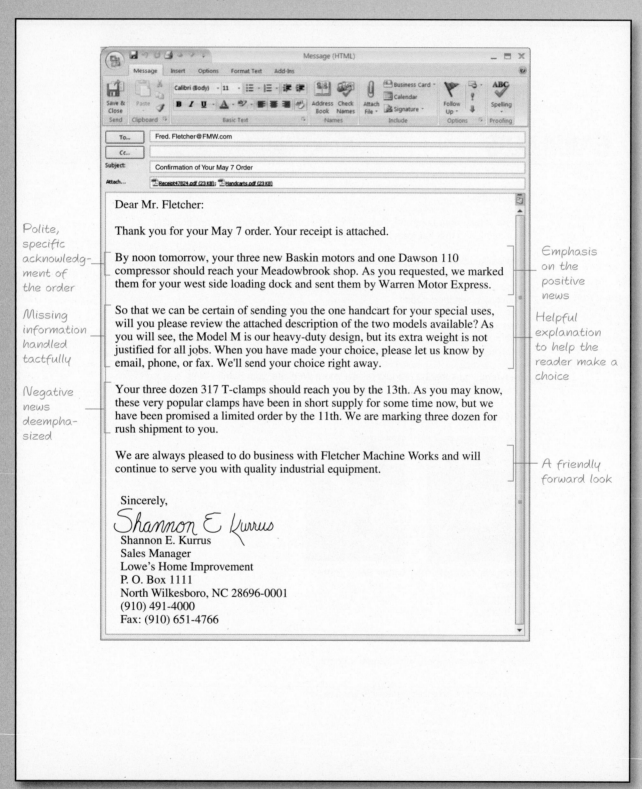

Polite, specific acknowledgment of the order

Missing information handled tactfully

Negative news deemphasized

Emphasis on the positive news

Helpful explanation to help the reader make a choice

A friendly forward look

Message (HTML)

To... Fred.Fletcher@FMW.com

Cc...

Subject: Confirmation of Your May 7 Order

Attach... Receipt47824.pdf (23 KB); Handcarts.pdf (23 KB)

Dear Mr. Fletcher:

Thank you for your May 7 order. Your receipt is attached.

By noon tomorrow, your three new Baskin motors and one Dawson 110 compressor should reach your Meadowbrook shop. As you requested, we marked them for your west side loading dock and sent them by Warren Motor Express.

So that we can be certain of sending you the one handcart for your special uses, will you please review the attached description of the two models available? As you will see, the Model M is our heavy-duty design, but its extra weight is not justified for all jobs. When you have made your choice, please let us know by email, phone, or fax. We'll send your choice right away.

Your three dozen 317 T-clamps should reach you by the 13th. As you may know, these very popular clamps have been in short supply for some time now, but we have been promised a limited order by the 11th. We are marking three dozen for rush shipment to you.

We are always pleased to do business with Fletcher Machine Works and will continue to serve you with quality industrial equipment.

Sincerely,

Shannon E. Kurrus

Shannon E. Kurrus
Sales Manager
Lowe's Home Improvement
P. O. Box 1111
North Wilkesboro, NC 28696-0001
(910) 491-4000
Fax: (910) 651-4766

Thank-You Message
(A Follow-Up to a Meeting)

This email from a representative of a telecommunications equipment company thanks a potential customer in Germany for a recent meeting. She also tactfully reminds him of his offer to help.

Hello, Herman:

Meeting you and the other members of the product selection team last Friday was a pleasure. We are honored not only that you took the time to explain the current dynamics and structure at General Telekom but also that you gave us your entire day. Thank you for your generosity.

Uses the reader's first name, which indicates a close business relationship

Incorporates specifics to add sincerity

I understand that you have graciously offered to share some of United Plorcon's key product qualities with other executives in your company. We are extremely appreciative that you have offered to help in this way. Based on our discussions at the meeting, we will summarize the main points of interest and send them to you.

Politely thanks the reader for something specific (while also reminding him of his promise)

To make the cost-benefit charts more self-explanatory, I will slightly condense the material and add some notes. They will not be confidential, so please share them if you wish. I will send those condensed charts to you before the end of the week. After we send them, I will follow up with you to make sure that you have all the material you need to explain who we are and what we offer.

Prepares for future communications between the two parties

Finally, thank you for your hospitality. Staying in an ideally located hotel, having coffee at a beautiful castle, viewing some of the world's best art, and sharing an exquisite meal made my first visit to Stuttgart a most memorable one. I hope that between our two companies, we can create many opportunities for our United Plorcon team to return to your beautiful city.

Ends on a positive note; looks to the future

I look forward to talking with you again soon.

Respectfully yours,

workplace scenario

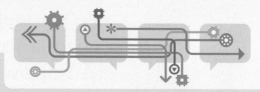

Requesting a Correct Shipment and Revised Invoice

Continue in your role with White Label Industries (WLI). As the assistant to the vice president of operations, you manage the supervisors on the paint production line. Today, one of the team leaders came to you for some feedback on his writing. Last week, he ordered some safety equipment (goggles and face masks) for employees on the production line; however, he received gloves instead of goggles and only half the masks he ordered, as well as an invoice for $50 more than his order should have cost. He is fairly certain he received someone else's order. He just wants the correct order shipped and his bill adjusted accordingly.

He called his sales representative but got her voice mail. He left her a quick message to call him about the order, but he also wants to send her an email explaining the situation. He asks for your feedback on the email, and you are surprised at the indirect language and unnecessarily negative and harsh tone. You need to use what you know about writing a direct claim to help him write a message that not only resolves the issue but also builds goodwill.

Organizing the Direct Claim

Because you anticipate that the reader will willingly grant your request, a direct claim begins with the claim, moves to an explanation, and ends with a goodwill closing.

beginning a direct claim The direct claim should open with just that—the direct claim. This should be a polite but direct statement of what you need. If the statement sounds too direct, you may soften it with a little bit of explanation, but the direct claim should be at the beginning of your message, as in this example:

> Please adjust the invoice (# 6379) for our May 10 order to remove the $7.50 shipping charge.

explaining the issue The body of the direct claim should provide the reader with any information he or she

This claim is insulting, too indirect, and too long.

Subject: Problem with Our Order #2478

Beth,

As you know, White Label Industries has been ordering our safety supplies from you for over 15 years. We have always depended on you for quick and accurate service, which, unfortunately, looks like it didn't happen this time. When our orders are not accurate and our safety gear is not what we ordered, you put our employees in jeopardy, and WLI loses money if employees don't have safety gear and can't work.

You can imagine how shocked I was when I opened the order expecting face masks and goggles but found gloves and half the face masks I ordered. I was also surprised to see a bill that was $50 more than I planned.

I tried to call you, but you didn't answer, so I left a voice message. I'm guessing you'll want to fix this quickly, so please call me or email me and let me know what you are going do. If my employees do not have the masks and goggles by tomorrow, employees can't work, and we will have to shut down our production line, which will cost us a lot of money. This was really disappointing service, but I'm sure it won't happen again, as you have always been accurate in the past. We would hate to think that we need to go with a different supplier. Thank you.

Ken

from the tech desk

Quick Parts Makes Quick Work for Business Writers

If you don't already do so, you may someday find yourself frequently using the same text in multiple email messages. For example, your company may require that you include a privacy statement or promotional message with every email sent outside the company. Or perhaps you have a standard message you send to customers who request information from you. Or maybe you frequently send reminder emails to your staff.

Whatever the scenario, if you find yourself typing the same information repeatedly, you may want to take advantage of Outlook's Quick Parts feature. Its use is simple:

1. Type your text in the message area of your email.
2. Select your text.
3. Go to Insert » Quick Parts » Save Selection to Auto Parts Gallery.

From there you can label and save the text you selected. As you create a library of Quick Parts features, you can also organize your texts by category. Then, whenever you need to use that text in an email, just go to the Quick Parts feature and select your text from your library. It will be automatically inserted in your message.

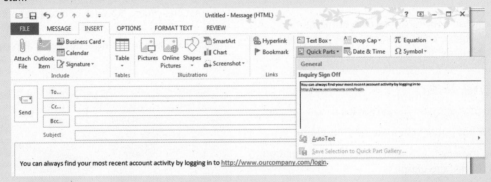

Subject: Need Correct Order Shipped (Invoice 6750)

Beth,

Please send 50 safety goggles and 100 face masks, as well as a new invoice, to replace the incorrect order that arrived this morning.

On Monday, I placed the order for the safety goggles and face masks, but today I received only 50 face masks and 15 boxes of safety gloves. The invoice indicated that this order was supposed to go to J&M Medical Supplies.

If you send the order today, we should receive it tomorrow. Our employees will need the safety equipment in order for us to keep the production line running.

Please let me know how you want to handle the return of J&M's order.

Thanks,

Ken

This direct claim is clear and efficient while also maintaining goodwill.

might need to understand your claim. To continue with the same example, we might write the following brief middle paragraph:

> Because our order totaled $73.50, we were able to take advantage of your offer for free shipping on orders of $50 or more and should not have been charged a shipping fee.

providing a goodwill closing Your close should end with an expression of goodwill. A simple ending like the following can suffice:

> Please send a corrected copy of the invoice to me at jsmith@americanmortgage.com. We look forward to continued business with National Office Supplies.

Reviewing the Plan

To write a direct claim message, you should use the following plan:

- Begin with a direct but polite statement of your claim.
- In the body of the message, give the reader the information he or she needs to adjust the claim.
- Close with an expression of goodwill.

Contrasting Examples of a Claim Message

The two email messages on page 120 and page 121 show contrasting ways of handling the erroneous shipment described in the Workplace Scenario on page 120. The first is slow and harsh. The second is courteous, yet to the point and firm.

an indirect and harsh message The first message starts slowly with a long explanation of the situation. Some of the details in the beginning sentence are helpful, but they do not deserve the emphasis that this position gives them. The problem is not described until the second paragraph. The wording here is clear but much too strong. The words are angry and insulting, and the writer talks down to the reader. Such words are more likely to produce resistance than acceptance. The negative writing continues into the close, leaving a bad final impression.

a firm yet courteous message The second message follows the plan suggested in the preceding paragraphs. A subject line quickly and neutrally identifies the situation. The message begins with a clear statement of the claim. Next, it uses objective language to tell what went wrong. The ending is

rational and shows that the writer is interested in resolving the issue, not placing blame.

LO 5-7 Compose adjustment grants that regain any lost confidence.

ADJUSTMENT GRANTS

When you can grant an adjustment, the situation is a happy one for your customer. You are correcting an error. You are doing what you were asked to do. As in other positive situations, a message written in the direct order is appropriate.

In most face-to-face business relations, people communicate with courteous directness. You should write most business messages this way.

workplace scenario

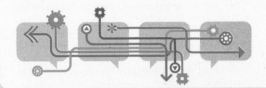

Dealing with the Unexpected

Continuing in your role with White Label Industries (WLI), this time you find on your computer an email message from an unhappy customer. It seems that Ms. Bernice Watson, owner of Tri-Cities Hardware, is upset because some of the 30 Old London lampposts she ordered from WLI arrived in damaged condition. "The glass is broken in 17 of the units," she writes, "obviously because of poor packing." She had ordered the lights for a special sale. In fact, she had even featured them in her advertising. The sale begins next Friday. She wants a fast adjustment—either the lamps by sale time or her money back.

Of course, you will grant Ms. Watson's request. You will send her an email message saying that the goods are on the way. And because you want to keep this good customer, you will try to regain any lost confidence with an honest explanation of the problem. This message is classified as an adjustment grant.

Considering Special Needs

The adjustment-grant message has much in common with the message types previously discussed. You begin directly with the most important point—here, the good news that you are granting the adjustment. You refer to the message you are answering, and you close on a friendly note. Because the situation stems from an unhappy experience, though, you have two special needs. One is the need to overcome any negative impressions caused by the experience. The other is the need to regain any confidence in your company, its products, or its service that the reader may have lost.

need to overcome negative impressions

To understand the first need, just place yourself in the reader's shoes. As the reader sees it, something bad has happened—goods have been damaged, equipment has failed, or sales have been lost. The experience has not been pleasant. Granting the claim will take care of much of the problem, but some negative thoughts may remain. You need to work to overcome any such thoughts.

You can attempt to do this using words that produce positive effects. For example, in the opening you can do more than just give the affirmative answer. You can add goodwill, as in this example:

> The enclosed check for $89.77 is our way of showing you that your satisfaction is our priority.

Throughout the message you should avoid words that unnecessarily recall the bad situation you are correcting. You especially want to avoid the negative words that could be used to describe what went wrong—words such as *mistake, trouble, damage, broken*, and *loss*. Even general words such as *problem, difficulty*, and *misunderstanding* can create unpleasant effects. Negative language makes the customer's complaint the focus of your message. Your goal is to move the customer beyond the problem and to the solution—that the customer is going to have his or her claim granted. You can only do this if you use positive, reader-centered language.

Also negative are the apologies often included in these messages. Even though well intended, the somewhat conventional "we sincerely regret the inconvenience . . ." type of comment is of questionable value because it emphasizes the negative happenings for which the apology is made. If you sincerely believe that you owe an apology or that one is expected, you can apologize early in the message and risk the negative effect. Don't repeat the apology at the end of the message or include an apology in the conclusion. In most instances, however, your efforts to correct the problem will show adequate concern for your reader's interests.

need to regain lost confidence

Except in cases in which the cause of the difficulty is routine or incidental, you also will need to regain the reader's lost confidence. Just what you must do and how you must do it depend on the situation. If something can be done to correct a bad procedure or a product defect, you should do it. Then you should tell your reader what has been done as convincingly and positively as you can. If what went wrong was a rare, unavoidable event, you should explain this. Sometimes you will need to explain how a product should be used or cared for. Sometimes you will need to resell the product. Whatever you say should be truthful, professional, and reader focused.

Reviewing the Plan

To organize a message granting an adjustment, writers should use the following plan:

- Begin directly—with the good news.

- Incidentally identify the correspondence that you are answering.

- Avoid negatives that recall the problem.

- Regain lost confidence through explanation or corrective action.

- End with a friendly, positive comment.

Contrasting Examples of an Adjustment Grant

The two messages on this page and the next illustrate an ineffective and effective way to write adjustment messages. The first, with its indirect order and grudging tone, is ineffective. The directness and positiveness of the second clearly make it the better message.

a slow and negative approach The ineffective message begins with an obvious comment about receiving the claim. It recalls vividly what went wrong and then painfully explains what happened. As a result, the good news is delayed for an additional paragraph. Finally, after two delaying paragraphs, the message gets to the good news. Though well intended, the close leaves the reader with a reminder of the trouble.

the direct and positive technique The better message uses the subject line to identify the transaction. The opening words tell the reader what she most wants to hear in a positive way that adds to the goodwill tone of the message.

With a you-viewpoint explanation, the message then reviews what happened. Without a single negative word, it makes clear what caused the problem and what has been done to prevent its recurrence. After handling the essential matter of picking up the broken lamps, the message closes with positive talk.

LO 5-8 Write clear and effective internal-operational communications.

INTERNAL-OPERATIONAL MESSAGES

As Chapter 1 explained, **internal-operational communications** are those messages that stay within a business. They are messages to and from employees that get the work of the organization done. The memorandums discussed in Chapter 2 are one form of operational communication. Internal email messages are another, and so are the various documents posted on bulletin boards, mailed to employees, uploaded on intranets, or distributed as handouts.

The formality of such messages ranges widely. At one extreme are the casual memorandum and email exchanges between employees concerning work matters. At the other are formal documents communicating company policies, directives, and procedures. Then, of course, there are the various stages of formality in between.

This adjustment grant is indirect and negative.

Subject: Your Broken Old London Lights

Ms. Watson,

We have received your May 1 claim reporting that our shipment of Old London lamppost lights reached you with 17 broken units. We regret the inconvenience and can understand your unhappiness.

Following our standard practice, we investigated the situation thoroughly. Apparently the fault is the result of an inexperienced temporary employee's negligence. We have taken corrective measures to assure that future shipments will be packed more carefully.

I am pleased to report that we are sending replacements today. They should reach you before your sale begins. Our driver will pick up the broken units when he makes delivery.

Again, we regret all the trouble we caused you.

Stephanie King

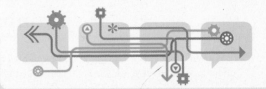

workplace scenario

Reminding Employees of the Shipping Policy

As the administrative assistant for the vice president of operations at White Label Industries (WLI), you have been asked by your boss to send a note on her behalf to all employees reminding them of the company's shipping policy. Whether customers pay shipping charges depends on the products they order. However, some customers who repeatedly order the same product are sometimes charged for shipping and sometimes are not, which, of course, leads to unhappy customers and is costly for WLI. Your challenge is to write a note that clearly and concisely explains the policy.

Casual Operational Messages

The documents at the bottom of the formality range typically resemble casual conversation. Usually they are quick responses to work needs. Rarely is there time or need for careful construction and wording. The goal is simply to exchange the information needed to conduct the company's work.

Frankness describes the tone of these casual operational messages as well as many of the messages at more formal levels. The participants exchange information, views, and recommendations forthrightly. They write with the understanding that all participants are working for a common goal—what is best for the company—and that people working together in business situations want and need straightforward communication.

Still, remember that being frank doesn't mean being impolite. Even in quick messages, you should build goodwill with a positive, courteous tone.

Moderately Formal Messages

Moderately formal messages tend to resemble the messages discussed earlier in this chapter. Usually they require more care in construction, and usually they follow a direct pattern. The most common arrangement begins with the most important point and works down. Thus, a typical beginning sentence is a topic (theme) statement. In messages in memorandum form, the opening repeats the subject-line information and includes the additional information needed to identify the situation. The

Subject: Your May 1 Report on Invoice 1248

Ms. Watson:

Seventeen carefully packed Old London lamppost lamps should reach your sales floor in time for your Saturday promotion. Our driver left our warehouse today with instructions to special deliver them to you on Friday.

Because your satisfaction with our service and products is important, we have thoroughly checked our shipping procedures. It appears that the shipment to you was packed by a temporary employee who was filling in for a hospitalized veteran employee. We now have our experienced packer back at work and have taken measures to ensure better performance by our temporary staff.

As you know, the Old London lamppost lights have become one of the hottest products in the lighting field. We are confident they will contribute to the success of your sale.

Stephanie King

This adjustment grant is direct and positive.

case illustration
Adjustment Grant (Explaining a Human Error)

This email message grants the action requested in the claim of a customer who received a leather computer case that was monogrammed incorrectly. The writer has no excuse because human error was to blame. His explanation is positive and convincing.

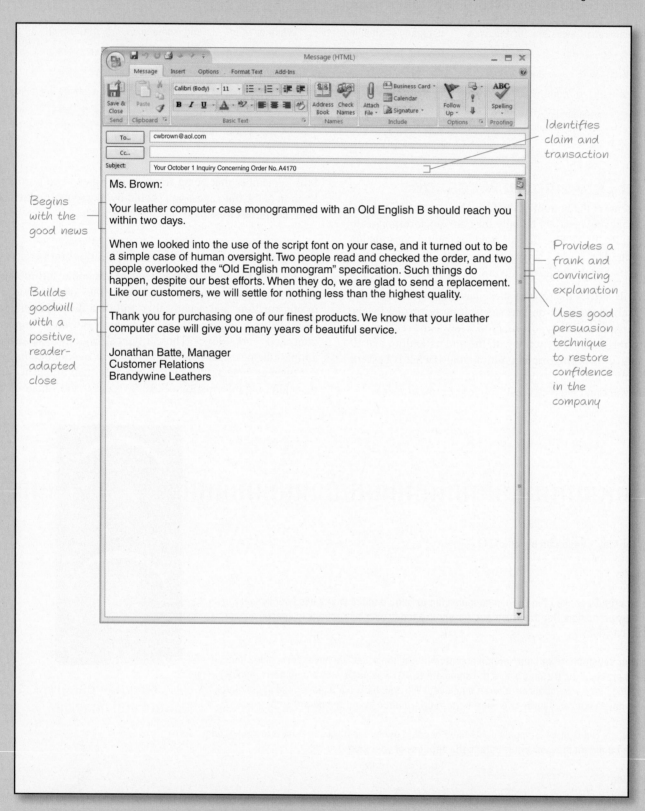

Identifies claim and transaction

Begins with the good news

Builds goodwill with a positive, reader-adapted close

Provides a frank and convincing explanation

Uses good persuasion technique to restore confidence in the company

Subject: Your October 1 Inquiry Concerning Order No. A4170

Ms. Brown:

Your leather computer case monogrammed with an Old English B should reach you within two days.

When we looked into the use of the script font on your case, and it turned out to be a simple case of human oversight. Two people read and checked the order, and two people overlooked the "Old English monogram" specification. Such things do happen, despite our best efforts. When they do, we are glad to send a replacement. Like our customers, we will settle for nothing less than the highest quality.

Thank you for purchasing one of our finest products. We know that your leather computer case will give you many years of beautiful service.

Jonathan Batte, Manager
Customer Relations
Brandywine Leathers

remainder of the message consists of a logical, orderly arrangement of the information covered. When the message consists of items in sequence, the items can be numbered and presented in this sequence.

Suggestions for writing moderately formal internal messages are much the same as those for writing the messages covered previously. Clarity, correctness, and courtesy should guide your efforts. The following example of a hard-copy memorandum illustrates these qualities. It is moderately formal, yet it is conversational. It is organized in the direct order, beginning with the objective and then systematically and clearly covering the vital bits of information. It is straightforward yet courteous.

DATE: April 1, 2014
TO: Remigo Ruiz
FROM: Becky Pharr
SUBJECT: Request for Cost Information Concerning
           Meeting at Timber Creek Lodge

As we discussed in my office today, please get the necessary cost information for conducting our annual sales meeting at the Timber Creek Lodge, Timber Creek Village, Colorado. Our meeting will begin on the morning of Monday, June 5; we should arrange to arrive on the 4th. We will leave after a brief morning session on June 9.

Specifically, we'll need the following information:

- Travel costs for all 43 participants, including air travel to Denver and ground travel between the airport and the lodge. I have listed the names and home stations of the 43 participants on the attached sheet.

- Room and board costs for the five-day period, including cost with and without dinner at the lodge. As you know, we are considering the possibility of allowing participants to purchase dinners at nearby restaurants.

- Costs for recreational facilities at the lodge.

- Costs for meeting rooms and meeting equipment (e.g., laptops, projectors). We will need a room large enough to accommodate our 43 participants.

I'd like to have the information by April 15. If you have any questions, please let me know.

Formal Messages

The most formal operational messages are those presenting policies, directives, and procedures. Usually written by executives for their subordinates, these administrative messages are often compiled in manuals, perhaps kept in loose-leaf form and updated as new material is developed or stored on a company intranet and updated as needed. Their official status accounts for their formal tone.

Formal operational messages usually follow a direct order, although the nature of their contents can require variations. The goal should be to arrange the information in the most logical order for quick understanding. Since the information frequently involves a sequence of information bits, numbering these bits can be helpful. And since these documents must be clearly understood and followed, the writing must be clear to all, including those with low verbal skills. The following example illustrates these qualities:

DATE: June 10, 2014
TO: All Employees
FROM: Terry Boedeker,
           President
SUBJECT: Energy Conservation

To help keep costs low, the following conservation measures are effective immediately:

- Thermostats will be set to maintain temperatures of 72 degrees Fahrenheit throughout the air-conditioning season.

- Air conditioners will be shut off in all buildings at 4 PM Monday through Friday.

- Air conditioners will be started as late as possible each morning so that buildings are at the appropriate temperature within 30 minutes after the start of the workday.

- Lighting levels will be reduced to approximately 80 to 100 watts in all work areas. Corridor lighting will be reduced to 50 watts.

- Outside lighting levels will be reduced as much as possible without compromising safety and security.

In addition, I ask that you do the following:

- Turn off lights not required in performing your work.

- Keep windows closed when the cooling system is operating.

- Turn off all computer monitors and printers at the end of the day.

I am confident that these measures will reduce our energy use significantly. Your efforts to follow them will be greatly appreciated.

Even though this message is straightforward, note the writer's courtesy and his use of *us* and *our*. When writing direct messages, skillful managers make use of such strategies for maintaining good relations with employees. Remembering this goal becomes especially important in situations where managers have news to convey or requests to make that employees may not be ready to accept. In fact, in these situations an indirect order will be more appropriate, as Chapters 6 and 7 will discuss. For most internal-operational communication, however, the direct order will be both expected and appreciated.

Summarizing the Structure of Internal-Operational Messages

To write an internal-operational message, writers should do the following:

- Organize in the direct order.

- Choose the appropriate tone (casual, moderately formal, or formal).

This indirect message wastes time and dwells on the negative.

Subject: Inconsistent Shipping Policies

WLI has been incurring increasing freight expenses and a decline in freight revenue over the last two years, impacting our ability to achieve our financial goals. The warehouse team has done a lot of research into the reasons behind this increase, and it has come to our attention that a very considerable number of shipments are going out of Cedar Rapids (1) as unbillable to the customer and/or (2) as overnight shipments rather than ground.

WLI has only one product for which shipping is not billed to the customer—the Chem-Treat paint. In all other cases, product shipments are supposed to be billed to the customer. ***Therefore, effective immediately, except for Chem-Treat shipments, which by contract provide for free overnight (weekday delivery) shipping, WLI will bill the customer for all shipments of products. Finance will screen all orders to ensure that they indicate billable shipping terms.***

WLI's overnight shipping falls into a few categories, including shipments of products to customers and shipments of marketing materials to prospects and customers. There are no customer programs or marketing programs for which WLI offers overnight shipping (except Chem-Treat). **Therefore, effective immediately, except for Chem-Treat shipments, which by contract provide for overnight (weekday delivery) shipping, WLI will not ship products overnight to customers unless the overnight shipping is billed to the customer. Also effective immediately, shipments of sales/marketing materials are to be shipped ground, not overnight.**

This policy change will impact some of your work processes, requiring you to be more planful in getting products shipped to customers in a businesslike and timely manner, and challenging you to prevent last-minute rush situations. I suspect that much of the freight performance situation, from a financial point of view, is an awareness issue for our Cedar Rapids team. I thank each of you in advance for adherence to this policy. We are fortunate to have an excellent distribution team in Cedar Rapids. That team needs all of our help so that their high-quality shipping and inventory control performance becomes matched by strong financial performance.

Exceptions to the billable shipping-only and no overnight shipping policies must be brought to me for approval prior to entering the order.

Dean Young
VP Operations

- Be clear and courteous.
- Order the information logically.
- Close in a way that builds goodwill.

Contrasting Examples of an Internal-Operational Message

The messages on pages 128 and 129 show contrasting ways in which the operational message regarding WLI's inconsistent shipping policies (see the Workplace Scenario on page 125) may be addressed.

a wordy, confusing, and indirect message
The reader really has to search for the writer's purpose and intent in the message on page 128. In addition, it is wordy, long, and disorganized and lacks visual appeal.

a direct, concise, and visually appealing message The message below is written directly and is more accurate because it communicates the point that this is a reminder of an existing policy, not an announcement of a new policy. This message is also more concise and gives the reader only the information he or she needs to know to comply. In addition, headings and bulleted lists make for much easier reading.

This direct message will be easy to read and reference. Its tone is straightforward but courteous.

Subject: Refresher on Our Shipping Policy

Please remember that our shipping policy is as follows:

Shipping Charges:
- *Chem-Treat paint* is the only product for which shipping is **not** billed to the customer.
- *All other product shipments* (including sales/marketing materials) **are** billed to the customer.

Overnight Shipping:
- Sales/marketing materials are to be shipped ground, not overnight.
- *Chem-Treat paint* may be shipped overnight at **no charge** to the customer, as provided by contract.
- *All other overnight product shipments* **are billed** to the customer.

Billing our customers accurately and consistently for shipping improves customer satisfaction with our service. In addition, the increased freight revenue will help us achieve our financial goals and control our shipping and inventory costs.

To ensure that your customers receive their products quickly, refer to the shipping and mailing timeline on WLI's intranet.

The Finance Department will be screening all shipment invoices to make sure that shipments are billed accurately. If you have questions regarding the shipping policy or require an exception, please contact me at Ext. 555.

Dean Young
VP Operations

OTHER DIRECT MESSAGE SITUATIONS

In the preceding pages, we have covered the most common direct message situations. Others occur, of course. You should be able to handle them with the techniques that have been explained and illustrated.

In handling such situations, remember that whenever possible you should get to the goal of the message right away. You should cover any other information needed in good logical order. You should carefully choose words that convey just the right meaning. More specifically, you should consider the value of using the you-viewpoint, and you should weigh carefully the differences in meaning conveyed by the positiveness or negativeness of your words. As in the good examples discussed in this chapter, you should end your message with appropriate and friendly goodwill words.

make your point and get results!

- What are best practices regarding email etiquette?
- Would you like more tips on writing routine business messages?
- Are you looking for the right words for a thank you note?

Scan the QR code with your smartphone or use your Web browser to find out at www.mhhe.com/RentzM3e. Choose Chapter 5 > Bizcom Tools & Tips. While you're there, you can view a chapter summary, exercises, PPT slides, and learn more about writing direct messages.

www.mhhe.com/RentzM3e

Writing **Bad-News** Messages

chapter six

Like all human resources professionals, Joan McCarthy, Senior Director of Human Resources Communication for Comcast Cable, sometimes has to deliver negative news to employees, whether it's about health care coverage, organizational change, or other issues. Her advice? "Balance, not spin, is the key. Frequent, candid communication that balances the good with the bad will go much further toward restoring and maintaining employee trust than the most creative 'spin.'"

Sometimes McCarthy will state negative news directly, while other times she takes a more gradual approach. Whichever pattern you use, "it's important to communicate openly and honestly," she advises. But you should also balance out the negative by "reinforcing the positive, putting the news in perspective, and showing what the organization is doing to help." In these ways you can "communicate bad news in a way that preserves your company's credibility and keeps employee trust and morale intact."

This chapter contains additional strategies for minimizing the negative impact of bad news, whether you're delivering it to internal or external readers. ∎

LEARNING OBJECTIVES

LO 6-1 Determine which situations require using the indirect order for the most effective response.

LO 6-2 Write indirect-order messages following the general plan.

LO 6-3 Use tact and courtesy in refusals of requests.

LO 6-4 Compose tactful, yet clear, claim messages using an indirect approach.

LO 6-5 Write adjustment refusals that minimize the negative and overcome bad impressions.

LO 6-6 Write negative announcements that maintain goodwill.

APPROACHES TO WRITING BAD-NEWS MESSAGES

The indirect order is especially effective when you must say "no" or convey other disappointing news. Several research studies indicate that negative news is received more positively when an explanation precedes it.[1] An explanation can convince the reader that the writer's position is correct or at least that the writer is taking a logical and reasonable position, even if the news is bad for the reader. In addition, an explanation cushions the shock of bad news. Not cushioning the shock makes the message unnecessarily harsh, and harshness destroys goodwill.

However, research also indicates that the direct approach is warranted for communicating negative news in some contexts.[2] In one study of "data breach notification letters" (letters a company uses to alert readers when the security of their personal information has been compromised), the researcher concluded that when "writers must convince readers that a potential problem exists and encourage them to act," a direct approach may be more appropriate.[3]

In addition, if you think that your negative news will be accepted routinely, you might choose directness. For example, in many buyer–seller relationships in business, both parties expect back orders and order errors to occur now and then. Thus, messages reporting this negative information would not really require indirectness. You also might choose directness if you know your reader well and feel that he or she will appreciate frankness. Although such instances are less common than those in which indirectness is the preferable strategy, you should always analyze your audience and business goals to choose the most appropriate organizational approach to delivering negative news.

As in the preceding chapter, we first describe a general plan. Then we adapt this plan to specific business situations—four in this case. First is the refusal of a request, a common task in business. Next we cover two related types of negative messages: indirect claims and adjustment refusals. Finally, we cover negative announcements, which are bad-news messages with unique characteristics.

THE GENERAL INDIRECT PLAN

The following plan will be helpful for most negative-news situations.

Using a Strategic Buffer

Indirect messages presenting bad news often begin with a strategic buffer. By **buffer** we mean an opening that identifies the subject of the message but does not indicate that negative news is coming. That is, the buffer is relevant to the topic of the message but does not state what the rest of the message will say about it.

A buffer can be neutral or positive. A neutral buffer might simply acknowledge your receipt of the reader's earlier message

> You do need to use care when opening on a positive note. You do not in any way want to raise the reader's hopes that you are about to deliver the news that he or she may be hoping for.

and indicate your awareness of what it said. A positive buffer might thank the reader for bringing a situation to your attention or for being a valued customer or employee. You do need to use care when opening on a positive note. You do not in any way want to raise the reader's hopes that you are about to deliver the news that he or she may be hoping for. That would only make your task of maintaining good relations more difficult.

Some may argue that not starting with the good news is, for savvy readers, a clear tip-off that bad news is coming. If this is the case, then why not just start with the bad news? The answer is that most readers appreciate a more gradual introduction to the message's main negative point even when they know it is coming. A buffer gives them a chance to prepare for the news—and even if they suspect that it will be negative, the use of a buffer indicates consideration for their feelings.

Setting Up the Negative News

For each case, you will have thought through the facts involved and decided that you will have to say "no" or present some other kind of negative news. You then have to figure out how you will present your reasons in such a way that your reader will accept the news as positively as possible. Your strategy might be to explain the fairness of a certain action. It might be to present facts that clearly make the decision necessary. Or you might cite the expert opinion of authorities whom both you and your reader respect. It might even be possible to show that your reasons for the negative decision will benefit the reader in the long run.

Whatever explanatory strategy you choose, these reasons should follow your buffer and precede the negative news itself. In other words, the paragraph after the buffer should start explaining the situation in such a way that by the time the negative news comes, the reader is prepared to receive it in the most favorable light possible.

Presenting the Bad News Positively

Next, you present the bad news. If you have developed your reasoning convincingly, this bad news should appear as a logical outcome. You should present it as positively as the situation will permit. In doing so, you must make certain that the negative message is clear—that your approach has not given the wrong impression.

One useful technique is to present your reasoning in first and third person, avoiding second person. To illustrate, in a message refusing a request for a refund for a returned product, you could write these negative words: "Since you have broken the seal, state law prohibits us from returning the product to stock." Or you could write these words emphasizing first and third person: "State law prohibits us from returning to stock all products with broken seals."

It is sometimes possible to take the sting out of negative news by linking it to a reader benefit. For example, if you preface a company policy with "in the interest of fairness" or "for the safety of our guests," you are indicating that all of your patrons, including the reader, get an important benefit from your policy.

Setting up your bad news indirectly helps your reader receive the news more easily.

Your efforts to present this part of the message positively should employ the positive word emphasis described in Chapter 4. In using positive words, however, you must make certain your words truthfully and accurately convey your message. Your goal is to present the facts in a positive way, not to confuse or mislead.

Offering an Alternative Solution

For almost any negative-news situation that you can think of, there is something you can do to help the reader with his or her problem.

If someone seeks to hold an event on your company grounds and you must say "no," you may be able to suggest other sites. If someone wants information that you do not have, you might know of another way that he or she could get similar information. If you cannot volunteer your time and services, perhaps you know someone who might, or perhaps you could invite the reader to make the request again at a later, better time. If you have to announce a cutback on an employee benefit, you might be able to suggest ways that employees can supplement this benefit on their own. Taking the time to help the readers in this way is a sincere show of concern for their situation. For this reason, it is one of your most powerful strategies for maintaining goodwill.

Ending on a Positive Note

Since even a skillfully handled bad-news message can be disappointing to the reader, you should end the message on a forward-looking note. Your goal here is to shift the reader's

When you deliver an apology, make sure the apology is sincere.

> If you do apologize in a bad-news message, do so early in the message as you explain the reasons and deliver the bad news.

thoughts to happier things—perhaps what you would say if you were in face-to-face conversation with the person. Your comments should fit the topic of your message, and they should not recall the negative message. They should make clear that you value your relationship with the reader and still regard it as a positive one.

Apologizing

Many times when a writer must deliver bad news, the first thought is to apologize. After all, if a customer or co-worker is unhappy—for any reason—somehow apologizing seems a good strategy for making a situation better.

Sometimes an apology can make a bad situation better, but other times it can make a bad situation worse. For example, if a customer incurs finance charges because you forgot to credit a payment to the customer's account, an apology, along with a credit to the account and removal of the finance charge, may help restore goodwill. On the other hand, if the bad news is something you had no control over (e.g., a customer didn't follow instructions for using a product and the item broke), apologizing can make you appear in the wrong even when you're not. A reader may also wonder why, if you're so sorry, you cannot do what the reader wants you to do. Apologies may even have legal implications if they can be construed as admissions of guilt.

If you do apologize in a bad-news message, do so early in the message as you explain the reasons and deliver the bad news.

Then move beyond the apology just as you move beyond the bad news and toward your forward-looking conclusion. If you think your apology may have legal implications, you can have your message reviewed by a supervisor or your company's legal department before sending it.

Following are adaptations of this general plan to four of the more common negative business message situations. From these applications you should be able to see how to adapt this general plan to almost any other situation requiring you to convey bad news.

LO 6-3 Use tact and courtesy in refusals of requests.

REFUSED REQUESTS

The **refusal of a request** is definitely bad news. Your reader has asked for something, and you must say no. Your primary goal, of course, is to present this bad news. You could do this easily with a direct refusal; however, opening with the bad news that you are refusing the reader's request could make you and your company appear insensitive. As a courteous and caring businessperson, you have the secondary goal of maintaining goodwill. To achieve this second goal, you must convince your reader that the refusal is fair and reasonable before you break the bad news.

workplace scenario

Denying a Request for a Donation

As in Chapter 5, assume the role of assistant to the White Label Industries (WLI) vice president. Today your boss assigned you the task of responding to a request from the local chapter of the National Association of Peace Officers. This worthy organization has asked WLI to contribute to a scholarship fund for certain children.

The request is persuasive. It points out that the scholarship fund is terribly short. As a result, the association is not able to take care of all the eligible children. Many of them are the children of officers who were killed in the line of duty. You have been moved by the persuasion and would like to comply, but you cannot.

You cannot contribute now because WLI policy does not permit it. Even though you do not like the effects of the policy in this case, you think the policy is good. Each year WLI earmarks a fixed amount—all it can afford—for contributions. Then it donates this amount to the causes that a committee of its executives considers the most worthy.

Unfortunately, all the money earmarked for this year has already been given away. You will have to say no to the request, at least for now. You can offer to consider the association's cause next year.

Your response must report the bad news, though it can hold out hope for the future. Because you like the association and because you want it to like WLI, you will try to handle the situation delicately. The task will require your best strategy and your best writing skills.

Developing the Strategy

Finding a fair and reasonable explanation involves carefully thinking through the facts of the situation. First, consider why you are refusing. Then, assuming that your reasons are just, try to find the best way of explaining them to your reader. To do so, you might well place yourself in your reader's shoes. Try to imagine how the explanation will be received. What comes out of this thinking is the strategy you should use in your message.

One often-used explanation is that company policy forbids compliance. This explanation may work but only if the company policy is defensible and clearly explained. Often you must refuse simply because the facts of the case justify a refusal—that is, you are right and the reader is wrong. In such cases, your best course is to review the facts and to appeal to the reader's sense of fair play.

In any situation, you may have multiple ways to offer a fair and reasonable explanation. Your job is to analyze your audience and communication goals and select the one that best fits your case.

Setting Up the Explanation in the Opening

Having determined the explanation, you begin the message with a buffer that sets up the discussion. For example, in the case of WLI's refusal to donate to the National Association of Peace Officers' worthy cause (see the Workplace Scenario above), the following opening meets this case's requirements well:

> Your organization is doing a commendable job of educating needy children. Like many other worthy efforts, it well deserves the support of our community.

This beginning, on-subject comment clearly marks the message as a response to the inquiry. It implies neither a yes nor a no answer. The second statement sets up the explanation, which will point out that the company has already given its allotted donation money to other worthy organizations. This buffer puts the reader in an agreeable or open frame of mind—ready to accept the explanation that follows.

Presenting the Explanation Convincingly

As with the general plan, you next present your reasoning. To do this, you use your best persuasion techniques: positive wording, proper emphasis, sound logic, and convincing details.

Handling the Refusal Positively

Your handling of the refusal follows logically from your reasoning. If you have built the groundwork of explanation and fact convincingly, the refusal comes as a logical conclusion and as no surprise. If you have done your job well, your reader may even support the refusal. Even so, because the refusal is the most negative part of your message, you should not give it much emphasis. You should state it quickly, clearly, and positively; and you should keep it away from positions of emphasis, such as paragraph endings.

You might even be able to make the message clear without stating the negative news explicitly. For example, if you are refusing a community member's request to use your company's retreat facility for a fundraiser, you will convey "no" clearly if you say that you must restrict the use of the facility to employees only

> "In any situation, you may have multiple ways to offer a fair and reasonable explanation."

Telling people news they don't want to hear requires your most careful communication effort.

and then go on to offer alternative locations. You must be sure, though, that your message leaves no doubt about your answer. Being unclear the first time will leave you in the position of writing an even more difficult, more negative message later.

To state the refusal positively, you should carefully consider the effects of your words. Such harsh words as *refuse, will not*, and *cannot* stand out. So do such apologies as "I deeply regret to inform you . . ." and "I am sorry to say . . ." You can usually phrase your refusal in terms of a positive statement of policy. For example, instead of writing "your insurance does not cover damage to buildings not connected to the house," write "your insurance covers damage to the house only." Or instead of writing "We must refuse," a wholesaler could deny a discount by writing "We can grant discounts only when. . . ." In some cases, your job may be to educate the reader. Not only will this be your explanation for the refusal, but it will also build goodwill.

Offering an Alternative When Possible

If the situation justifies an alternative, you can use it in making the refusal positive. More specifically, by saying what you can do (the alternative), you can clearly imply what you cannot do. For example, if you write "What we can do is to (the

> ## To state the refusal positively, you should carefully consider the effects of your words.

This refusal of a request email is harsh because of its directness and negative language.

Subject: Your request for a donation

Ms. Cangelosi:

We regret to inform you that we cannot grant your request for a donation to the association's scholarship fund.

So many requests for contributions are made of us that we have found it necessary to budget a definite amount each year for this purpose. Unfortunately, our budgeted funds for this year have been exhausted, so we simply cannot consider additional requests. We won't be able to consider your request until next year.

We deeply regret our inability to help you now and trust that you understand our position.

Mark Stephens

compromise), . . ." you clearly imply that you cannot do what the reader requested and avoid negative words.

Closing with Goodwill

Even a skillfully handled refusal is the most negative part of your message. Because the news is disappointing, it is likely to put your reader in an unhappy frame of mind. That frame of mind works against your goodwill goal. To leave your reader with a feeling of goodwill, you must shift his or her thoughts to more pleasant matters.

The best closing subject matter depends on the facts of the case, but it should be positive talk that fits the one situation. For example, if your refusal involves a counterproposal, you could say more about the counterproposal. Or you could make some friendly remark about the subject of the request as long as it does not remind the reader of the bad news. In fact, your closing subject matter could be almost any friendly remark that would be appropriate if you were handling the case face to face. The major requirement is that your ending words have a goodwill effect.

Ruled out are negative apologies, such as "Again, may I say that I regret that we must refuse." Also ruled out are the equally timeworn appeals for understanding, such as "I sincerely hope that you understand why we must make this decision." Such words sound selfish and emphasize the bad news.

Adapting the General Plan to Refused Requests

Adapting the general plan to refusals of requests, we arrive at the following outline:

- Begin with words that indicate a response to the request, are neutral about the answer, and set up the strategy.

- Present your justification or explanation, using positive language and you-viewpoint.

- Refuse clearly and positively.

- Include a counterproposal or compromise when appropriate.

- End with an adapted goodwill comment.

Contrasting Examples of a Refused Request

The advantage of the indirect order in refusal messages is illustrated by contrasting examples of WLI's possible response to the request from the National Association of Peace Officers described in the Workplace Scenario on page 137. Both the example on page 138 and the example on this page refuse clearly. But only the one that uses the indirect order is likely to maintain the reader's goodwill.

harshness in the direct refusal The first example states the bad news right away. This blunt treatment puts the reader in an unreceptive frame of mind. The result is that the reader is less likely to accept the explanation that follows. The explanation is clear, but note the unnecessary use of negative words (*exhausted, regret, cannot consider*). Note also how the closing words leave the reader with a strong reminder of the bad news.

tact and courtesy in an indirect refusal The second example skillfully handles the negative message. Its opening words are on subject and neutral. They set up the explanation that follows. The clear and logical explanation ties in with the opening. Using no negative words, the explanation leads smoothly to the refusal. Note that the refusal is also handled without negative words and yet is clear. The friendly close fits the one case.

Subject: Your Scholarship Fund Request

Ms. Cangelosi:

Your efforts to build the scholarship fund for the association's needy children are commendable.

White Label Industries assists worthy causes whenever we can. That is why every January we budget in the upcoming year the maximum amount we believe we are able to contribute to such causes. Then we distribute that amount among the various deserving groups as far as it will go. Since our budgeted contributions for this year have already been made, we are placing your organization on our list for consideration next year.

We wish you success in your efforts to improve the lives of the children in our city.

Mark Stephens

This refusal of a request using the indirect approach builds goodwill.

communication matters

LO 6-4 Compose tactful, yet clear, claim messages using an indirect approach.

INDIRECT CLAIMS

When something goes wrong between a business and its customers, usually someone begins an effort to correct the situation. Typically, the offended party calls the matter to the attention of those responsible. This claim can be made in person, by phone, or by written message (email or letter).

Our concern here is how to make it in a written message. You would likely choose a written medium if you wanted a record of the interchange, were not on personal terms with the recipient, or knew that writing to the recipient would be quicker and more efficient than contacting the reader by phone. While some claim messages are written directly (see Chapter 5), many are also written indirectly when the writer anticipates resistance or a strong negative reaction on the part of the reader. In this chapter, we present an approach for writing **indirect claims**.

Choosing the Right Tone

Your goal in a claim message is to convince your recipient that you deserve some kind of compensation or remedy for a situation that has occurred. But even if you are completely in the right, you will not advance your cause with accusatory, one-sided language.

When writing this kind of message, project an image of yourself as a reasonable person. Just as importantly, project an image of the reader as a reasonable person. Give him or her a chance to show that, if presented with the facts, he or she will do the right thing. Do not blame or whine. Keep your tone as objective as you can while also making sure that the reader understands the problems caused by the situation. Focus as much as possible on facts, not feelings.

Leading into the Problem in the Beginning

A claim message needs to identify the transactions involved. This you can do early in the message as a part of the beginning. One way is to put the identification in the subject line of an email message or in the subject line of a letter, as in this example:

Subject: Fire Extinguishers: Your Invoice C13144

Another way is just to include a neutral but relevant buffer:

Today we received via FedEx Ground the fire extinguishers we ordered on 5 May 2009 (invoice # C13144).

Whether you use a subject line and your first paragraph or the first paragraph alone to introduce the problem, choose your words with care. Such negatively charged words as *complaint* or *disappointment* can put your readers on the defensive before you've even had a chance to make your case.

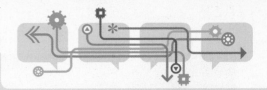
Seeking an Adjustment for a Subpar Experience

Play the role of Jeff Sutton, owner and president of Sutton Creative Services. You've just received a bill from Regal Banquet Center for the winter-holiday party that your company held there last week. It's for $1,410, which you had agreed to pay for an elegant three-course meal, plus drinks, for your 27 employees.

The food was as good as its reputation, but there were two problems. First, the room for the party was much too warm. You complained to the servers but to no avail. You would have opened windows to correct the problem yourself, but the room you were given did not have any windows (something you weren't happy about either). Second, there was apparently a shortage of servers on the night of your event. Some of your employees had to wait a long time for their food, while those who had their food first either had to start eating before the others or let their food get cold while waiting for all to be served. This ragged timing ruined the dinner, and it also threw off the timing of the program you had planned.

You were embarrassed by these problems. They reflected poorly on you and your efforts to thank your employees for their work. While you understand that unexpected problems can arise, you just don't think you should have to pay the full amount for a subpar experience. You'll need to write a claim message asking for an adjustment to your bill.

Describing the Problem Clearly

In the body of your message, explain what happened. The words describing the problem should be courteous yet firm. They should cover the problem completely, giving enough information to permit the reader to judge the matter. Present your case using facts and logic. If there were consequences to what happened, include them. This beginning sentence illustrates the point:

> When we purchased a Quick Time microwave (Serial No. 713129), we were told that because of our light use and the quality of the microwave, we needed only the six-month warranty rather than the three-year extended warranty. We have had the microwave for only seven months, but it suddenly quit working.

from the tech desk

Customizing Your Word or Outlook Toolbar

As a writer, you repeatedly use some features in Word and Outlook (e.g., print, save, undo). Word and Outlook let you customize your toolbar (the space above the ribbon) so that you can more quickly access these options you use frequently. As shown here in Word, you can choose File > Options > Quick Access Toolbar. Once there, you simply click a command in the "Choose commands from" list and then click "Add." You can see that the items in the "Customize Quick Access Toolbar" list now appear in the Word document toolbar (circled). Whenever you need to use one of those commands, just select it from your toolbar.

Since unanticipated problems occur in business, writing a clear, complete, and fair-minded claim will usually solve them.

Notice that this example uses the passive voice ("were told") to avoid accusing or blaming language. You should follow these statements with any other evidence that supports your eventual request to replace the microwave.

Requesting the Correction

The facts you present should prove your claim, so your next step is to follow logically with making the claim. How you handle the claim, however, is a matter for you to decide. You have two choices: You can state what you want (money back, replacement), or you can leave the decision to the reader. You choose which, based on the situation.

Building Goodwill with a Fair-Minded Close

Your final friendly words should leave no doubt that you are trying to maintain a positive relationship. You could express appreciation for what you seek. However, you want to avoid the cliché "Thanking you in advance." Instead, say something like "I would be grateful if you could get the new merchandise to me in time for my Friday sale." Whatever final words you choose, they should clearly show that yours is a firm yet cordial and fair request.

This blunt and accusing claim is unlikely to lead to a cooperative reply or further business with the reader.

Subject: Bill Adjustment

To Whom It May Concern:

I just received a bill for $1,410 for the winter party that I held for my employees at the Regal Banquet Center. I absolutely refuse to pay this amount for the subpar job you did of hosting this event.

First, you put us in an unpleasant room with no windows even though we had made our reservations weeks in advance. The room was also much too warm. I asked your staff to adjust the temperature, but apparently they never did. Since the room didn't have any windows, we just had to sit there and swelter in our dress clothes. As if this weren't bad enough, it took the servers so long to bring all our food out that some people had finished eating before others were even served. This made a complete mess of the nice dinner and the scheduled program.

I had heard good things about your center but now regret that I chose it for this important company event. The uncomfortable and chaotic experience reflected poorly on me and on my appreciation for my employees. Enclosed is my payment for $1,000, which I feel is more than fair.

Sincerely,
Jeff Sutton, Owner and President
Sutton Creative Services

Outlining the Indirect Claim Message

Summarizing the preceding points, we arrive at this outline for the indirect claim message:

- Identify the situation (invoice number, product information, etc.) and lead into the problem.
- Present enough facts to be convincing.
- Seek corrective action.
- End positively—friendly but firm.

Contrasting Examples of an Indirect Claim

The two messages on this page and page 142 show contrasting ways of handling Jeff Sutton's problem with the Regal Banquet Center. The first is blunt and harsh. The second is courteous, yet clear and firm.

a blunt and harsh message From the very beginning, the first message (on page 142) includes the writer's refusal to pay and is insulting. "To whom it may concern" is impersonal, generic, and outdated. The opening paragraph is a further affront, blurting out the writer's stance in angry language. The middle of the message continues in this negative vein, accusing the reader with *you* and *your* and using emotional language. The negative writing continues into the close, leaving a bad final impression. Such wording is more likely to produce resistance than acceptance.

a firm yet courteous message The second message (below) follows the plan recommended. A subject line quickly identifies the situation. The first paragraph leads into the problem. Next, in a tone that shows firmness without anger, it tells what went wrong. Then it requests a specific remedy. The ending uses subtle persuasion by implying confidence in the reader. The words used here leave no doubt about the writer's interest in a continued relationship.

Subject: Invoice #3712 (for Sutton Party on December 12, 2014)

Dear Ms. Sanchez:

As you know, Sutton Creative Services held its winter-holiday party at Regal Banquet Center on December 12. While the food was exceptional, I have some concerns regarding our experience.

When I booked the party last August, I requested that we have the party in Salon A because of its size and view of the city. The room we were given for the event was Salon C. As you know, the room is small and has no windows. In addition, the location also had the drawback of making the temperature hard to control. The servers were sympathetic but were unable to keep the room from getting too warm for my 27 employees. I know that you book many parties during the holiday season; however, as the attached copy of our contract shows, we agreed that Sutton Creative Services would be in Salon A.

It also appeared that more servers were needed for our party. The fare was elegant, but with only two servers, some guests had finished eating before others had even started. As a result, we had to start the after-dinner program in the middle of the meal, requiring the speaker to talk while people were eating. This made it difficult for people to pay attention to his presentation.

Overall, the event was not the impressive "thank-you" to my hard-working employees I had in mind when we drew up the contract. In light of these circumstances, I am requesting a revised invoice of $1,000. I believe this is a fair amount for an experience that I am sure did not represent the Regal's typical level of customer service.

I would be grateful for your response by the end of the month so that I can forward the adjusted bill to my accountant for payment.

Sincerely yours,
Jeff Sutton, President and Owner
Sutton Creative Services

This more tactful but honest claim invites the reader to do what is fair and retains goodwill.

Refused Request Message to an External Audience (Denying an Artist's Request)

A regional medical facility displays local artists' work at its various satellite locations. Artists submit applications to have their work displayed. This message shows a good strategy for denying a request to an artist who applied to have her work displayed in the Lake Superior Family Medicine Clinic's reception area.

Lake Superior
Family Medicine Clinic

Visit us: Web: https://www.lsfm.org

Lake Superior Family Medicine Clinic
4546 Burger Lane
North Concord, WI 54746
Web: www.lsfm.org
phone: 715-987-4958
fax: 715-567-7684

June 15, 2014

Ms. Jane Burroughs
2942 County Highway J
North Concord, WI 54746

Dear Ms. Burroughs:

Relevant, neutral buffer— gains the reader's favor by thanking her

Thank you for submitting your artwork for display at Lake Superior Family Medicine Clinic. The jury's deliberation process took more time than expected due to the number of submissions. Such a delay is a rare occurrence, so your patience was appreciated.

Provides a reasonable, convincing explanation supported by a fact

The Medical Center's art wall and case is a free service open to local artists like you. This exhibit area has been embraced by not only artists but also community members because of the beauty it showcases. In fact, it's so popular that we had 75 requests from local artists last month. Due to the limited wall space available and the large number of art submissions, your artwork was not chosen for display at this time.

Offers an alternative

The jury enjoyed your pieces and noted that your art "personified light." We encourage you to submit up to 10 pieces from your collection once again in 120 days. As outlined in the initial request letter dated May 22, 2014, artists can submit up to 10 submissions every 120 days.

Provides a relevant, forward-looking conclusion that builds goodwill

If you have any questions, please contact me. Again, thank you for submitting your artwork. We look forward to your next submission.

Sincerely,

Samantha Kennedy

Director, Marketing/Community Relations
Lake Superior Family Medicine Clinic
Email: sakennedy@lsfmc.org
Phone: 715-456-7890

Refused Request Message to an Internal Audience (Saying "No" to an Employee)

This message denies a hard-working employee. Showing appreciation for his work and citing the CEO's directive are likely to keep the reader's goodwill. In addition, the writer shows respect for the reader's request by remaining logical and objective in his explanation and offers a positive alternative.

Positive, relevant buffer highlights points the reader and writer can agree upon—presents the writer as a reasonable person

Logically explains the reasons for rejecting the reader's request

Provides good news and a positive alternative for the reader

Offers a goodwill close and moves beyond the bad news

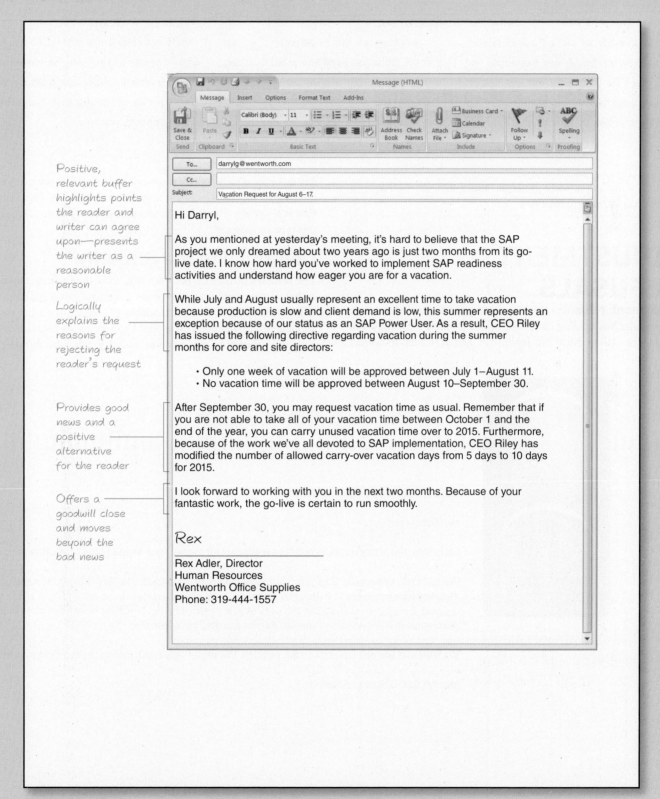

To... darrylg@wentworth.com

Cc...

Subject: Vacation Request for August 6–17.

Hi Darryl,

As you mentioned at yesterday's meeting, it's hard to believe that the SAP project we only dreamed about two years ago is just two months from its go-live date. I know how hard you've worked to implement SAP readiness activities and understand how eager you are for a vacation.

While July and August usually represent an excellent time to take vacation because production is slow and client demand is low, this summer represents an exception because of our status as an SAP Power User. As a result, CEO Riley has issued the following directive regarding vacation during the summer months for core and site directors:

- Only one week of vacation will be approved between July 1–August 11.
- No vacation time will be approved between August 10–September 30.

After September 30, you may request vacation time as usual. Remember that if you are not able to take all of your vacation time between October 1 and the end of the year, you can carry unused vacation time over to 2015. Furthermore, because of the work we've all devoted to SAP implementation, CEO Riley has modified the number of allowed carry-over vacation days from 5 days to 10 days for 2015.

I look forward to working with you in the next two months. Because of your fantastic work, the go-live is certain to run smoothly.

Rex

Rex Adler, Director
Human Resources
Wentworth Office Supplies
Phone: 319-444-1557

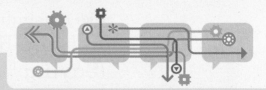
Denying a Customer's Claim

Sometimes your job at White Label Industries (WLI) involves handling a complaint. Today that is one of your tasks because the morning email has brought a strong claim for adjustment on an order for WLI's Do-Craft fabrics. The claim writer, Ms. Arlene Sanderson, explains that a Do-Craft fabric her upholstering company used on some outdoor furniture has faded badly in less than 10 months. She even includes photographs of the fabric to prove her point. She contends that the product is defective, and she wants her money back—all $2,517 of it.

Inspection of the photographs reveals that the fabric has been subjected to strong sunlight for long periods. Do-Craft fabrics are for indoor use only. Both the WLI brochures on the product and the catalog description stress this point. In fact, you have difficulty understanding how Ms. Sanderson missed it when she ordered from the catalog. Anyway, as you see it, WLI is not responsible and should not refund the money. At the same time, it wants to keep Ms. Sanderson as a repeat customer. Now you must write the message that will do just that. The following discussion tells you how.

LO 6-5 Write adjustment refusals that minimize the negative and overcome bad impressions.

ADJUSTMENT REFUSALS

Adjustment refusals are a special type of refused request. Your reader has made a claim asking for a remedy. Usually you grant these claims. Most are legitimate, and you want to correct any error for which you are responsible. But such is sometimes not the case as in Ms. Sanderson's situation above. The facts require that you say no. The following section shows you how to handle this type of message.

Determining the Strategy

The primary difference between this and other refusal messages is that in these situations, as we are defining them, your company will probably have clear, reasonable guidelines for what should and should not be regarded as legitimate requests for adjustment. You will, therefore, not have to spend much time figuring out

The adjustment refusal shows little concern for the reader's feelings.

Subject: Your May 3 claim for damages

Ms. Sanderson,

I regret to report that we must reject your request for money back on the faded Do-Craft fabric.

We must refuse because Do-Craft fabrics are not made for outside use. It is difficult for me to understand how you failed to notice this limitation. It was clearly stated in the catalog from which you ordered. It was even stamped on the back of every yard of fabric. Since we have been more than reasonable in trying to inform you, we cannot possibly be responsible.

We trust that you will understand our position. We regret very much having to deny your request.

Marilyn Cox, Customer Relations

why you cannot grant the reader's request. You will have good reasons to refuse. The challenge will be to do so while still making possible an ongoing, positive relationship with the reader.

Setting Up Your Reasoning

With your strategy in mind, you begin with words that set it up. Since this message is a response to one the reader has sent, you also acknowledge this message. You can do this by a date reference early in the message. Or you can do it with words that clearly show you are writing about the specific situation.

One good way of setting up your strategy is to begin on a **point of common agreement** and then to explain how the case at hand is an exception. To illustrate, a case involving a claim for adjustment for failure of an air conditioner to perform properly might begin this way:

> You are correct in believing that an 18,000 BTU Whirlpool window unit should cool the ordinary three-room apartment.

The explanation that follows this sentence will show that the apartment in question is not an ordinary apartment.

Another strategy is to build the case that the claim for adjustment goes beyond what can reasonably be expected. A beginning such as this one sets it up:

> Assisting families to enjoy beautifully decorated homes at budget prices is one of our most satisfying goals. We do all we reasonably can to reach it.

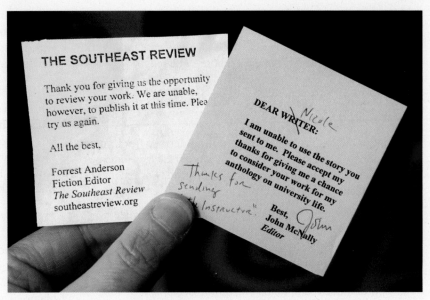

When refusing a request, remember how disappointing "no" can be and do all you reasonably can to spare your reader's feelings.

The explanation that follows this sentence will show that the requested adjustment goes beyond what can be reasonably expected.

Making Your Case

In presenting your reasons for refusal, explain your company's relevant policy or practice. Without accusing the reader, call attention to facts that bear on the case—for example, that the item

This adjustment refusal is indirect, tactful, and helpful.

Subject: Your May 3 Message about Do-Craft Fabric

Ms. Sanderson:

Certainly, you have a right to expect the best possible service from Do-Craft fabrics. Every Do-Craft product is the result of years of experimentation. And we manufacture each yard under the most careful controls. We are determined that our products will do for you what we say they will do.

We carefully inspected the photos of Do-Craft Fabric 103 that you sent us. It appears that each sample has been subjected to long periods in extreme sunlight. Because Do-Craft fabrics cannot withstand exposure to sunlight, our advertising, the catalog, and a stamped reminder on the back of every yard of the fabric advise customers that the fabric is meant for indoor use only.

As you can see from our catalog, the fabrics in the 200 series are recommended for outdoor use. You may also be interested in the new Duck Back cotton fabrics listed in our 500 series. These plastic-coated cotton fabrics are economical, and they resist sun and rain remarkably well.

If we can help you further in your selection, please contact us at service@wli.com.

Marilyn Cox, Consumer Relations

in question has been submerged in water, that the printed material warned against certain uses, or that the warranty has expired. Putting together the policy and the facts should lead logically to the conclusion that the adjustment cannot be granted.

Refusing Positively and Closing Courteously

As in other refusal messages, your refusal derives from your explanation. It is the logical result. You word it clearly, and you make it as positive as the circumstances permit. For example, this one is clear, and it contains no negative words:

> For these reasons, we can pay only when our employees pack the goods.

If a compromise is in order, you might present it in positive language like this:

> In view of these facts, we can repair the equipment at cost.

As in all bad-news messages, you should end this one with some appropriate, positive comment. You could reinforce the

illustrated by the messages on pages 146 and 147. The bad one, which is blunt and insulting, destroys goodwill. The good one, which uses the techniques described in the preceding paragraphs, stands a fair chance of keeping goodwill.

bluntness in a direct refusal The bad adjustment refusal on page 146 begins bluntly with a direct statement of the refusal. The language is negative (*regret, must reject, claim, refuse, damage, inconvenience*). The explanation is equally blunt. In addition, it is insulting ("It is difficult for me to understand how you failed . . ."). It uses little tact, little you-viewpoint. Because the close is negative, it recalls the bad news.

tact and indirect order in a courteous refusal The good adjustment refusal (page 147) begins with friendly talk on a point of agreement that also sets up the explanation. Without accusations, anger, or negative words, it reviews the facts of the case, which free the company from blame. The refusal is clear, even though it is implied rather than stated. It uses no negatives, and it does not receive undue emphasis. The close shifts to helpful suggestions

> " **Putting together the policy and the facts should lead logically to the conclusion that the adjustment cannot be granted.** "

message that you care about the reader's business or the quality of your products. In cases where it would not seem selfish, you could write about new products or services that the reader might be interested in. Neither negative apologies nor words that recall the problem are appropriate here.

Adapting the General Plan

When we apply these special considerations to the general plan, we come up with the following outline for adjustment refusals:

- Begin with words that are on subject, are neutral about the decision, and set up your strategy.
- Present the strategy that explains or justifies, being factual and positive.
- Refuse clearly and positively, perhaps including a counterproposal.
- End with positive, forward-looking, friendly words.

Contrasting Examples of an Adjustment Refusal

Bad and good treatment of WLI's refusal to refund the money for the faded fabric (see the Workplace Scenario on page 146) are

that fit the one case—suggestions that may actually result in a future sale.

LO 6-6 Write negative announcements that maintain goodwill.

NEGATIVE ANNOUNCEMENTS

Occasionally, businesses must announce bad news to their customers or employees. For example, a company might need to announce that prices are going up, that a service or product line is being discontinued, or that a branch of the business is closing. Or a company might need to tell its employees that the company is in some kind of trouble, that people will need to be laid off, or, as in the Workplace Scenario (page 150), that employees will contribute more to the cost of their health insurance. Such **negative announcements** generally follow the instructions previously given in this chapter.

Determining the Strategy

When faced with the problem of making a negative announcement, your first step should be to determine your overall strategy. Will you use direct or indirect organization?

case illustration
Adjustment Refusal Letter
(Refusing a Refund)

An out-of-town customer bought an expensive dress from the writer and mailed it back three weeks later asking for a refund. The customer explained that the dress was not a good fit and that she did not like it anymore. But perspiration stains on the dress proved that she had worn it. This letter skillfully presents the refusal.

MARIE'S
Fashions

103 BREAKER RD. HOUSTON, TX 77015 713-454-6778 Fax: 713-454-6771

On-subject opening

February 19, 2014

Ms. Cherie Ranney
117 Kyle Avenue E
College Station, TX 77840-2415

Dear Ms. Ranney:

We understand your concern about the elegant St. John's dress you returned February 15. As always, we are willing to do as much as we reasonably can to make things right.

Set-up for the explanation

Review of the facts—supports the writer's position

What we can do in each instance is determined by the circumstances. With returned clothing, we generally give refunds. Of course, to meet our obligations to our customers for quality merchandise, all returned clothing must be unquestionably new. As you know, our customers expect only the best from us, and we insist that they get it. Thus, because the perspiration stains on your dress would prevent its resale,

Good restraint—no accusations, no anger

Negative language minimized in the refusal

we must consider the sale final. We are returning the dress to you. With the proper alterations, it can be an elegant addition to your wardrobe.

Emphasis on what can be done—helps restore goodwill

Friendly goodwill close

Please visit us again when you are in the Houston area. It would be our pleasure to serve you.

Sincerely,

Marie O. Mitchell

Marie O. Mitchell
President

dm

workplace scenario

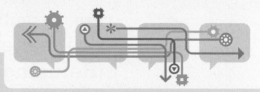

Announcing an Increase in Health Insurance Costs

As the assistant to the human resources director at National Window Systems, you have been given the difficult assignment of writing a bad-news message for your boss. She has just returned from a meeting of the company's top executives in which the decision was made to deduct 25 percent of the employees' medical insurance premiums from their paychecks. Until now, National Window Systems has paid it all. But declining profits are forcing the company to cut back on these benefits. Something has to give if National Window Systems is to remain competitive while also avoiding lay-offs. The administrators decided on a number of cost-cutting measures including this reduction in the company's payment for medical insurance. The message you will write to National Window Systems employees is a negative announcement.

In most cases the indirect arrangement will be better. This route is especially recommended when it is reasonable to expect that the readers would be surprised, particularly disappointed, or even angered by a direct presentation. When planning an indirect announcement, you will need to think about what kind of buffer opening to use, what kind of explanation to give, how to word the news itself, and how to leave your readers feeling that you have considered their interests.

As in other negative situations, you should use positive words and avoid unnecessary negative comments when presenting the news itself. Since this is an announcement, however, you must make certain that you cover all the factual details involved. People may not be expecting this news. They will therefore want to know the whys and whats of the situation. And if you want them to believe that you have done all you can to prevent the negative situation, you will need to provide evidence that this

> **When planning an indirect announcement, you will need to think about what kind of buffer opening to use, what kind of explanation to give, how to word the news itself, and how to leave your readers feeling that you have considered their interests.**

Setting Up the Bad News

As with the preceding negative message types, you should plan your indirect beginning (buffer) carefully. You should think through the situation and select a strategy that will set up or begin the explanation that justifies the announcement. Perhaps you will begin by presenting justifying information. Or maybe you will start with complimentary or cordial talk focusing on the good relationship that you and your readers have developed. Choose the option that will most likely prepare your reader to accept the coming bad news.

Positively Presenting the Bad News

In most cases, the opening paragraph will enable you to continue with background reasons or explanations in the next paragraph, before you present the negative news. Such explaining will help you put the negative news in the middle of the paragraph rather than at the beginning where it would be emphasized.

is true. If there are actions the readers must take, these should be covered clearly as well. All questions that may come to the readers' minds should be anticipated and covered.

Focusing on Next Steps or Remaining Benefits

In many cases negative news will mean that things have changed. Customers may no longer be able to get a product that they have relied upon, or employees may have to find a way to pay for something that they have been getting for free. For this reason, a skillful handling of a negative announcement will often need to include an effort to help people solve the problem that your news just created for them. In situations where you have no further help to offer—for example, when announcing certain price increases—you can still help people feel better about your news by calling attention to the benefits that they will continue to enjoy. You can focus on the good things that have not changed and perhaps even look ahead to something positive or exciting on the horizon.

When making a negative announcement, remember that an indirect, tactful approach is usually better than a blunt or aggressive approach.

[The ending words . . . can be whatever is appropriate for this one situation—a positive look forward, a sincere expression of gratitude, or an affirmation of your positive relationship with your readers.]

Closing on a Positive or Encouraging Note

The ending words should cement your effort to cover the matter positively. They can be whatever is appropriate for this one situation—a positive look forward, a sincere expression of gratitude, or an affirmation of your positive relationship with your readers.

Reviewing the Plan

Applying the preceding instructions to the general plan, we arrive at this plan for negative announcements written in indirect order:

- Start with a buffer that begins or sets up justification for the bad news.
- Present the justification material.

- Give the bad news positively and clearly.
- Help solve the problem that the news may have created for the reader.
- End with appropriate goodwill talk.

Contrasting Examples of a Negative Announcement

Good and bad techniques in negative announcements are illustrated in the sample messages on pages 152 and 153 that announce WLI's plan to have employees contribute more to the cost of their health insurance (see the Workplace Scenario on page 150). The bad example is written in a direct pattern, which in some circumstances may be acceptable but clearly is not in this case. The good one follows the pattern just discussed.

directness here alarms the readers The bad example below clearly will upset the readers with its abrupt announcement in the beginning. The readers aren't prepared to receive the negative message. They probably don't understand the reasons behind the negative news. The explanation comes later, but the readers are not likely to be in a receptive mood when they see it. The message ends with a repetition of the bad news.

convincing explanation begins a courteous message The better example on page 153 follows the recommended indirect pattern. Its opening words begin the task of convincing the readers of the appropriateness of the action to be taken. After more convincing explanation, the announcement flows logically. Perhaps it will not be received positively by all recipients, but it represents a reasonable position given the facts presented. After the announcement, comes an offer of assistance to help readers deal with their new situation. The last paragraph reminds readers of remaining benefits and reassures them that management understands their interests. It ends on an appreciative, goodwill note.

Using Directness in Some Cases

As we mentioned at the beginning of this chapter, in some cases it is likely that the reader will not have a strong negative reaction to the bad news. If, for example, the negative news is expected (as when the news media have already revealed it), its impact may be viewed as negligible. There is also a

Directness in this negative announcement sends an abrupt, upsetting message.

To our employees:

National Window Systems management sincerely regrets that effective February 1 you must begin contributing 25 percent of the cost of your medical insurance. As you know, in the past the company has paid the full amount.

This decision is primarily the result of the continued high cost of medical insurance and declining profits over the last several quarters. Given this tight financial picture, we needed to find ways to reduce expenses.

We trust that you will understand why we must ask for your help with cutting costs to the company.

Sincerely,

good case for directness when the company's announcement will contain a remedy or announce new benefits that are designed to offset the effects of the bad news. As in all announcements with some negative element, this part must be worded as positively as possible. Also, the message should end on a goodwill note. The sample message below, announcing the end of a store's customer reward program, illustrates this situation.

Dear Ms. Cato:

Effective January 1 Frontier Designs is discontinuing our Preferred Customer program so that we may offer several new promotions.

Your accumulated points will be converted to a savings coupon worth as much as or more than your points total. Your new points total is on the coupon enclosed with this letter. You may apply this coupon in these ways:

- When shopping in our stores, present your coupon at the register.

- When shopping from our catalogs, give the coupon number to the telephone service agent, enclose your coupon with your mail order, or enter it with your online order at www.frontierdesigns.com/catalog.

In all these cases we will deduct your coupon value from your purchase total. If you have any questions, please call us at 1-800-343-4111.

We thank you very much for your loyalty. You'll soon hear about exciting new opportunities to shop and save with us.

Sincerely,

To All Employees:

Even though the Affordable Care Act has increased competition among insurance providers, the cost of medical coverage has not gone down for most companies.

Such is the case at our company. The premiums that we pay to cover our health benefits have increased by 34 percent over the last two years, and they now represent a huge percentage of our expenditures. Meanwhile, as you know, our sales have been lower than usual for the past several quarters.

For the short term, we must find a way to cut overall costs. Your management has considered many options and rejected such measures as cutting salaries and reducing personnel. Of the solutions that will be implemented, the only change that affects you directly concerns your medical insurance. On **March 1** we will begin deducting 25 percent of the cost of the premium.

Jim Taylor in the Personnel Office will soon be announcing an informational meeting about your insurance options. Switching to spousal coverage, choosing a less expensive plan with lighter deductibles, or setting up a flexible spending account may be right for you. You can also see Jim after the meeting to arrange a personal consultation. He is well versed in the many solutions available and can give you expert advice for your situation.

Our health care benefits are some of the best in our city and in our industry, and those who continue with the current plan will not see any change in their medical coverage or their co-pays. Your management regards a strong benefits program as critical to the company's success, and we will do all we can to maintain these benefits while keeping your company financially viable.

Sincerely,

This indirect bad-news announcement provides a well-reasoned explanation for the bad news.

case illustration
Negative Announcement
(Decreasing Work Hours)

Shop employees are told of the effects a slow economy will have on their work hours. The message is friendly and empathetic but clearly conveys the negative news. The goodwill close looks forward to better economic times.

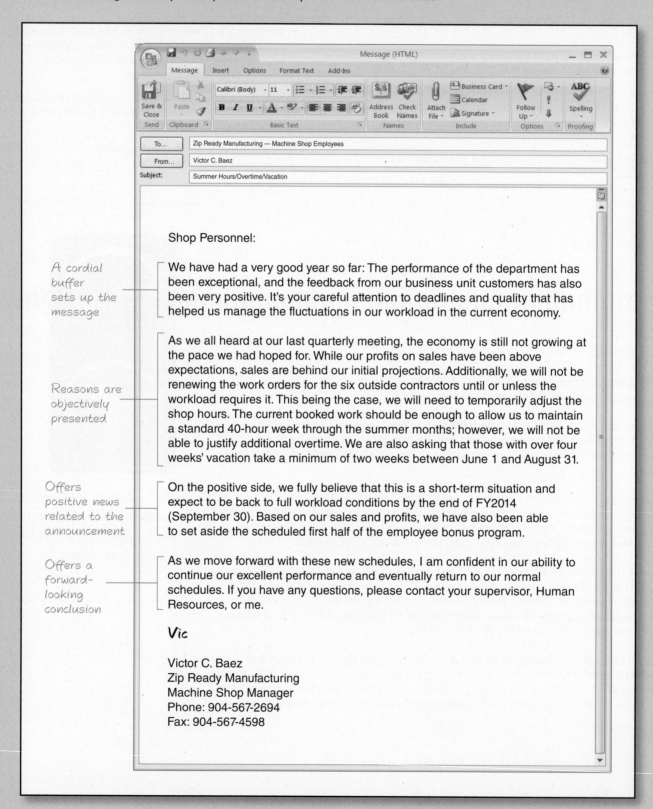

Message (HTML)

Message | Insert | Options | Format Text | Add-Ins

Save & Close | Paste | Calibri (Body) · 11 · | B I U · A · | Address Book | Check Names | Attach File · | Business Card · | Calendar | Signature · | Follow Up · | Spelling

Send | Clipboard | Basic Text | Names | Include | Options | Proofing

To... | Zip Ready Manufacturing — Machine Shop Employees
From... | Victor C. Baez
Subject: | Summer Hours/Overtime/Vacation

Shop Personnel:

A cordial buffer sets up the message
We have had a very good year so far: The performance of the department has been exceptional, and the feedback from our business unit customers has also been very positive. It's your careful attention to deadlines and quality that has helped us manage the fluctuations in our workload in the current economy.

Reasons are objectively presented
As we all heard at our last quarterly meeting, the economy is still not growing at the pace we had hoped for. While our profits on sales have been above expectations, sales are behind our initial projections. Additionally, we will not be renewing the work orders for the six outside contractors until or unless the workload requires it. This being the case, we will need to temporarily adjust the shop hours. The current booked work should be enough to allow us to maintain a standard 40-hour week through the summer months; however, we will not be able to justify additional overtime. We are also asking that those with over four weeks' vacation take a minimum of two weeks between June 1 and August 31.

Offers positive news related to the announcement
On the positive side, we fully believe that this is a short-term situation and expect to be back to full workload conditions by the end of FY2014 (September 30). Based on our sales and profits, we have also been able to set aside the scheduled first half of the employee bonus program.

Offers a forward-looking conclusion
As we move forward with these new schedules, I am confident in our ability to continue our excellent performance and eventually return to our normal schedules. If you have any questions, please contact your supervisor, Human Resources, or me.

Vic

Victor C. Baez
Zip Ready Manufacturing
Machine Shop Manager
Phone: 904-567-2694
Fax: 904-567-4598

© Chris Ryan/age fotostock

master the art of breaking bad news!

- Do you want to learn more about tactfully breaking bad news?
- Would you like to learn how to apologize in a message?
- Would you like help in developing the appropriate tone for your bad-news messages?

Scan the QR code with your smartphone or use your Web browser to find out at www.mhhe.com/RentzM3e. Choose Chapter 6 > Bizcom Tools & Tips. While you're there, you can view a chapter summary, exercises, PPT slides, and more to learn techniques that will help you effectively deliver bad news.

www.mhhe.com/RentzM3e

Writing Persuasive
Messages and Proposals

chapter seven

chapter

Everything you write on the job will have some kind of persuasive purpose—to convince the reader of your professionalism, convey an appealing company image, promote good relations, or all of these. But in some situations, persuasion will be your central goal. In these cases, your readers will hold a certain position, and your task will be to move them from this position to one that is more favorable to you and/or your company. Meeting this challenge requires careful analysis, strategic thinking, and skillful writing.

But the rewards of persuasive communication are many. Read on to see how to boost your persuasion savvy. ■

LEARNING OBJECTIVES

LO 7-1 Describe important strategies for writing any persuasive message.

LO 7-2 Write skillful persuasive requests that begin indirectly, develop convincing reasoning, make a call to action, and close with goodwill.

LO 7-3 Discuss ethical concerns regarding sales messages.

LO 7-4 Describe the planning steps for direct mail or email sales messages.

LO 7-5 Compose sales messages that gain attention, present persuasive appeals, use appropriate visual elements, and effectively drive for action.

LO 7-6 Write well-organized and persuasive proposals.

THE PREDOMINANCE OF INDIRECTNESS IN PERSUASIVE MESSAGES

By definition, persuasive messages are written to potentially uncooperative readers; that is why persuasion is necessary. For this reason, it is often best to organize persuasive messages in an **indirect order**. Preparing the reader to accept your idea is a much better strategy than announcing the idea from the start and then having to argue uphill through the rest of the message. Ideally, you should organize each persuasive message so that, from the title or subject line to the end, your readers will agree with you. If you try to have them on your side from start to finish, you'll have your best chance of success.

Although indirectness works for most persuasive messages, sometimes you will want to use a **direct approach**. For example, if you know your reader prefers directness or if you believe your readers will discard your message unless you get to the point early, then directness is in order. As we discuss later in this chapter, proposals in response to specific requests may also use directness.

In the following pages we first provide general advice for effective persuasion using the indirect approach. We then explain how the indirect order is used in two kinds of persuasive messages: the persuasive request and the sales message. Finally, we cover another important category of persuasive writing: proposals. These, as you will see, can use either the direct or indirect pattern, depending on whether they are invited or uninvited.

LO 7-1 Describe important strategies for writing any persuasive message.

GENERAL ADVICE ABOUT PERSUASION

All our previous advice about adapting your messages to your readers comes into play with persuasive messages—only more so. Moving your reader from an uninterested or even resistant position to an interested, cooperative one is a major accomplishment. To achieve it, keep the following advice in mind.

Know Your Readers

For any kind of persuasive message, thinking about your subject from your readers' point of view is critical. To know what kind of appeals will succeed with your readers, you need to know as much as you can about their values, interests, and

Knowing your readers enables you to target their interests.

needs. Companies specializing in email and direct-mail campaigns spend a great deal of money to acquire this kind of information. Using a variety of research techniques, they gather **demographic information** (such as age, gender, income, and geographic location) and **psychographic information** (such as social, political, and personal preferences) about their target audience. They also develop mailing lists based on prior shows of interest from consumers and purchase mailing lists from other organizations that have had success with certain audiences.

If you don't have these resources, you can use other means to learn as much as possible about the intended readers. You can talk with customer service about the kinds of calls they're getting, study the company's customer database, chat with people around the water cooler or online, and run ideas past colleagues. Good persuasion depends on knowledge as well as on imagination and logic.

Choose and Develop Targeted Reader Benefits

No one is persuaded to do something for no reason. Sometimes their reasons for acting are related to **tangible** or measurable rewards. For example, they will save money, save time, or acquire some kind of desired object. But often, the rewards that persuade are **intangible**. People may want to make their lives easier, gain prestige, or have more freedom. Or perhaps they want to identify with a larger cause, feel that they are helping others, or do the right thing. In your quest for the appeals that will win your

Extrinsic benefits like giveaways can add incentive, but intrinsic benefits are usually stronger.

readers over, do not underestimate the power of intangible benefits, especially when you can pair them with tangible rewards.

When selecting the **reader benefits** to feature in your persuasive messages, bear in mind that such benefits can be intrinsic, extrinsic, or a combination. **Intrinsic benefits** are benefits that readers will get automatically by complying with your request. For example, if you are trying to persuade people to attend your company's awards dinner, the pleasure of sharing in their colleagues' successes will be intrinsic to the event. Door prizes would be an **extrinsic benefit**. We might classify the meal itself as a combination—not really the main feature of the event but definitely central to it. Intrinsic benefits are tightly linked to what you're asking people to do, while extrinsic ones are added on and more short-lived. Let intrinsic benefits do the main persuasive work. Focusing too much on extrinsic benefits can actually cheapen your product or service in the readers' eyes.

When presenting your reader benefits, be sure the readers can see exactly how the benefits will help them. The literature on selling makes a useful distinction between **product features** and reader benefits. If you say that a wireless service uses a certain kind of technology, you're describing a feature. If you say that the technology results in fewer missed or dropped calls, you're describing a benefit. Benefits persuade by enabling readers to envision the features of the recommended product or action in their own worlds.

One common technique for achieving this goal is to use what we call **scenario painting**—a description that pictures the reader in a sample situation enjoying the promised benefits. Here is an example of scenario painting from a Web site promoting a tour of Warner Brothers Studio:

> Led by our top tour guides, this exclusive tour takes you into the craft shops where artisans create sets, props, costumes and more. You'll visit working production sets to talk to key crew people, watch as Foley artists create sound effects for film and TV, and get an up close and personal look at the magic of movie making. In addition, your private group will dine in the studio commissary where you have the real opportunity to dine with the stars.

Scenario painting is common in sales messages, but you can also use it to good advantage in other persuasive messages, even internal ones. Whatever your persuasive situation or strategy, be sure to provide enough detail for readers to see how they will benefit from what you are asking them to do.

Make Good Use of Three Kinds of Appeals

The first acknowledged expert on persuasion, the Greek philosopher Aristotle, lived almost 2,500 years ago, but many of his core concepts are still widely taught and used. Of particular value is his famous categorizing of persuasive appeals into three kinds, summarized in Exhibit 7-1 (next page): those based on **logic** (*logos*), those based on **emotion** (*pathos*), and those based on the **character** of the speaker (*ethos*). All three kinds come into play in every persuasive message—in fact, one might say, in every kind of message. But as the writer of a persuasive message, you will need to think especially carefully about how to manage these appeals and which ones to emphasize given your intended audience.

Painting a vivid scene showing the benefits can help you persuade.

workplace scenario

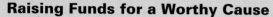

Raising Funds for a Worthy Cause

You're a mid-level manager for Arslan, one of the largest custodial services companies in your city. Like many others in the company, you devote some of your personal time to serving the community. Arslan wants you to do this volunteer work for the sake of good public relations. You want to do it because it is personally and professionally rewarding.

Currently, as chair of the fundraising committee of the city's Junior Achievement program, you head all efforts to get financial support for the program from local business-people. The committee can contact some of these potential donors by phone, but there are too many for you to be able to reach all of them this way.

At its meeting today, the Junior Achievement board of directors discussed various solutions. One director suggested using a fundraising letter. The board accepted the idea with enthusiasm. With just as much enthusiasm, it gave you the assignment of writing the letter.

As you view the assignment, it is not a routine letter-writing problem. Although the local businesspeople are probably generous, they are not likely to part with money without good reason. In fact, their first reaction to a request for money is likely to be negative. So you will need to overcome their resistance in order to persuade them. Your task is indeed challenging.

> **Whether written to internal or external readers, requests that are likely to be resisted require a slow, deliberate approach.**

As you plan your message, consider what kind of logical appeals you might use. Saved money? Saved time? A more dependable or effective product? How about emotional appeals? Higher status? More sex appeal? Increased popularity? And don't neglect appeals based on character. What kind of image of yourself and your company will resonate with the reader? Should you get a celebrity or expert to endorse your product or to serve as the spokesperson? Not only when planning but also when revising your persuasive message, assess your appeals. Be sure to choose and develop the ones most likely to persuade your audience.

Make It Easy for Your Readers to Comply

Sometimes writers focus so much on creating persuasive appeals that they put insufficient thought into making the requested action as clear and easy to perform as possible. If you want people to give money or buy your product, tell them where and how to do it, and supply a preaddressed mailing envelope or a Web address if applicable. If you want employees to give suggestions for improving products or operations, tell them exactly where and how to submit their ideas, and make it easy for them to do so. If you want people to remember to work more safely or conserve on supplies, give them specific techniques for achieving these goals and include reminders at the actual locations where they need to remember what to do. Making the desired action specific and easy to perform is a key part of moving your readers from resistance to compliance with your request.

With this general advice in mind, we now turn to the three main types of persuasive writing in business: persuasive requests, sales messages, and proposals.

LO 7-2 Write skillful persuasive requests that begin indirectly, develop convincing reasoning, make a call to action, and close with goodwill.

PERSUASIVE REQUESTS

At many points in your career—starting with your job search—you will need to make **persuasive requests**. Perhaps, as in the Workplace Scenario above, you will be asked to write a fundraising message. Perhaps you will need to ask your management for another staff position or for special equipment. You may need to persuade a potential client to join you in a meeting so that you can demonstrate the benefits of your products. Or maybe you will be trying to persuade your employees to change their behavior in some way.

▼ EXHIBIT 7-1   The Three Main Types of Appeals

- Logical (*logos*)—makes a rational argument
- Emotional (*pathos*)—arouses certain feelings and values
- Character based (*ethos*)—projects appealing traits of the writer or speaker

Whether written to internal or external readers, requests that are likely to be resisted require a slow, deliberate approach. In essence, you must persuade the reader that he or she should grant your request before you actually state it. Such an achievement requires that you carefully plan your persuasive strategy.

Determining Your Strategy

Figuring out the best persuasive approach involves three interrelated tasks: determining what you want, figuring out your readers' likely reactions, and deciding upon a persuasive strategy that will overcome reader objections and evoke a positive response.

Think carefully about your actual goals for your persuasive request. A request for a one-time-only donation might be written very differently from a request that is intended to create a long-time, multiple donor. If you were convincing employees to leave the parking places next to the building for customers' use, you would write a very different message if it were the third rather than the first request. Your goals, considered in the context of your organization's goals and your relationship with your readers, are key shapers of your persuasive message.

To anticipate how your readers will react to your request, consider everything you know about them and then put yourself in their shoes. Look at the request as they are likely to see it. Figure out what's in it for them and how to overcome any objections they may have. From this thinking and imagining, your plan should emerge.

The specific plan you develop will depend on the facts of the case. You may be able to show that your reader stands to gain in time, money, or other tangible benefits. Or you may be able to show that your reader will benefit in goodwill or prestige. In some cases, you may persuade readers by appealing to their love of beauty, excitement, serenity, or other emotions. In different cases, you may be able to persuade readers by appealing to the pleasant feeling that comes from doing a good turn. You decide on the benefits that will be most likely to win over your readers.

A special kind of persuasive request is one that casts the request as a problem–solution message. With this strategy, you first present a problem that you and the readers share—called the **common-ground persuasion technique**—and then show how doing as you propose will solve the problem for all concerned. Many fundraising letters use this technique, giving us striking facts about the current economic climate, the environment, or living conditions in a certain area of the world. But this strategy can also be a powerful one for internal audiences who

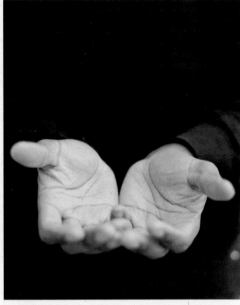

Many times in your career you will need to ask for something. This chapter tells you how.

might not be receptive to a straightforward proposal for action but who share your opinion that something needs to be done.

A persuasive request situation is a special opportunity for analysis, creativity, and judgment. With careful use of all three, you can plan messages that will change your readers' minds and move them to action.

Gaining Attention in the Opening

In the indirect messages discussed in Chapter 6, the goal of the opening is to set up the explanation for the negative news. The opening of a persuasive request has a similar goal: to lead into your central strategy. But the opening of a persuasive request has an additional goal: to gain attention.

Many persuasive messages arrive uninvited, and they compete with many other messages. Unless they gain the reader's attention at the beginning, they are likely to end up in the recycle bin.

You need to draw your reader in with the opening of your persuasive message because you are writing to a person who has not invited your message and may not agree with your goal. An interesting beginning is a good step toward getting this person in a receptive mood.

Determine what your reader will find compelling. It might be a statement that arouses curiosity, or it might be a statement offering or implying a reader benefit. Because questions get people thinking, they are often effective openings. The following examples indicate the possibilities.

From the cover letter of a questionnaire seeking the opinions of medical doctors:

> What, in your opinion as a medical doctor, is the future of the private practice of medicine?

From a message requesting contributions for orphaned children:

> While you and I dined heartily last night, 31 orphans at San Pablo Mission had only dried beans to eat.

From a message seeking the cooperation of business leaders in promoting a fair:

> What would your profits be if 300,000 free-spending visitors came to our town during a single week?

If writing your request in the form of a problem–solution message, you should start with a goal that you and the readers share. For example, let's say that a project manager in your company has retired and that you want to recommend his capable administrative assistant as his replacement. Since no member of the support staff has ever broken into the managerial ranks, any direct proposal to promote your candidate will likely be resisted. To get readers on your side from the beginning, you could start your message with facts that everyone can agree upon: that someone has retired, that his or her duties are important, and that someone capable needs to be found quickly. Your subject line for an email along these lines might be something like, "Reassigning Jim Martin's Duties" (which everyone supports), not "Promoting Kathy Pearson" (which your readers will resist).

Whatever the case, the form of indirectness that you choose for your opening should engage your readers right away and get

This direct, bland request is not likely to persuade.

Dear Mr. Williams:

Will you please donate to the local Junior Achievement program? We have set $50 as a fair minimum for businesses to give. But larger amounts would be appreciated.

The organization badly needs your support. Currently, about 900 young people will not get to participate in Junior Achievement activities unless more money is raised.

If you do not already know about Junior Achievement, let me explain. Junior Achievement is an organization for high school students. They work with local business executives to form small businesses and then operate the businesses. In the process, they learn about our economic system. This is a good thing, and it deserves our help.

Hoping to receive your generous donation,

Jane Monroe

> ## If writing your request in the form of a problem–solution message, you should start with a goal that you and the readers share.

them thinking along the lines that will lead to their approval of your request.

Developing the Appeal

Following the opening, you should proceed with your goal of persuading. Your task here is a logical and orderly presentation of the reasoning you have selected.

As with any argument intended to convince, you should do more than merely list points. You should develop the points with convincing details. Since you are trying to penetrate a neutral or resistant mind, you need to make good use of the you-viewpoint. You need to pay careful attention to the meanings of your words and the clarity of your expression. You need to use logic and emotion appropriately and project an appealing image (see the Communication Matters feature on the next page for a list of persuasive personal qualities). And because your reader may become impatient if your appeal is not clear, you need to make every word count.

Making the Request Clearly and Positively

After you have done your persuading, move to the action you seek. You have tried to prepare the reader for what you want. If you have done that well, the reader should be ready to accept your request.

This indirect, interesting request has a much greater chance of success.

Dear Mr. Williams:

Right now—right here in our city—620 teenagers are running 37 corporations. The kids run the whole show; their only adult help comes from business professionals who work with them.

Last September these young people applied for charters and elected officers. They created plans for business operations. For example, one group planned to build Web sites for local small businesses. Another elected to conduct a rock concert. Yet another planned to publish electronic newsletters for area corporations. After determining their plans, the kids issued stock—and sold it, too. With the proceeds from stock sales, they began their operations. This May they will liquidate their companies and account to their stockholders for their profits or losses.

What's behind these impressive accomplishments? As you've probably guessed, it's Junior Achievement. Since 1919, this nonprofit organization has been teaching school kids of all ages about business, economics, and entrepreneurship. Thanks to partnerships between volunteers and teachers, these students gain hands-on experience with real business operations while learning the fundamentals of economics and financial responsibility. They also learn cooperation and problem solving. It's a win–win situation for all involved.

To continue to succeed, Junior Achievement needs all of us behind it. During the 13 years the program has been in our city, it has had enthusiastic support from local business leaders. But with over 900 students on the waiting list, our plans for next year call for expansion. That's why, as a volunteer myself, I ask that you help make the program available to more youngsters by contributing $50 (it's deductible). By helping to cover the cost of materials, special events, and scholarships, you'll be preparing more students for a bright future in business.

Please make your donation now by completing our online contribution form at www.juniorachievement .org. You will be doing a good service—for our kids, for our schools, and for our community.

Best regards,

Jane Monroe

As with negative messages, your request requires careful word choice. You should avoid words that bring to mind images and ideas that might work against you. Words that suggest reasons for refusing are especially harmful, as in this example:

> I am aware that businesspeople in your position have little free time to give, but will you please consider accepting an assignment to the board of directors of the Children's Fund?

The following positive tie-in with a major point in the persuasion strategy does a much better job:

> Your organizing skills and stature in the community would make you an ideal board member for the Children's Fund.

Whether your request should end your message will depend on the situation. In some cases, you will profit by following the request with additional persuasive material. This technique is especially effective when your reader needs a lot of convincing. In cases where your request is relatively simple, won't cost the reader much, and isn't likely to be resisted, you can end your message with the request. Even here, though, you combine or follow the request with wording that makes the reader feel good about doing as you ask (see the sample messages on pages 165–166).

Summarizing the Plan for Requests

From the preceding discussion, the general plan for persuasive requests can be summarized as follows:

- Open with words that (1) gain attention and (2) set up the strategy.

- Develop the strategy using persuasive language and the you-viewpoint.

- Make the request clearly and without negatives (1) either at the end of the message or (2) followed by words that continue the persuasive appeal.

Contrasting Examples of a Persuasive Request

The persuasive request is illustrated by contrasting letters that ask businesspeople to donate to Junior Achievement. The first message is direct and bland. The second message, which follows the indirect approach and provides convincing details, is much more likely to succeed.

a selfish blunt approach The weaker letter (page 162) begins with the request. Because the requested action is something the reader probably doesn't want to do, the direct beginning is likely to get a negative reaction. In addition, the comments about how much to give tend to lecture rather than suggest. Some explanation follows, but it is weak and scant. In general, the letter is poorly written. It makes little use of the you-viewpoint. Perhaps its greatest fault is that the persuasion comes too late. The selfish close is a weak reminder of the action requested.

skillful persuasion using the indirect order The message on page 163 follows the recommended indirect pattern. Its opening generates interest and sets up the persuasive strategy. Notice the effective use of the you-viewpoint throughout. Not until the reader has been sold on the merits of the request does the message ask the question. It does this clearly and directly. The final words leave the reader thinking about the benefits that a *yes* answer will give.

SALES MESSAGES

One of the most widely disseminated forms of business communication is the **sales message**. It is such an important component of most businesses' sales strategies that it has become an elaborate, highly professionalized genre, shaped by extensive consumer research. Think about the typical sales letter that you receive. Careful attention has been paid to the message on the envelope, to the kinds of pieces inside, and to the visual appeal of those pieces, as well as to the text of the letter itself. Clearly, advertising professionals produce many of these mailings, as well as much of the fundraising literature that we receive. You can also see a professional's hand in many of the sales emails that appear in your in-box, as well as in other sales communications, such as blogs, Facebook pages, and white papers. Why, then, should you study sales writing?

As a businessperson, you will often find yourself in the position of helping to shape a major sales campaign or contribute to its success through social networking. You may have valuable insight into your product's benefits and your potential customers. You need to be familiar with the conventions for sales messages and to be able to offer your own good ideas for their success.

A Persuasive Internal Request (Using a Central Emotional Appeal Supported by Logical and Character-Based Appeals)

The writer wants employees to participate in the company's annual blood drive. He needs to convince them of the importance of the drive and overcome their likely objections. This message will be distributed to employees' mailboxes.

AMBERLY
Engineering & Construction

Department of Community Relations
Mail Location 12
123 Jackson Street
Edison, Colorado 80864
(719) 777-4444
CommunityRelations@Amberly.com

February 27, 2014

Opens with an attention-getting, you-focused question

Did you help save Brad Meyer's life?

Uses a character-based appeal; invites the reader to identify with these "lifesavers"

A few years ago, an employee of Amberly was driving to a friend's wedding when an oncoming car, operated by a drunk driver, swerved across the center line. Brad doesn't remember the crash. But he does remember two months spent in the hospital, two months of surgery and therapy.

Tells an engaging story with specific details

Without the help of people like us, Brad would not have lived. Some Amberly employees save lives regularly. We're blood donors. Please be a lifesaver and join us on Friday, March 19, for Amberly's annual blood drive.

Your help is needed for a successful drive.

Avoids words such as "draw blood" or "needle" that would bring unpleasant thoughts to mind

Giving blood is simple. The entire process will take less than 45 minutes.

Giving blood is safe. Experienced health professionals from the Steinmetz Blood Center will be on site to conduct the procedure exactly as they would in a clinical setting.

Giving blood is convenient. The Steinmetz staff will be in Room 401, Building B, between 9:00 a.m. and 3:00 p.m. To save time, make an appointment to donate. Call the Steinmetz Blood Center at 569-1170.

Addresses likely reader objections

Giving blood is important. Nobody knows who will need blood next, but one thing is certain—it will be available only if healthy, caring people take time to give it. Brad's accident required 110 units—more than 12 gallons—of blood. Because 110 people set aside 45 minutes, Brad Meyer has a lifetime of minutes to be grateful.

Recalls the emotion-based opening and links it to a logical appeal: You or someone in your family might benefit

Take a few moments now to make your pledge on the reverse side of this letter. Then return it to the Community Relations department, Mail Location 12, by March 15.

Makes the requested action clear and easy

From Brad and from other families—like yours and mine—who might need it in the days to come,

Thank you,

John M. Piper

John M. Piper
Director, Community Relations

case illustration

A Persuasive Email to Members of a Professional Organization

The writer of this message, the president of the American Society for Training and Development (ASTD), uses character appeal as well as logic and emotion to persuade readers to participate in a survey.

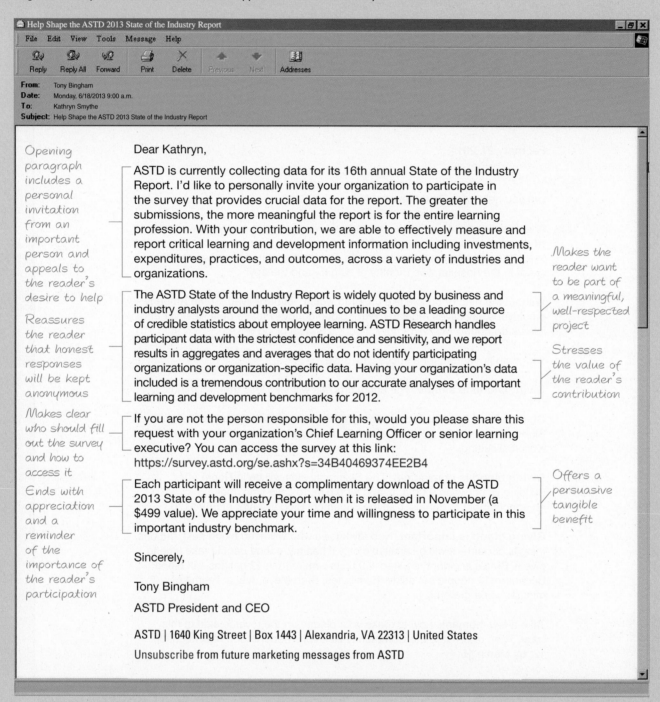

Help Shape the ASTD 2013 State of the Industry Report

File   Edit   View   Tools   Message   Help

Reply   Reply All   Forward   Print   Delete   Previous   Next   Addresses

From: Tony Bingham
Date: Monday, 6/18/2013 9:00 a.m.
To: Kathryn Smythe
Subject: Help Shape the ASTD 2013 State of the Industry Report

Dear Kathryn,

Opening paragraph includes a personal invitation from an important person and appeals to the reader's desire to help

ASTD is currently collecting data for its 16th annual State of the Industry Report. I'd like to personally invite your organization to participate in the survey that provides crucial data for the report. The greater the submissions, the more meaningful the report is for the entire learning profession. With your contribution, we are able to effectively measure and report critical learning and development information including investments, expenditures, practices, and outcomes, across a variety of industries and organizations.

Makes the reader want to be part of a meaningful, well-respected project

Reassures the reader that honest responses will be kept anonymous

The ASTD State of the Industry Report is widely quoted by business and industry analysts around the world, and continues to be a leading source of credible statistics about employee learning. ASTD Research handles participant data with the strictest confidence and sensitivity, and we report results in aggregates and averages that do not identify participating organizations or organization-specific data. Having your organization's data included is a tremendous contribution to our accurate analyses of important learning and development benchmarks for 2012.

Stresses the value of the reader's contribution

Makes clear who should fill out the survey and how to access it

If you are not the person responsible for this, would you please share this request with your organization's Chief Learning Officer or senior learning executive? You can access the survey at this link:
https://survey.astd.org/se.ashx?s=34B40469374EE2B4

Ends with appreciation and a reminder of the importance of the reader's participation

Each participant will receive a complimentary download of the ASTD 2013 State of the Industry Report when it is released in November (a $499 value). We appreciate your time and willingness to participate in this important industry benchmark.

Offers a persuasive tangible benefit

Sincerely,

Tony Bingham

ASTD President and CEO

ASTD | 1640 King Street | Box 1443 | Alexandria, VA 22313 | United States

Unsubscribe from future marketing messages from ASTD

workplace scenario

Generating More Customers for Your Business

Play the role of Zach Miller, a student in your university's college of business. You've had your own house-painting business ever since you started college. So far, you've gained your customers through word of mouth only, but you think it's time to expand.

You've got two buddies willing to come on board if you can generate enough business for them to work full time in the summer months and part time in the spring and fall.

You've decided that your first step in growing the business will be to advertise your services to the faculty and staff of your school. After getting permission to post your message to the listservs of several offices and academic departments, you sit down to think through what you want to say.

Considering your audience, you decide to reveal to them that you're a student. Given the line of work they're in, your readers obviously

like the feeling of helping young people succeed. On the other hand, you'll need to overcome any concern that the job you'll do will be less than professional. As you start figuring out your persuasive details, you realize that it may be a good idea to include links to other information. Hmmm . . . this message is going to take some careful planning in order to be effective. The following sections will help you think through your options.

In addition, knowledge of selling techniques can help you in many of your other activities, especially the writing of other kinds of business messages. As we've said, most of them involve selling something—an idea, a line of reasoning, your company, yourself. Sales techniques are more valuable to you than you might think. After you have studied the remainder of this chapter, you should see why.

LO 7-3 Discuss ethical concerns regarding sales messages.

Questioning the Acceptability of Sales Messages

Sales messages are a controversial area of business communication, for two main reasons: They are often unwanted, and they sometimes use ethically dubious persuasive tactics. You probably know from your own experience that sales letters are not always received happily. Called "junk" mail, these mailings often go into the wastebasket or recycle bin without being read. They must be profitable, though, because they are still heavily used.

Sales messages sent by email may create even more hostility among intended customers. Angrily referred to as **spam**, unsolicited email sales messages have generated strong resistance among email users. Perhaps it is because these messages clutter in-boxes. Maybe it is because mass mailings place a heavy burden on Internet providers, driving up costs to the users. Or perhaps the fact that they invade the reader's privacy is to blame. Whatever the explanation, the resistance is real. You will need to consider these objections any time you use this sales medium.

Fortunately, a more acceptable form of email selling has developed. Called **permission-based email**, it permits potential customers to sign up for email promotions on a company's Web site or provide their email addresses to an email or phone

Face-to-face selling is only part of the picture. Many sales occur through mail, email, Web-based media, and mobile messaging.

marketer. The potential customers may be asked to indicate the products, services, and specific topics of their interest, or the company may be able to track those interests based on what the customer has searched for or ordered in the past. The marketers can then tailor their messages to the customer, and the customer will receive only what he or she wants. According to a recent report by *eMarketer*, 93 percent of U.S. online consumers have at least one email subscription, and it is by far their preferred method for hearing about sales or promotions (64 percent of those surveyed ranked this method first, as opposed to 25 percent for postal mail and 8 percent for social media sites). Permission-based email is thus a powerful sales practice, and it has helped address the problem of unwanted sales messages.[1]

As for the charge that persuasive messages use unfair persuasive tactics, this is, unfortunately, sometimes the case. The unfair tactics could range from deceptive wording and visuals to the omission of important information to the use of emotional elements that impair good judgment.

In a Missouri court case, Publishers Clearing House was found guilty of deception for direct mail stating that the recipients were already winners, when in fact they were not.[2] To consider a different example, one linen supply company sent a letter to parents of first-year students at a university telling them that the students would need to purchase extra-long sheets, offered by this company, to fit the extra-long beds on campus—but omitted the fact that only one dorm out of four had such beds. And it is well documented that images, because they work on an emotional level, persuade in ways that tend to bypass the viewers' reasoned judgment, leading some to question the ethics of such elements.[3]

Certain kinds of online sales messages are particularly obnoxious, such as giant pop-ups that obscure the screen. Some of these trap readers in a loop that they can't exit without restarting their computers or, in extreme cases, having to remove spyware or viruses.

Any persuasive message is, by its very nature, biased. The writer has a favored point of view and wants to persuade the reader to adopt it. Therefore, considering the ethical dimension of your communication, while important for all types of messages, is especially critical for persuasive messages. Let your conscience and your ability to put yourself in the readers' shoes guide you as you consider how to represent your subject and win others to your cause.

LO 7-4   Describe the planning steps for direct mail or email sales messages.

Preparing to Write a Sales Message

Before you can begin writing a sales message, you must know all you can about the product or service you are selling. You simply cannot sell most goods and services unless you know them well and can tell the prospects what they need to know. Before prospects buy a product, they may want to know how it is made, how it works, what it will do, and what it will not do. Clearly, a first step in sales writing is careful study of your product or service.

As we have stressed, you must also study your readers. You should gather demographic, psychographic, and any other kind

from the tech desk

Learn about e-Selling from Chief Marketer and MailChimp

The use of digital media for sales campaigns has grown enormously over the last several years because of the increased popularity of these media and the devices people use to access them. If you'd like to learn more about the complex art and science of e-marketing, read on.

Chief Marketer's Web site (www. chiefmarketer.com) has great information about using technology to conduct effective email, Web, social, and mobile campaigns. Their researchers have gathered statistics about consumer behavior regarding each medium and developed marketing strategies based on that behavior. If you've been wanting to go behind the scenes and see how

the pros design electronic messages to sell through different channels, Chief Marketer is for you.

Chief Marketer

Source: www.chiefmarketer.com

If you're particularly interested in writing effective sales emails, let MailChimp show you how. This company specializes in conducting email sales campaigns. You can access excellent material on their "Resources" page (mailchimp.com/resources), such as their guide to designing for mobile devices, mobile-friendly templates, and an "Email Marketing Benchmarks by Industry" report. Check

out their "Subject Line Comparison," which examined the open rates of 40 million emails and compared the subject lines of the best and worst performers.

Source: mailchimp.com

communication matters

Are Sales Letters Becoming Extinct? Absolutely Not!

You might be wondering if digital media are making sales letters obsolete. While it's true that the volume of sales mailings has declined, print sales messages still have a lot going for them, as these sources attest:

- According to Tristan Loo of the marketing firm ProfitFuzion.com, direct mail has the highest rate of success in new customer acquisition as compared to other channels. Also, 65 percent of people who received direct mail made a purchase in 2012, and 22- to 24-year-olds were the consumers most likely to respond to a mail piece ("Direct Mail Statistics," 22 Nov. 2012, Web, 17 May 2013).

- An article in *Forbes* magazine reports that physical mail has certain advantages over electronic mail. It "leaves a 'deeper footprint' in the brain because it involves more emotional processing, connects better with readers' feelings, and evokes more emotionally vivid memories" (Steve Olenski, "In the Land of Digital, Let's Not Forget the Physical," 18 Oct. 2011, Web, 17 May 2013).
- A study by the marketing firm Epsilon found that "a majority of consumers still prefer postal mail for a large portion of their multichannel communication." Interestingly, their preferences for physical mail are stronger in some industries than in others, as shown in the chart below.

To be effective, though, sales mailings should be part of an "integrated multichannel communication plan," advises marketer Beth Negus Viveiros. You also need a well-researched mailing list, writing that is targeted to your specific readers, and content that clearly differentiates you from the competition ("Make Mail Part of Your 2013 Integrated Marketing Plan," *Chief Marketer*, 15 Nov. 2012, Web, 17 May 2013).

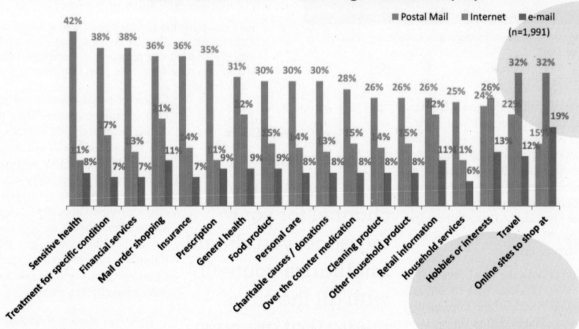

Stated Preference for Receiving Information (US)

Source: Epsilon, *Channel Preferences for Both the Mobile and Non-Mobile Consumer, Epsilon.com,* Epsilon Data Management, LLC, 2012, Web, 17 May 2013. Reprinted with permission.

of information that will help you understand why they might want or need your product. The more you know about your readers, the better you will be able to adapt your sales message to them. (Read about consumers' different media preferences by industry in the Communication Matters box above.)

In large businesses, a marketing research department or agency typically gathers information about prospective customers. If you do not have such help, you will need to gather this information on your own. If time does not permit you to do the necessary research, you may have to use

Begin work on a sales message by thoroughly studying the product or service to be sold.

Determining the Central Appeal

With your product or service, your prospects, and your medium or media in mind, you are ready to create the sales message. This involves selecting and presenting your persuasive appeals, whether emotional, logical, character based, or a combination. But for most sales messages, one appeal should stand out as the main one—mentioned in the beginning, recalled in the middle, and reiterated at the end. While other benefits can be brought in as appropriate, the message should emphasize your central, best appeal.

Emotional appeals—those based on our senses and emotions—can be found in almost any sales message, but they predominate in messages for goods and services that do not perform any discernable rational function. The following example illustrates:

> Linger in castle corridors on court nights in London. Dance on a Budapest balcony high above the blue Danube. Seek romance and youth and laughter in charming capitals on five continents. And there you'll find the beguiling perfume that is fragrance Jamais.

Logical appeals are useful for selling products that help readers save money, do a better job, or get better use from a product. Illustrating a rational appeal (saving money) are these words from a message selling magazine subscriptions:

> We're slashing the regular rate of $36 a year down to only $28, saving you a full 22 percent. That means you get 12 information-filled new issues of *Science Digest* for only $2.33 a copy. You save even more by subscribing for 2 or 3 years.

logic and imagination. For example, the nature of a product can tell you something about its likely buyers. Industrial equipment would probably be bought by people with technical backgrounds. Expensive French perfumes and cosmetics would probably be bought by people in high-income brackets. Long-term care insurance would appeal to older people in middle-income brackets.

To be able to choose the most persuasive channel for your message, you should familiarize yourself with all the sales media that are now available. Often times, several media will need to work together to have the best effect. These days, sales emails are often linked to Web sites, and Web sites often provide links to social networking sites. You may even decide that the best approach is to start with an email and follow up with a phone call, or speak face to face with a prospect and then hand him or her some sales literature. There have never been so many media available, and the more you know about them, the better your selling can be.

> " **To be able to choose the most persuasive channel for your message, you should familiarize yourself with all the sales media that are now available.** "

Character-based appeals can enhance any kind of sales message. They persuade by stating or implying "I use this product, so you should, too" or "I am an authority, so you should do what I recommend." Ads that employ sports figures, film stars, or experts to sell their products are relying heavily on character-based appeals. Companies themselves often project an appealing "character" in their sales campaigns. Note how the following excerpt from a sales letter for *Consumer Reports* magazine uses the company's identity to persuade:

> *Consumer Reports* is on your side. We're a nonprofit consumer protection organization with no commercial interests whatsoever. To put it bluntly, we don't sell out to big companies and private interest groups—we're accountable to no one except to consumers. And when you're not beholden to advertisers (like other so-called consumer protection publications), you can tell it like it is.

People may also buy a certain product because they want to identify with, and be identified with, a certain successful,

A well-known spokesperson can add character appeal.

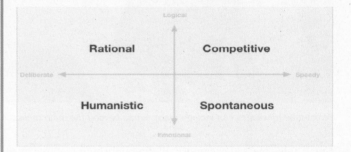

communication matters

What Type of Decision Maker Is Your Reader?

As a consumer, are you more emotional or more logical? And do you make your purchases quickly or deliberately?

According to Des Traynor, an expert on customer acquisition and management, how your potential customers would answer these questions will put them in one of the following four quadrants:

	Logical	
Rational		**Competitive**
Deliberate ←		→ Speedy
Humanistic		**Spontaneous**
	Emotional	

When you know your readers' likely decision-making style, you can choose the most appropriate sales materials for them, as shown here:

	Logical	
• Detailed Case Studies • Extensive Docs • Uptime & Support Details		• Fact sheet • Checklists • Discounts
Deliberate ←		→ Speedy
• Brand/Reputation • Recommendations • External activities (social media etc)		• Emotive one liner • Customer logos • Customer testimonial
	Emotional	

So select and design your sales materials with your readers' decision-making style(s) in mind.

Source: "Know Your Customers and How They Decide," *Inside Intercom*, Intercom, Inc., n. d., Web, 16 May 2013. Reprinted with permission.

socially responsible, or "cool" company as projected in the company's sales messages.

In any given case, many appeals are available to you. You should use those that fit your product or service and your readers best. Keep in mind that how the buyer will use the product may be a major basis for selecting a sales strategy. For example, cosmetics might well be sold to the final user through emotional appeals, but selling cosmetics to a retailer (who is primarily interested in reselling them) would require rational appeals. A retailer's main questions about the product would be "Will it sell? What turnover can I expect? How much money will it make for me?"

Determining the Makeup of the Mailing

When you write a sales message to be sent by mail or email, a part of your effort is to determine the makeup of the mailing. To know what you want to say in your main message, you'll need to decide what kinds of additional pieces will be included and how they will support the main piece.

For example, the letter featured in the Case Illustration on page 175 came with a second page that included a detailed agenda for the workshop, as well as a list of management skills the attendees would learn. Moving this material to a second

page enabled the writer to keep the letter itself relatively short and fast moving. Many sales and fundraising mailings are even more elaborate, including brochures, giveaways, order forms, and more, like the National Audubon Society mailing on page 172. You will need to plan what to include and in what form. Even if someone else, such as a graphic artist or desktop publishing expert, will be designing the pieces of your mailing, you will need to be able to explain what you want and plan how all parts of the sales package will work together, especially for a complex mailing like the one described here.

Direct-mail messages can include many extras beyond the main message.

Email sales messages can use all the publishing features available on the computer. Like the Case Illustration on page 179, the message can be presented creatively with color, font variations, box arrangements, artwork, and more. It may include links to the seller's Web site as well as to other supporting material and to the ordering procedure. And it may have attachments. Just as with a direct-mail package, the email sales package can use many elements to persuade and to provide all the information a reader will need in order to make a purchase.

LO 7-5 Compose sales messages that gain attention, present persuasive appeals, use appropriate visual elements, and effectively drive for action.

Gaining Attention Before the Message Begins

Sales messages must gain the reader's attention right away. Otherwise, they won't be read. For this reason, the sales effort often begins before the actual message does.

Many mailed messages have an attention getter on the envelope. It may be the offer of a gift ("Your gift is enclosed"). It may present a brief sales message ("12 months of *Time* at 60 percent off the newsstand price"). It may present a picture and a message (a picture of a cruise ship and "Tahiti and more at 2-for-1 prices"). An official-appearing envelope sometimes is used. So are brief and simple messages such as "Personal" or "Sensitive material enclosed."

With email, of course, there is no envelope. The effort to gain attention begins with the *From* and *Subject* information. To avoid having your message

regarded as spam, you should clearly tell who you are and identify your company. You should also address the reader by name. Though some readers will delete the message even with this clear identification, the specificity will induce some to read on.

The subject line in email messages is the main place for getting attention. Here honesty and simplicity should be your guide. The subject line should tell clearly what your message is about, and it should be short. It should avoid sensational wording, such as "How to earn $60,000 the first month." In addition, avoiding sensationalism involves limiting the use of solid caps, exclamation points, dollar signs, and "free" offers. In fact, you risk having spam filters block your message or send it to the recipient's junk folder if you use "free" or other words and phrases commonly used in spam. An email with the subject line "Making your restaurant more profitable" that is sent to a researched list of restaurant managers and owners is much more likely to be opened and read than a message with the subject line "You have to read this!" that is sent to thousands of readers.

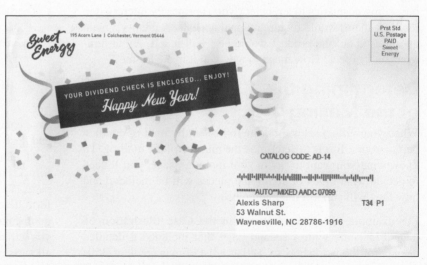

The envelope often begins the persuasive effort.

Gaining Attention in the Opening of the Message

The first words of your message must also gain attention and motivate the reader to keep reading. What you do here can be creative, but the method you use should help set up your strategy. It should not just gain attention for attention's sake. Attention is easy to gain if nothing else is needed. In a sales letter, a sensational statement such as "You can be a millionaire!" or "Free chocolate for the rest of your life!" would gain attention, but it wouldn't be likely to help sell most products or services, which can't live up to such extravagant claims.

One of the most effective attention-gaining openings is a statement or question that introduces a need that the product will satisfy. For example, this rational-appeal opening would be likely to tap into a retailer's main need:

> Here is a proven best-seller—and with a 12 percent greater profit.

This paragraph of a message selling a fishing vacation at a lake resort illustrates a need-fulfilling beginning of an emotional-appeal approach:

> Your line hums as it whirs through the air. Your line splashes and dances across the smooth surface of the clear water as you reel. From the depth you see the silver streak of a striking bass. You feel a sharp tug. The battle is on!

A different tack is illustrated by the following example. It attracts interest by telling a story and using character-based appeal:

> In 1984 three enterprising women met to do something about the lack of accessible health information for women.

Whatever opening strategy you choose, it should lead into your central selling point.

Building a Persuasive Case

With the reader's attention gained, you proceed with the sales strategy that you have planned. In general, you establish a need. Then you present your product or service as fulfilling that need.

The plan of your sales message will vary with each case. But it is likely to follow certain general patterns determined by your choice of appeals. If your main appeal is emotional, for example, your opening has probably established an emotional atmosphere that you will continue to develop. Thus, you will sell your product based on its effects on your reader's senses. You will describe the appearance, texture, aroma, and taste of your product so vividly that your reader will mentally see it, feel it—and want it. In general, you will seek to create an emotional need for your product.

If you select a rational appeal as your central theme, your sales description is likely to be based on factual material. You should describe your product based on what it can do for your reader rather than how it appeals to the senses. You should

communication matters

Gaining—and Keeping—Readers' Attention on Facebook and Twitter

Messages on Facebook and Twitter don't use envelopes or subject lines, so how do you gain attention in these media? Monica Clarke of The Wakeman Agency, a New York–based public relations firm serving nonprofits and small to medium-sized businesses, has some advice.

If you can afford it, hire a celebrity spokesperson; "co-branding or collaborating with a 'star' may give your cause the spark it needs." But if that's not feasible, use television ads, magazine ads, your website, and other outlets to start a buzz.

Once you have people logged in, "the major effort must be focused on not losing their attention." Keep the feeds coming, and provide content "that will get people talking to their friends about your social media marketing campaign." Find an interesting angle "and work it—don't just provide information about your organization, but provide information about the bigger picture." Just setting up a Facebook or Twitter account and hoping for followers is a surefire way to become "just another '@' symbol in the crowd."

Source: "Standing Out Amongst Millions—How to Get Attention on Twitter and Facebook," *The Wakeman Agency*, The Wakeman Agency, n.d., Web, 20 May 2013.

write matter-of-factly about such qualities as durability, savings, profits, and ease of operation.

When using character-based appeals, you will emphasize comments from a well-known, carefully selected spokesperson. Or, if the character being promoted is that of the company itself, you will provide evidence that your company is expert and dependable, understands customers like "you," and stands behind its service or product.

The writing that carries your sales message can be quite different from your normal business writing. Sales writing usually is highly conversational, fast moving, and aggressive. It even uses techniques that are incorrect or inappropriate in other forms of business writing, such as sentence fragments or catchy slang. As the Case Illustrations show, it also uses visual emphasis devices (underscore, capitalization, boldface, italics, exclamation marks, color) and a variety of type sizes and fonts. And its paragraphing often appears choppy. Any sales message is competing with many other messages for the intended reader's attention. In this environment of information overload, punchy writing and visual effects that enable quick processing of the message's main points have become the norm in professional sales writing.

A Direct-Mail Message
(Selling a Lawn Care Service)

With an emphasis on character-based appeals, this attractive letter builds a central argument: Only the experts at Scotts can give you the results you want. Inside the mailing were two pages of additional appeals and information—with pictures of luscious green grass, of course.

Scotts *LawnService®*

The attention getter is featured twice for emphasis.

> Make your lawn a Scotts® Lawn.
> **Get a Free Lawn Analysis.**
>
> ---
>
> FREE Lawn Analysis for
> Primary Address 2----------

The letter can be filled in with the reader's information for a personal touch.

The Lastname Household------------
Firm Name (Float Up) -------------
Sec Add (Float Up) ----------------
Pr Add (Float Up) ------------------
Csz (Float Up) ---------------------

Variable Date -----------------

Priority Code: XXXXXXX

The main theme of the letter is highlighted here.

All lawn services are not the same. Scotts LawnService® has new programs to help your lawn and your budget.

Dear (Insert Name),

The story and the scenario painting engage the reader and continue the letter's theme.

Here's a familiar story. At one point, another lawn service company promised you a perfect lawn without weeds. The lawn pictured in their ad looked nice and you signed up. Now, there are days when you drive away from your house, glance at your lawn, and wonder why you even bothered. You deserve better, your lawn deserves better—**and Scotts LawnService has new programs that can help.**

Look inside to see **the variety of programs we offer to give you a Scotts Lawn.** Everything from basic maintenance to "best on the block" quality, you'll find Scotts LawnService is more flexible—and more affordable—than you might expect.

Don't take chances with anyone else.

The main theme is repeated, with added emotional appeal.

This readable list of benefits uses all three types of appeals, with the emphasis on character-based appeal (Scott's unmatched expertise).

Only Scotts LawnService has:
- Access to **Scotts' research and development** team, the largest and most experienced in the industry.
- A system of **professional products from the Scotts Miracle-Gro® family of brands that no one else has**—including our competition.
- Exclusive **WaterSmart®** formulation, which improves your lawn's ability to absorb water, nutrients and weed controls by up to 25%, **giving you a greener lawn and fewer weeds.**
- Exclusive **Weed B Gon® EdgeGuard,™** which reinforces weed barriers to provide **up to 75% more weed control** along your sidewalks and driveway, where weeds seem to creep up the most.

Along with our industry-leading products and services, **only our lawn technicians have been trained at the Scotts LawnService Training Institute.™** So you'll never worry about an untrained part-timer working on your lawn.

Scotts LawnService offers a Complete Satisfaction Guarantee for every one of our customers: we're not satisfied until you are. And when you get started with Scotts LawnService, **we promise to never charge you a start-up fee—unlike some of our competitors.**

The call to action is clear and linked to the letter's theme.

Call 800-728-4940 or go online in the next 14 days to request your **Free Lawn Analysis**. Let us give you a genuine Scotts Lawn—and make the other guys' empty promises a thing of the past.

Sincerely,

Having the letter come from a local manager adds more personalization.

(Name)
(Insert Branch Manager)
(Insert Branch Name)

This visually appealing invitation leads to additional persuasive facts.

Learn more about
The Scotts Advantage®

inside

P.S. Go online to get your Free Lawn Analysis at mySLS.com

The P.S. repeats the call to action in "you" language.

A Direct-Mail Message
(Selling a Management Seminar)

This sales letter uses all three types of appeals (logical, emotional, and character based). The mailing included a second page that provided a detailed outline of the seminar and a longer list of likely reader benefits.

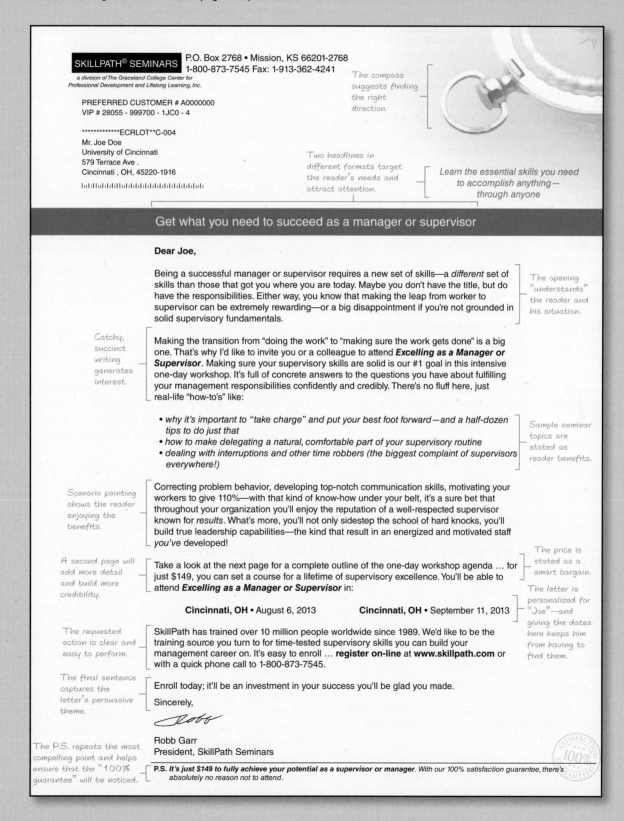

SKILLPATH® SEMINARS P.O. Box 2768 • Mission, KS 66201-2768
1-800-873-7545 Fax: 1-913-362-4241
a division of The Graceland College Center for Professional Development and Lifelong Learning, Inc.

PREFERRED CUSTOMER # A0000000
VIP # 28055 - 999700 - 1JC0 - 4

*************ECRLOT**C-004
Mr. Joe Doe
University of Cincinnati
579 Terrace Ave .
Cincinnati , OH, 45220-1916
Iulılıdılıdıldılıdıldıdıldıdıldıldıldıldıldıl

The compass suggests finding the right direction.

Two headlines in different formats target the reader's needs and attract attention.

Learn the essential skills you need to accomplish anything— through anyone

Get what you need to succeed as a manager or supervisor

Dear Joe,

Being a successful manager or supervisor requires a new set of skills—a *different* set of skills than those that got you where you are today. Maybe you don't have the title, but do have the responsibilities. Either way, you know that making the leap from worker to supervisor can be extremely rewarding—or a big disappointment if you're not grounded in solid supervisory fundamentals.

The opening "understands" the reader and his situation.

Catchy, succinct writing generates interest.

Making the transition from "doing the work" to "making sure the work gets done" is a big one. That's why I'd like to invite you or a colleague to attend **Excelling as a Manager or Supervisor**. Making sure your supervisory skills are solid is our #1 goal in this intensive one-day workshop. It's full of concrete answers to the questions you have about fulfilling your management responsibilities confidently and credibly. There's no fluff here, just real-life "how-to's" like:

- *why it's important to "take charge" and put your best foot forward—and a half-dozen tips to do just that*
- *how to make delegating a natural, comfortable part of your supervisory routine*
- *dealing with interruptions and other time robbers (the biggest complaint of supervisors everywhere!)*

Sample seminar topics are stated as reader benefits.

Scenario painting shows the reader enjoying the benefits.

Correcting problem behavior, developing top-notch communication skills, motivating your workers to give 110%—with that kind of know-how under your belt, it's a sure bet that throughout your organization you'll enjoy the reputation of a well-respected supervisor known for *results*. What's more, you'll not only sidestep the school of hard knocks, you'll build true leadership capabilities—the kind that result in an energized and motivated staff *you've* developed!

A second page will add more detail and build more credibility.

Take a look at the next page for a complete outline of the one-day workshop agenda … for just $149, you can set a course for a lifetime of supervisory excellence. You'll be able to attend **Excelling as a Manager or Supervisor** in:

The price is stated as a smart bargain.

    Cincinnati, OH • August 6, 2013   **Cincinnati, OH** • September 11, 2013

The letter is personalized for "Joe"--and giving the dates here keeps him from having to find them.

The requested action is clear and easy to perform.

SkillPath has trained over 10 million people worldwide since 1989. We'd like to be the training source you turn to for time-tested supervisory skills you can build your management career on. It's easy to enroll … **register on-line** at **www.skillpath.com** or with a quick phone call to 1-800-873-7545.

The final sentence captures the letter's persuasive theme.

Enroll today; it'll be an investment in your success you'll be glad you made.

Sincerely,

Robb

Robb Garr
President, SkillPath Seminars

The P.S. repeats the most compelling point and helps ensure that the "100% guarantee" will be noticed.

P.S. It's just $149 to fully achieve your potential as a supervisor or manager. With our 100% satisfaction guarantee, there's absolutely no reason not to attend.

Stressing the You-Viewpoint

In no area of business communication is the use of the you-viewpoint more important than in sales writing. A successful sales message bases its sales points on reader interest. You should liberally use and imply the pronoun *you* throughout the sales message as you present your well-chosen reader benefits.

For example, assume you are writing a sales message to a retailer and that one point you want to make is that the manufacturer will help sell the product with an advertising campaign. You could write this information in a matter-of-fact way: "HomeHealth products will be advertised in *Self* magazine for the next three issues." Or you could write it based on what the advertising means to the reader: "Your customers will read about HomeHealth products in the next three issues of *Self* magazine." As you can see, viewing things from the reader's perspective strengthens your persuasiveness. The following examples further illustrate the value of presenting facts as reader benefits:

Facts	You-Viewpoint Statements
We make Aristocrat hosiery in three colors.	You may choose from three lovely shades.
The Regal weighs only a few ounces.	The Regal's featherlight touch makes vacuuming easier than ever.
Lime-Fizz is a lime-flavored carbonated beverage.	Your customers will keep coming back for the refreshing citrus taste of Lime-Fizz.
Baker's Dozen is packaged in a rectangular box with a bright bull's-eye design.	Baker's Dozen's new rectangular package fits compactly on your shelf, and its bright bull's-eye design is sure to catch the eyes of your customers.

You may also want to make use of scenario painting, putting the reader in a simulated context that brings out the product's

Choosing Words Carefully

In persuasive messages, every word can influence whether the reader will act on your request. Try putting yourself in your reader's place as you select words for your message. Some words, while closely related in meaning, have different emotional effects. For example, the word *selection* implies a choice, while the word *preference* implies a first choice. Consider how changing a single adjective changes the effect of these sentences:

> The NuPhone's *small* size . . .
>
> The NuPhone's *compact* size . . .
>
> The NuPhone's *sleek* size . . .

Framing your requests in the positive is also a proven persuasive technique. Readers will tend to opt for solutions to problems that avoid negatives. Here are some examples:

Original Wording	Positive Wording
Tastee ice cream has nine grams of fat per serving.	Tastee ice cream is 95 percent fat free.
Our new laser paper keeps the wasted paper from smudged copies to less than 2 percent.	Our new laser paper ensures smudge-free copies over 98 percent of the time.

Enhancing Your Message with Visuals

The Web has made today's readers more visually oriented than any before in history. When preparing any kind of sales message, be sure to consider whether photos, tables, boxes, word art, borders, or other graphical elements would enhance your message's appeal.

> " In no area of business communication is the use of the you-viewpoint more important than in sales writing. "

appeal. The J. Peterman company is famous for this technique, exemplified in the following excerpt from an advertisement for a picnic backpack:

> It's a 7-mile hike to Snowmass Lake.
>
> Best do it in late spring, early summer.
>
> At sunset, the orange hues of Snowmass Mountain reflect off this glassy, ripple free water. Many stop here to rest before a summit run.
>
> Audrey and I are here for a picnic. No further quest required, less equipment, too. I have the overnight gear. She, this perfectly equipped lightweight picnic backpack.

As pointed out earlier, sales letters can contain elaborately designed materials. Often, the envelope will contain artful text and images. Inside, you'll find visual elements ranging from logos and creatively chosen fonts to photos of products and customers.

Email sales messages have become just as visually appealing, if not more so. Many that sell products are almost all pictures; others are at the oppositive end of the spectrum, such as some email messages designed to be read on mobile devices (see the contrasting Delta Airlines messages on page 178). Probably most fall somewhere in between these two extremes, like the Case Illustration on page 179.

communication matters

Kelly Gadd began working for the Eisen Agency, a PR firm in Cincinnati, Ohio, right after graduation. Here she shares her view of today's promotional writing landscape.

Q: How do current promotional strategies differ from those of, say, 10 years ago? What has changed?

A: Certainly one significant change is the move toward digital promotions. Now, many consumers view promotional material exclusively through social media channels, Web sites, and email. Many companies make certain offers, events, or products available only to their loyal digital audiences to entice them to engage with their brands more frequently over these channels. Brands need to be wherever their customers want them whenever they need them.

An increase in experiential promotional tactics has also become popular. With consumers receiving more than 5,000 marketing messages each day, it's hard for brands to stand out with traditional promotions. Whether through special events, street teams, or a traveling display, companies want to create an experience that consumers can't ignore.

Q: What persuasion principles and strategies are still important, no matter what the medium?

A: While the communication channels have changed, people still have the same basic needs, so persuasive appeals are still based on credibility, logic, and emotion. If anything, consumers' sensibilities have been heightened by the overwhelming access to information. In particular, the credibility of the communicator or company has become more important. Since anyone can post whatever he/she likes online and call it accurate, consumers have had to become more skeptical and discerning—but they also are quicker to make decisions and less likely to fact check.

Q: What advice would you give other young business professionals about promoting their companies' products or services?

A: Familiarity with digital media is certainly a strength. However, I would caution today's young employees not to discount promotional and marketing strategies that have worked well for decades. Well-planned strategy is still the key to successful execution, and professionalism and good, solid writing skills are still very important. But be aware that information about your products/company will be available quickly, good or bad, so be as prepared as possible and put problem-solving skills to work when things do not go as planned.

The visual literacy you read about in Chapter 1 definitely comes into play when designing sales messages. You need to understand, for example, that a photo is not just a literal representation of a person or a product; it conveys a mood, a set of values, and even an experience. Where to put the visual elements is also an important decision, requiring that you imagine the readers' likely response to the screen or page design. Whatever visual elements you choose and wherever you put them, be sure that they project the desired message and complement the textual content.

Including All Necessary Information

Of course, the information you present and how you present it are matters for your best judgment. But you must make sure that you present enough information to complete the sale. You should leave none of your readers' questions unanswered. Nor should you fail to overcome any likely objections. You must work to include all such information in your message, and you should make it clear and convincing.

You will also need to decide how to apportion your information across all the pieces in your mailing or the layout of a screen. With direct mail, you should use your letter to do most of the persuading, with any enclosures, attachments, or links providing supplementary information. These supplements might provide in-depth descriptions, price lists, diagrams, and pictures—in short, all the helpful information that does not fit easily into the letter. You may want to direct your readers' attention to these other pieces with such comments as "you'll find comments from your satisfied neighbors in the enclosed brochure," "as shown on page 7 of the enclosed catalog," or "you'll see testimonials of satisfied customers in our brochure."

When you send the sales message by email, the supporting information can be worked into the message, accessed via Web links, or provided in attachments that you invite the reader to view. Because people skim email quickly, be sure to keep the message itself relatively short. Skillfully chunking the message visually (see the Case Illustration on page 179) also helps reduce the impression of excessive length.

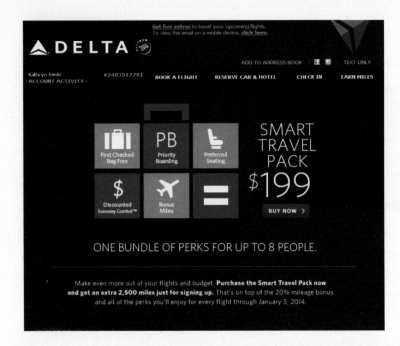

Get five extras to boost your upcoming flights.

DELTA AIR LINES

Comment/Complaint? | Add to Address Book

Kathryn Rentz: SkyMiles® #2483517781

Book a Flight | Reserve Car & Hotel | Check In | Earn Miles | Account Activity

ONE BUNDLE OF PERKS FOR UP TO 8 PEOPLE.

- First checked Bag Free
- Priority Boarding
- Preferred Seating
- Discounted Economy Comfort™
- 20% Bonus Miles

Make even more out of your flights and budget. **Purchase the Smart Travel Pack now and get an extra 2,500 miles just for signing up.** That's on top of the 20% mileage bonus and all of the perks you'll enjoy for every flight through January 5, 2014.

BUY NOW >

===

Terms & Conditions

Eligibility: Only SkyMiles General members are eligible for this offer. To participate in this offer eligible SkyMiles members

Many sales emails today are designed to be displayed in two formats, one for a computer screen and one for a mobile device. As you see here, the large-screen version can be graphically rich, but the mobile version might include text and links only.

Source: Delta Airlines. Reprinted with permission.

from the tech desk

Visuals Help Business Writers Add Interest to Sales Messages

Sales messages—both print and rich email—often include art and animation to increase their visual appeal as well as attract attention to the message.

Today's business writers need not be artists or professional photographers to use good visuals in their documents. Major software programs include bundled art, animation, photographs, and sounds; and scanners and easy-to-use programs are readily available to help writers create customized visuals. Additionally, on the Web, writers can find a vast assortment of specialists with products and services to help enhance their sales messages.

Here is a short list of helpful Web sites. You'll find more on the textbook Web site as well.

- *www.wpclipart.com*
A wide range of free artwork in the public domain (available for you to use without permission).

- *www.stockfreeimages.com*
The largest Web collection of free images, with 856,030 images, stock photos, and illustrations.

- *www.corbisimages.com/*
A gigantic collection of photos searchable by date, location, photographer, orientation, and more. Use of the images requires an up-front fee, but it is often quite low, and there's no royalty fee (a fee charged for each use of the image).

- *www.istockphoto.com*
Another popular source of royalty-free photos, but with illustrations, video, and audio available, too.

- *www.findsounds.com*
The Web's leading source of sound files. Some are in the public domain (e.g., animal sounds), but others require that you obtain permission.

But a note of caution: If your favorite way to find an image is to search **Google Images**, be very wary of using these images in your own work. Many are protected by copyright—and it can be difficult to trace the source in order to ask permission to use the image.

An Email Sales Message (Persuading Readers Who Used a Trial Version of an Application to Purchase It)

This message uses logical appeals and a variety of visual elements to make the product's benefits stand out.

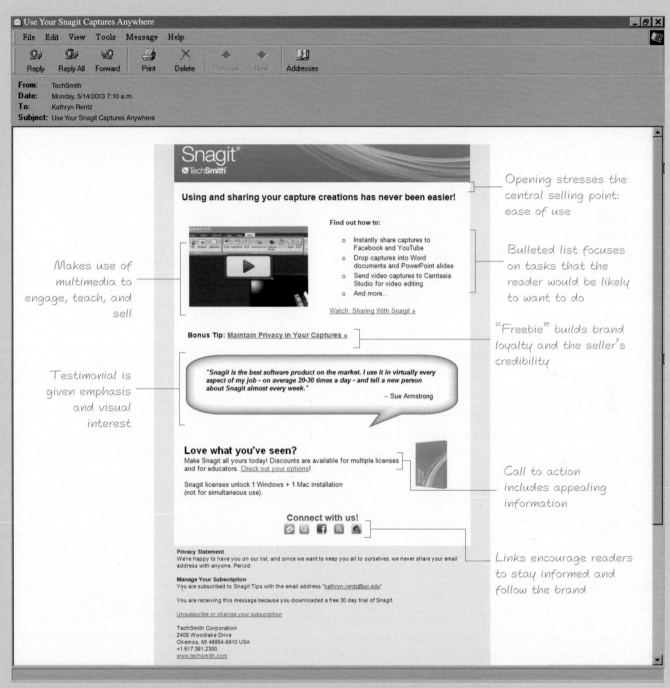

Opening stresses the central selling point: ease of use

Bulleted list focuses on tasks that the reader would be likely to want to do

"Freebie" builds brand loyalty and the seller's credibility

Makes use of multimedia to engage, teach, and sell

Testimonial is given emphasis and visual interest

Call to action includes appealing information

Links encourage readers to stay informed and follow the brand

Source: Reprinted with permission.

communication matters

Driving for the Sale

After you have developed your reader's interest in your product or service, the next logical step is to drive for the sale. After all, this is what you have been working for all along. It is the natural conclusion to the previous paragraphs.

How to word your drive for the sale depends on your strategy. If your selling effort is strong, your drive for action also may be strong. It may even be worded as a command ("Order your copy today—while it's on your mind."). If you use a milder selling effort, you could use a direct question ("May we send you your copy today?"). In any event, the drive for action should be specific and clear. For best effect, it should take the reader through the motions of whatever he or she must do. Here are some examples:

> Just check your preferences on the enclosed order card and drop it in the mail today.

> To start enjoying *House and Garden*, just call 1-888-755-5265. Be sure to have promo code 3626 handy to receive your 40 percent discount.

Similarly, in email selling you will need to make the action easy. Make it a simple click—a click to an order form or to the first part of the ordering process. For example, you might say "Just click the button below to order your customized iPhone case now!" or "You can download our free new catalog of business gifts at http://thankyoutoo.com." Many sales emails, such as the one shown on the next page, make the desired action easy by including multiple places for readers to perform it.

Because readers who have been persuaded sometimes put things off, you should urge immediate action. "Do it now" and "Act today" are versions of this technique, although some people dislike the commanding tone of such words. A milder and generally more acceptable way of urging action is to tie it in with a good reason for acting now. Here are some examples:

> . . . to take advantage of this three-day offer.

> . . . so that you can be ready for the Christmas rush.

> . . . so that you can immediately begin enjoying. . . .

Another effective technique for the close of a sales message is to use a few words that recall the main appeal. Associating the action with the benefits that the reader will gain by taking it adds strength to your sales effort. Illustrating this technique is a message selling a fishing resort vacation that follows its action words with a reminder of the joys described earlier:

> It's your reservation for a week of battle with the fightingest bass in the Southland.

Adding a Postscript

Unlike other business messages where a postscript (P.S.) appears to be an afterthought, a sales message can use a postscript as a part of its design. It can be used effectively in a number of ways: to urge the reader to act, to emphasize the major appeal, to invite attention to other enclosures, or to suggest that the reader pass along the sales message. Postscripts effectively used by professionals include the following:

> PS: Remember—if ever you think that *Action* is not for you, we'll give you every cent of your money back. We are that confident that *Action* will become one of your favorite magazines.

> PS: Hurry! Save while this special money-saving offer lasts.

> PS: Our little magazine makes a distinctive and appreciated gift. Know someone who's having a birthday soon?

> PS: Click now to order and you'll automatically be entered into a contest for a 4G Android smartphone.

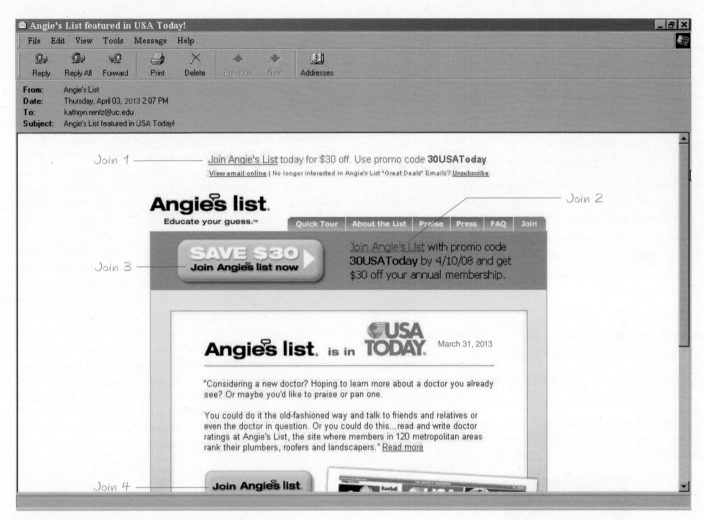

This email sales message makes the desired action easy.

Source: Reprinted with permission.

Offering Name Removal to Email Readers

Until January 1, 2004, it was a courtesy to offer the recipients of commercial email the option of receiving no further emails from the sender. Now, thanks to the so-called CAN-SPAM Act, it is a legal requirement as well (see Exhibit 7-3).[4] Consider placing this invitation in a prominent place—perhaps even before the main message. Wherever it is, the link should be easy to identify.

Ideally, it should also take only one or two clicks to work. Unfortunately, a 2010 report indicated that this trend was going in the wrong direction: The unsubscribe function of nearly half of the retail emails studied required three or more clicks, which was an increase from the year before.[5] On the other hand, providing readers the option of being emailed less frequently, of receiving emails about certain products only, or of following the brand through another channel (e.g., Facebook) seems to be regarded favorably by both consumers and businesses.[6]

▼ **EXHIBIT 7-3**   CAN-SPAM: It's the Law

The CAN-SPAM Act doesn't apply just to bulk email; it covers all commercial email messages whose purpose is to promote a product or service. Here are its main requirements:

1. Don't use false or misleading header information.

2. Don't use deceptive subject lines.

3. Identify the message as an ad.

4. Tell recipients where you're located (i.e., a physical postal address).

5. Tell recipients how to opt out of future email from you.

6. Honor opt-out requests promptly.

7. Be sure your email sales contractor (if you've hired one) is following the law.

Reviewing the General Sales Plan

From the preceding discussion, a general plan for the sales message emerges. This plan is similar to the classic AIDA (attention, interest, desire, action) model developed almost a century ago. It should be noted, however, that in actual practice, sales messages vary widely. Creativity and imagination are continually leading to innovative techniques. Even so, the most common plan is the following:

- Gain favorable attention.
- Create desire by presenting the appeals, emphasizing supporting facts, emphasizing the reader's viewpoint, and enhancing the message with appropriate visual elements.
- Include all necessary information—using a coordinated sales package (brochures, leaflets, links, and other appended parts).
- Drive for the sale by urging action now and recalling the main appeal.
- Possibly add a postscript.
- In email writing, offer to remove the reader from your email list to comply with legal requirements.

Contrasting Examples of a Sales Message

The following two email sales messages show bad and good efforts to promote Zach Miller's painting business (described in the Workplace Scenario on page 167).

a weak, self-centered message The ineffective example, below, begins with a dull, vague subject line. The first sentence talks only about the writer and delays answering the reader's obvious question, "What's the offer?" The second sentence begins to hint at the topic, but it is still writer focused. The middle paragraph contains some potentially good logical points, but the general, bland language won't generate much enthusiasm. Also, the character appeal is weak. Zach comes across as just another struggling student trying to pay his way through school; he does not make a convincing case that he and his friends would actually do a good job. The final sentence doesn't tell the reader what information to include in a response—and no final selling ends the message.

skillful use of character and rational appeals The better message (next page) follows the advice presented in the preceding pages. The subject line contains concrete reader benefits and states right away that the job will be professionally done. The opening sentence is an engaging question and is again focused on key benefits. The second paragraph builds confidence in the writer and his co-workers, and it ends by allaying any worries that the homeowner will be held liable if one of the students is injured. The third paragraph provides convincing logical details about the job. In the final paragraph, the writer uses a Web link to make it easy for readers to submit an estimate request, and he directs them to an attachment that contains further information and one last reader benefit.

This me-focused message is short on appealing reader benefits.

Subject: Offer from a Student

Hello,

My name is Zach Miller, and I'm a business student here at North Rapids University. I've been learning about business not only in the classroom but on the job, by running my own house-painting operation.

If you'd like to have your house painted this summer, I can do it for considerably less than what you'd pay a professional service. I would carefully prepare your house for painting, and I use good-quality paint. My two friends and I will also guarantee to complete the job within three days (weather permitting).

Please email me if you'd like an estimate.

Sincerely,
Zach Miller

PROPOSALS

Proposals share certain characteristics with reports. Both genres require that information be carefully gathered and presented. Visually, they can seem quite similar; at their most formal, they use the same kinds of prefatory material (e.g., title page, letter of transmittal, table of contents). And proposals frequently use the direct pattern that most reports use. But proposals differ from reports in one essential way: Proposals are intentionally *persuasive*. Proposal writers are not just providing information in an orderly, useful form. They are writing to get a particular result, and they have a vested interest in that result. Whether they use the direct or indirect approach, their purpose is to persuade. The following sections provide an introduction to the main types of proposals and offer guidelines for preparing them.

Types of Proposals

Proposals can vary widely in purpose, length, and format. Their purpose can be anything from acquiring a major client to getting a new copier for your department. They can range from one page to hundreds of pages. They can take the form of an email, a memo, a letter, or a report. They are usually written, but they can be presented orally or delivered in both oral and written form. As with other kinds of business communication, the context will determine the specific traits of a given proposal. But all proposals can be categorized as either internal or external and either

EXHIBIT 7-4   The Main Categories of Proposals

Answer these two questions to start your proposal planning:

- Is it an *internal or external* proposal? That is, will it go to readers inside or outside my organization?

- Is it a *solicited or unsolicited* proposal? That is, am I responding to a request for proposal or "selling" my idea to someone who doesn't expect it?

How you answer will help shape your proposal's contents, format, and tone.

solicited or unsolicited (see Exhibit 7-4). It is with the unsolicited type that indirect organization most often comes into play.

internal or external   Proposals can be either **internal** or **external**. That is, they may be written for others within your organization or for readers outside your organization.

The reasons for internal proposals differ, but you will almost surely find yourself having to write them. They are a major means by which you will get what you need in order to do your job better or change your organization. Whether you want a computer upgrade, an improved physical environment, specialized training, travel money, or additional staff members, you will usually need to make your case to management. Of course,

The you-viewpoint and better details give this message strong appeal.

Subject: A Professional-Quality Paint Job for Less

Would you like a great deal on having your house painted this summer while supporting North Rapids University students?

My name is Zach Miller, and I'm a junior business major here at NRU, in the Honors Plus program. I'm a graduate of St. Xavier High School and have been running my own house-painting business for three years. I employ two other NRU students, Jeff Barnes and Alex Wilson, who are also business students. We are all fully licensed and insured.

We use only 100% acrylic paints from top manufacturers like Benjamin Moore, Porter, and Sherwin Williams. Our specialty is careful preparation. We will scrape or power wash all surfaces of your house and do any needed minor repairs before we paint. The work comes with a two-year guarantee and, combined with our low rates, is an excellent value.

If you'd like a free estimate, just fill out the online estimate request form. The attached flyer tells more about me and also includes a coupon for 10% off if you schedule your paint job the day you receive your estimate.

Zach Miller
millerzs@mail.nru.edu
cellular: (431) 445-5560

workplace scenario

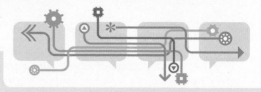

Selling Your Services through Proposal Writing

Play the role of Evan Lockley, vice president of account management at Whitfield Organizational Research. Your company collects internal information for businesses that want to improve their management techniques, information flow, employee morale, work processes, or other parts of their operations. To keep a steady stream of clients coming in, Whitfield must write numerous proposals for performing this kind of research.

As the manager of client accounts and the lead proposal writer at Whitfield, you now sit down to write a proposal to submit to RT Industries. This company is about to implement an enterprise resource planning (ERP) system. This implementation will require employees in every functional area of the business—from purchasing to inventory to design, manufacturing, and shipping—to learn the system and enter the data for their area. If the implementation is successful, the management at RT Industries will be able to tell, with the click of a few buttons, exactly how every facet of the business is doing. But implementing such a system is a major and potentially disastrous organizational change, and RT knows it. That's why they want to pay an organizational research firm to track the implementation and make sure it's as successful as possible. RT has invited Whitfield, along with other firms, to bid on this job.

You and one of your principal researchers have visited with the implementation team at RT Industries to learn more about the system they've chosen and their particular concerns. Whitfield has experience tracking such organizational changes, so you feel your odds of winning this client are good. But now you need to make your case. How can you craft a proposal that will make as positive an impression as possible? How can you make sure the readers at RT Industries will choose you over the competition? Read on to see how to write a persuasive proposal.

much of what you need as an employee will already be provided by your company. But when resources are tight, as they almost always are, you will have to persuade your superiors to give you the money rather than allocating it to another employee or department. Even if your idea is to enhance company operations in some way—for example, to make a procedure more efficient or cost effective—you may find yourself having to persuade. Companies tend to be conservative in terms of change. The management wants good evidence that the trouble and expense of making a change will pay off.

In addition, as the practice of outsourcing has grown, many companies have adopted a system in which departments have to compete with external vendors for projects. As the director

from the tech desk

Web Resources for Proposal Writing

You can find excellent free advice about proposal writing—along with examples and templates—on the Internet.

Here are just a few useful sites to check out:

- CapturePlanning at **CapturePlanning.com** is perhaps the best all-around site for learning about proposal writing. The site is full of freebies—from examples to articles to templates—that you can access just by registering.

- The team at **learntowriteproposals.com** offers such useful articles as "Our Top Ten Proposal Writing Tips," "10 Ways to Make Your Proposal Easier to Understand," and "11 Things Not to Do When You Write a Business Proposal."

- Debra Klug, author of **ProposalWriter.com**, discusses a wide range of relatively specific topics, such as getting a government grant to start a business, and provides a huge assortment of links to proposal-related resources.

- At **fedmarket.com**, Richard White provides links to his free webinars and publications about how to win government contracts.

- The **YouTube.com** videos by Julia King Tamang ("How to Write a Winning Proposal") and Doug Stern ("Six Keys to Writing a Great Proposal") offer sound, specific advice that supplements the advice in this chapter.

of technical publications for a company, for example, you may find yourself bidding against a technical-writing consulting firm for the opportunity, and the funding, to write the company's online documentation. If you are not persuasive, you may find yourself with a smaller and smaller staff and, eventually, no job yourself. Clearly, the ability to write a persuasive internal proposal is an important skill.

External proposals are also written for a variety of reasons, but the most common purpose is to acquire business for a company or money from a grant-awarding organization. Every consulting firm—whether in training, financial services, information technology, or virtually any other business specialty—depends upon external proposals for its livelihood. If such firms cannot persuade companies to choose their services, they will not be in business for long. Companies that supply other companies with goods they need, such as uniforms, computers, or raw materials, may also need to prepare proposals to win clients. Business-to-business selling is a major arena for external proposals.

But external proposals are also central to other efforts. A company might propose to merge with another company; a city government might propose that a major department store choose the city for its new location; a university professor might write a proposal to acquire research funding. Many nonprofit and community organizations depend upon proposals for grant money to support their work. They might write such proposals to philanthropic foundations, to wealthy individuals, to businesses, or to government funding agencies. Depending on the nature of the organization that you work for, proficiency in external proposal writing could be critical.

solicited or unsolicited Another way to categorize proposals is **solicited** versus **unsolicited**. A solicited proposal is written in response to an explicit invitation offered by a company, foundation, or government agency that has certain needs to meet. An unsolicited proposal, as you can probably guess, is one that you submit without an official invitation to do so.

The primary means by which organizations solicit proposals is the **request for proposals,** or **RFP** (variations are requests for quotes—RFQs—and invitations for/to bid—IFBs or

Whether solicited or unsolicited, proposals must do an effective job of presenting what is being proposed. You must work hard to meet the needs of your audience to win their business.

ITBs—both of which tend to focus only on price). These can range from brief announcements to documents of 50, 100, or more pages, depending upon the scope and complexity of the given project. As you might expect, their contents can also vary. But a lot of thought and research go into a good RFP. In fact, some RFPs—for instance, a company's request for proposals from IT firms to design and implement its technology infrastructure—need to be just as elaborately researched as the proposals being requested. Whatever its topic or purpose, the RFP needs to include a clear statement of the organization's need, the proposal guidelines (due date and time, submission process, and proposal format and contents), and the approval process, in addition to such helpful information as background about the organization.

When responding to an RFP, you should be careful to heed its guidelines. With some firms, your proposal gets eliminated if it arrives even one minute late or omits a required section. This is particularly true for proposals to the federal government, whose proposal guidelines are notoriously, and perhaps understandably, regimented (see the example on page 187). On the other hand, most RFPs give you some latitude to craft your proposal in such a way that your organization can put its best foot forward. You will want to take advantage of this maneuvering room to make your proposal the most persuasive of those submitted. Of course, you will decide in the first place to respond only to those RFPs that give your organization (or, if it is an internal RFP, your department) a good chance to win.

In business situations, solicited proposals usually follow preliminary meetings between the parties involved. For example, if a business has a need for certain production equipment, its buyers might first identify likely suppliers by considering those they already know, by looking at industry material, or by asking around in their professional networks. Next they would initiate meetings with these potential suppliers to discuss the business's needs. Some or all of these suppliers would then be invited to submit a proposal for filling the need with its particular equipment. As you can see, the more relationships you have with companies that might use your goods or services, the more likely it is that they will invite you to a preliminary meeting and then invite you to bid. One expert, in fact, asserts that "winning starts with pre-RFP relationships," which "provide the

Many nonprofit organizations depend upon grant writing for their survival.

intelligence you need to design and execute a winning plan."[7] Another advises that "proposals can be won (or lost) before the RFP hits the streets."[8]

Even if you are preparing a proposal for a government or foundation grant, it is wise—unless the RFP specifically forbids it—to call the funding source's office and discuss your ideas with a representative.

When writing unsolicited proposals, your job is harder than with solicited proposals. After all, in these scenarios, the intended reader has not asked for your ideas or services. For this reason, your proposal should resemble a sales message. It should quickly get the readers' attention and bring a need of theirs vividly to mind. It should then show how your product or services will answer the need. And from beginning to end, it should build your credibility. For example, if you want to provide training for a company's workforce or persuade a company to replace its current insurance provider with your company, you will need to target your readers' need in the opening, use further details to prepare them to receive your plan, lay out the benefits of your proposal quickly and clearly, and get the readers to believe that yours is the best company for the job. Careful and strategic preparation of unsolicited proposals can result in much success.

As with solicited proposals, you should try, if at all possible, to make prior contact with a person in the organization who has some power to initiate your plan. All other things being equal,

a proposal to someone you know is preferable to a "cold" proposal. It is best to view the unsolicited proposal as part of a larger relationship that you are trying to create or maintain.

Proposal Format and Contents

Every proposal is unique, but some generalizations can be made. To succeed, proposals must be designed with the key decision makers in mind, emphasize the most persuasive elements, and present the contents in a readable format and style.

format and formality The simplest proposals are often email messages. Internal proposals (those written for and by people in the same organization) usually fall into this category. The more complex proposals may take the form of long reports, including prefatory pages (title pages, letter of transmittal, table of contents, executive summary), extensive text, and an assortment of appended parts. Most proposals have arrangements that fall somewhere between these extremes.

Because of the wide variations in the makeup of proposals, you need to investigate the situation carefully before designing a particular proposal. Try to find out what format the readers will expect and what other proposal writers have submitted in similar situations. In the case of an invited proposal, review the request thoroughly, looking for clues concerning the preferences of the inviting organization. If you are unable to follow any of these courses, design a format based on your analysis of the audience and your knowledge of formatting strategies. Your design should be the one that you think is best for the one situation.

The same advice applies to your decisions about formality. Let your reader and the circumstances be your guide. Internal proposals tend to be less formal than external ones because the parties are often familiar with each other and because internal documents, in general, are less formal than external ones. If you

When your proposal is uninvited, the challenge is harder.

case illustration
First Page of a Government RFP

This first page of an RFP posted on Vermont's "Buildings and General Services" Web site (www.bgs.vermont.gov) shows the beginning of a long list of vendor requirements. The complete RFP for this relatively simple project is 23 pages long. To have a chance of winning the contract, you would need to study all 23 pages carefully and follow the instructions to the letter.

STATE OF VERMONT
OFFICE OF PURCHASING & CONTRACTING
RFP – BGS SNOW REMOVAL SERVICES 2012
PAGE 1

1. **OVERVIEW:**

   1.1. **SCOPE AND BACKGROUND:** The Office of Purchasing & Contracting is seeking to establish purchasing agreements with one or more companies that can provide BGS Snow Removal Services in the Burlington Area, Burlington, Vermont.

   1.2. **CONTRACT PERIOD:** Contracts arising from this request for proposal will be for a period of **12 <u>months</u>** with an option to renew for two (2) additional 12-**<u>month</u>** periods. Proposed start date will be **October 1, 2012.**

   1.3. **CONTRACT VALUE/QUANTITY:** The estimated annual value of this contract is $80,000.00. The annual value and quantities are estimated only based on prior usage; actual purchases may be higher or lower depending on the state's needs.

   1.4. **SINGLE POINT OF CONTACT:** All communications concerning this Request for Proposal (RFP) are to be addressed in writing to the attention of: **Robert Pierce**, Purchasing Agent, State of Vermont, Office of Purchasing & Contracting, 10 Baldwin St - Montpelier, Montpelier, VT 05633-7501. **Robert Pierce**, Purchasing Agent is the sole contact for this proposal. Actual contact with any other party or attempts by bidders to contact any other party could result in the rejection of their proposal.

   1.5. **BIDDERS' CONFERENCE:** A bidder's conference will be held on **July 23, 2012 at 11:00 AM**. Meeting will start at the BGS maintenance shop at 1705 Hedgeman Avenue in Colchester, Vermont.

   1.6. **QUESTION AND ANSWER PERIOD:** Any vendor requiring clarification of any section of this proposal or wishing to comment or take exception to any requirements or other portion of the RFP must submit specific questions in writing no later than **July 30, 2012 4:30PM**. Questions may be e-mailed to robert.pierce@state.vt.us . Any objection to the RFP or to any provision of the RFP, that is not raised in writing on or before the last day of the question period is waived. At the close of the question period a copy of all questions or comments and the State's responses will be posted on the State's web site http://bgs.vermont.gov/purchasing/bids . Every effort will be made to have these available as soon after the question period ends, contingent on the number and complexity of the questions.

   1.7. **INSTRUCTIONS FOR BIDDERS**: see sections 5 and 6.

2. **DETAILED REQUIREMENTS:**

   1.0 Without notification from the State, the Contractor will plow all snowfall of two inches or more, or at the end of each snow fall.

   2.0 Snow removal services for this site will be (24) hours a day, (7) days a week.

   3.0 Contractor shall take care not to damage buildings, guardrails, curbs and automobiles, Contractor will make repairs to material damaged during performing these services. Determination of the need for and extent of repair is the sole discretion of the State of Vermont Contract Coordinator.

   4.0 Contractor shall be responsible for incidental and emergency calls for removing snow that may interfere with State operations in the complex.

   5.0 Where applicable, driver/operator must be CDL certified.

   6.0 The contractor shall at all times provide adequate supervision of his/her employees to ensure complete and satisfactory performance of all work in accordance with the terms of the contract. The contractor will have a responsible supervisor on the job at all times when the work of the contract is being carried out.

   7.0 The contractor and his/her employees will be subject to all applicable State and Federal regulations for the conduct of personnel.

   8.0 Contractor's employees shall not utilize or operate State owned equipment of any type without specific authorization of the contract coordinator.

   9.0 The contractor will screen all personnel to assure the State that all employees are capable of performing the services in accordance with the RFP. A background check will be required to perform services as part of this contract.

A proposal or grant is the beginning of a relationship. Essentially, the readers are interviewing your company or organization, trying to determine whether a basis for a positive, constructive alliance exists. Your proposal is the face you are presenting to the client or funding source. If they feel comfortable with your proposal, they will feel comfortable with your company or organization.

Source: Richard Johnson-Sheehan, *Writing Proposals*, 2nd ed. (New York: Pearson/Longman, 2008) 232–33, print.

are proposing a major initiative or change, however, using a formal presentation—whether oral, written, or both—may be in order. Likewise, external proposals, while they tend to be formal, can be quite informal if they are short and the two parties know each other well. Many successful business proposals are pitched in letter format. As with every other kind of message, knowledge of and adaptation to your reader are key.

content Whether you are writing an external or internal proposal or a solicited or unsolicited one, your primary goal is the same: to make a persuasive argument. Every element of your proposal—from the title to the cover letter to the headings and organization of your content to the way you say things—needs to contribute to your central argument.

To be able to design your proposal according to this principle, you need to know your readers and their needs (which may be represented in an RFP). You also need to know how you can meet those needs. From these two sets of facts, you can develop your central argument. What is your competitive edge? Value for the money? Convenience? Reliability? Fit of your reader's needs or mission with what you have to offer? Some or all of the above? How you frame your argument will depend on how you think your proposal will be evaluated.

The reader of a business proposal will bring three basic criteria to the evaluation process:

- Desirability of the solution (Do we need this? Will it solve our problem?)
- Qualifications of the proposer (Can the author of the proposal, whether

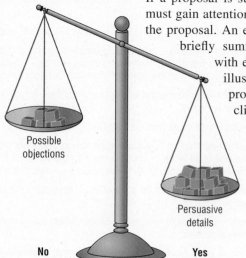

Possible objections

Persuasive details

No   **Yes**

Give your readers enough information to tip them toward yes.

an individual or company, really deliver, and on time and on budget?)

- Return on investment (Is the expense, whether time or money, justified?)

If you can answer these questions affirmatively from the point of view of your intended recipient, you have a good chance of winning the contract or your management's approval.

When you have figured out what to propose and why, you need to figure out how to propose it. If the RFP provides strict guidelines for contents and organization, follow them. Otherwise, you have considerable latitude when determining your proposal's components. Your reader is likely to expect some version of the following eight topics, but you should adapt them as needed to fit the facts of your case. (See page 190 and pages 191–196 for two very different examples.)

1. *The writer's purpose and the reader's need.* An appropriate beginning is a statement of your purpose (to present a proposal) and the reader's need (such as reducing the turnover of sales staff). If the report is in response to an invitation, that statement should tie in with the invitation (for example, "as described in your July 10 announcement"). The problem and your purpose should be stated clearly. This proposal beginning illustrates these recommendations:

> As requested at the July 10 meeting with Alice Burton, Thomas Cheny, and Victor Petrui in your Calgary office, Murchison and Associates present the following proposal for studying the high rate of turnover among your field representatives. We will assess the job satisfaction of the current sales force, analyze exit interview records, and compare company compensation and human resource practices with industry norms to identify the causes of this drain on your resources.

If a proposal is submitted without invitation, its beginning must gain attention in order to motivate the recipient to read the proposal. An effective way of doing this is to begin by briefly summarizing the highlights of the proposal with emphasis on its benefits. This technique is illustrated by the beginning of an unsolicited proposal that a consultant sent to prospective clients:

> Is your social marketing strategy working?
>
> Twitter, blogs, Facebook, LinkedIn—such tools have become essential to building your brand. Are you making the best use of them?
>
> Using a three-step social media audit, Mattox and Associates can find out. With access to your media and just one day of on-site interviews, our experts will tell you . . . [the rest of the proposal follows].

Your clear statement of the purpose and problem may be the most important aspect of the proposal. If you do not show right away that you understand what needs to be done and have a good plan for doing it, you may well have written the rest of your proposal in vain.

2. *The background.* A review of background information promotes an understanding of the problem. Thus, a college's proposal for an educational grant might benefit from a review of the relevant parts of the college's history. A company's proposal of a merger with another company might review industry developments that make the merger desirable. Or a chief executive officer's proposal to the board of directors that a company be reorganized might discuss related industry trends.

3. *The need.* Closely related to the background information is the need for what is being proposed. In fact, background information may well be used to establish need. Your goal in this important section is to paint a picture of the problem or goal in such a way that the reader feels a keen need for what you are proposing.

You might wonder if this section applies in situations where an RFP has been issued. In such cases, won't readers already know what they need? In many cases the answer is no, not exactly. They may think they know, but you may see factors that they've overlooked. Plus, restating their problem in ways that lead to your proposed solution helps your persuasive effort. And whatever the situation, elaborating on the receiving organization's needs enables your readers to see that *you* understand those needs.

4. *The description of your plan.* The heart of a proposal is the description of what the writer proposes to do. This is the primary message of the proposal. It should be concisely presented in a clear and orderly manner, with headings and subheadings as needed. It should give sufficient detail to convince the reader of the plan's logic, feasibility, and appropriateness. It should also identify the "deliverables," or tangible products, of the proposal.

5. *The benefits of the proposal.* Your proposal should make it easy for your readers to see how your proposed action will benefit them. A brief statement of the benefits should appear at the front of your proposal, whether in the letter of transmittal, executive summary, opening paragraph, or all of the above. But you should elaborate on those benefits in the body of your proposal. You might do so in the section describing your plan, showing how each part will yield a benefit. Or, you might have a separate section detailing the benefits. As with sales writing, the greater the need to persuade, the more you should stress the benefits.

As an example of benefits logically covered in proposals, a college's request for funding to establish a program for retraining the older worker could point to the profitability that such funding would give local businesses. A proposal

offering a consulting service to restaurants could stress such benefits as improved efficiency, reduced employee theft, savings in food costs, and increased profits.

6. *Cost and other particulars.* Once you have pitched your plan, you need to state clearly what it will cost. You may also need to cover such other particulars as time schedules, performance standards, means of appraising performance,

"We need something to come after this part. Any ideas?"

case illustration
An Internal Unsolicited Proposal

This email proposal asks a company to sponsor an employee's membership in a professional organization. Starting with the subject line, the writer tries to avoid saying anything that the reader—in this case, the head of a corporate communications department—would disagree with. When enough background and benefits are given, the writer states the request and then describes the cost in the most positive terms. Offering to try the membership for one year helps the proposal seem relatively modest.

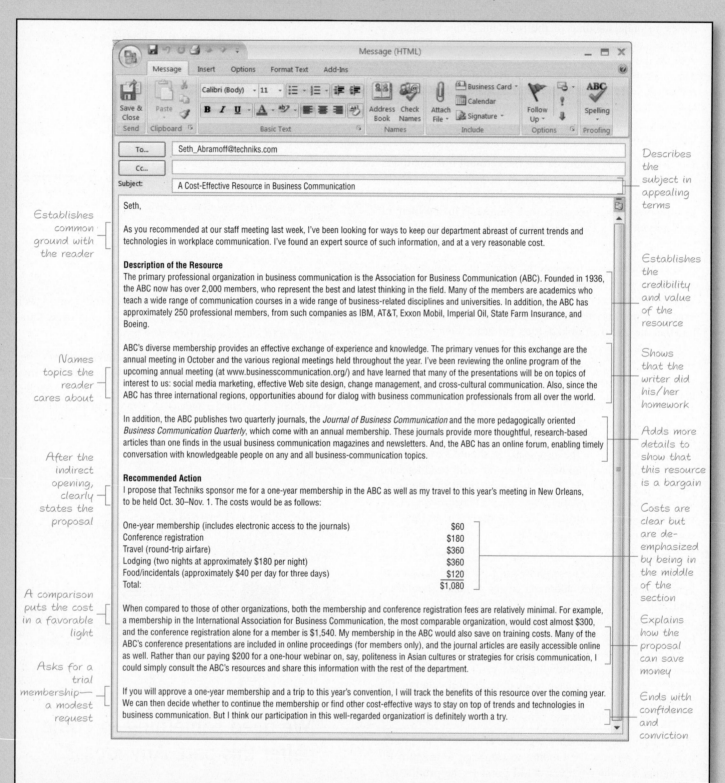

Describes the subject in appealing terms

Establishes common ground with the reader

Names topics the reader cares about

After the indirect opening, clearly states the proposal

A comparison puts the cost in a favorable light

Asks for a trial membership—a modest request

Establishes the credibility and value of the resource

Shows that the writer did his/her homework

Adds more details to show that this resource is a bargain

Costs are clear but are de-emphasized by being in the middle of the section

Explains how the proposal can save money

Ends with confidence and conviction

Message (HTML)

Message   Insert   Options   Format Text   Add-Ins

To... Seth_Abramoff@techniks.com

Cc...

Subject: A Cost-Effective Resource in Business Communication

Seth,

As you recommended at our staff meeting last week, I've been looking for ways to keep our department abreast of current trends and technologies in workplace communication. I've found an expert source of such information, and at a very reasonable cost.

Description of the Resource
The primary professional organization in business communication is the Association for Business Communication (ABC). Founded in 1936, the ABC now has over 2,000 members, who represent the best and latest thinking in the field. Many of the members are academics who teach a wide range of communication courses in a wide range of business-related disciplines and universities. In addition, the ABC has approximately 250 professional members, from such companies as IBM, AT&T, Exxon Mobil, Imperial Oil, State Farm Insurance, and Boeing.

ABC's diverse membership provides an effective exchange of experience and knowledge. The primary venues for this exchange are the annual meeting in October and the various regional meetings held throughout the year. I've been reviewing the online program of the upcoming annual meeting (at www.businesscommunication.org/) and have learned that many of the presentations will be on topics of interest to us: social media marketing, effective Web site design, change management, and cross-cultural communication. Also, since the ABC has three international regions, opportunities abound for dialog with business communication professionals from all over the world.

In addition, the ABC publishes two quarterly journals, the *Journal of Business Communication* and the more pedagogically oriented *Business Communication Quarterly*, which come with an annual membership. These journals provide more thoughtful, research-based articles than one finds in the usual business communication magazines and newsletters. And, the ABC has an online forum, enabling timely conversation with knowledgeable people on any and all business-communication topics.

Recommended Action
I propose that Techniks sponsor me for a one-year membership in the ABC as well as my travel to this year's meeting in New Orleans, to be held Oct. 30–Nov. 1. The costs would be as follows:

One-year membership (includes electronic access to the journals)	$60
Conference registration	$180
Travel (round-trip airfare)	$360
Lodging (two nights at approximately $180 per night)	$360
Food/incidentals (approximately $40 per day for three days)	$120
Total:	$1,080

When compared to those of other organizations, both the membership and conference registration fees are relatively minimal. For example, a membership in the International Association for Business Communication, the most comparable organization, would cost almost $300, and the conference registration alone for a member is $1,540. My membership in the ABC would also save on training costs. Many of the ABC's conference presentations are included in online proceedings (for members only), and the journal articles are easily accessible online as well. Rather than our paying $200 for a one-hour webinar on, say, politeness in Asian cultures or strategies for crisis communication, I could simply consult the ABC's resources and share this information with the rest of the department.

If you will approve a one-year membership and a trip to this year's convention, I will track the benefits of this resource over the coming year. We can then decide whether to continue the membership or find other cost-effective ways to stay on top of trends and technologies in business communication. But I think our participation in this well-regarded organization is definitely worth a try.

A design and manufacturing company has invited research firms to propose plans for tracking its implementation of an enterprise resource planning (ERP) system—information technology that integrates all functions of the company, from job orders to delivery and from accounting to customer management. The midlevel formality of this proposal responding to the RFP is appropriate given the proposal's relative brevity and the two parties' prior meeting.

WHITFIELD
Organizational Research

7 Research Parkway, Columbus, OH 45319 614-772-4000 Fax: 614-772-4001

February 3, 2014

Ms. Janice Spears
Chief Operations Officer
RT Industries
200 Midland Highway
Columbus, OH 45327

Dear Janice:

Identifies the context for the proposal and shows appreciation for being invited to submit

Thank you for inviting Whitfield Organizational Research to bid on RFP 046, "Study of InfoStream Implementation at RT Industries." Attached is our response.

Reminds the reader of the previous, pleasant meeting

Reinforces the need for the study

We enjoyed meeting with you to learn about your goals for this research. All expert advice supports the wisdom of your decision to track InfoStream's implementation. As you know, the road of ERP adoption is littered with failed, chaotic, or financially bloated implementations. Accurate and timely research will help make yours a success story.

Summarizes the proposing company's advantages

Whitfield Organizational Research is well qualified to assist you with this project. Our experienced staff can draw upon a variety of minimally invasive, cost-effective research techniques to acquire reliable information on your employees' reception and use of InfoStream. We are also well acquainted with ERP systems and can get a fast start on collecting the data you need. And because Whitfield is a local firm, we will save you travel and lodging costs.

Compliments the receiving company, shows the writer's knowledge of the company, and states the benefits of choosing the writer's company

RT's culture of employee involvement has earned you a place on *Business Ohio*'s list of the Best Ohio Workplaces for the last five years. The research we propose, performed by Whitfield's knowledgeable and respectful researchers, will help you maintain your productive culture through this period of dramatic change. It will also help you reap the full benefits of your investment.

Indirectly asks the reader for the desired action

We would welcome the opportunity to work with RT Industries on this exciting initiative.

Sincerely yours,

Evan Lockley

Evan Lockley
Vice President, Account Management

enclosure

case illustration (continued)

Response to RFP 046:
Study of InfoStream Implementation at RT Industries

Proposed by
Whitfield Organizational Research
February 3, 2014

Executive Summary

Provides a clear overview of the problem, purpose, and benefits

RT Industries has begun a major organizational change with its purchase of InfoStream enterprise resource planning (ERP) software. To track the effect of this change on personnel attitudes and work processes in the company, RT seeks the assistance of a research firm with expertise in organizational studies. Whitfield Organizational Research has extensive experience with personnel-based research, as well as familiarity with ERP software. We propose a four-part plan that will help ensure the success of your implementation.

Our methodology will be multifaceted, minimally disruptive, and cost effective. The results will yield a reliable picture of how InfoStream is being received and used among RT's workforce. With this information, RT's change leaders can intervene appropriately to effectively manage this companywide innovation.

Project Goals

RT Industries has so far invested over $1.6 million and over 1,000 employee hours in the purchase of and management's training on InfoStream's ERP system. As RT integrates the system fully into its company of 800+ employees over the next 12 months, it will invest many additional dollars and hours in the project, with the total investment likely to top $2 million. Adopting such a system is one of the most wide-ranging and expensive changes a company can make.

Shows knowledge of the company; reminds readers of the investment they want to protect

Reinforces the need for the study

As Jeri Dunn, Chief Information Officer of Nestle USA, commented in *CIO Magazine* about her company's troubles with its ERP adoption, "No major software implementation is really about the software. It's about change management." An ERP system affects the daily work of everyone in the company. The most common theme in ERP-adoption failure stories—of which there are many—is lack of attention to the employees' experience of the transition. Keeping a finger on the pulse of the organization during this profound organizational change is critical to maximizing the return on your investment.

Our research will determine

- How well employees are integrating InfoStream into their jobs.
- How the new system is changing employees' work processes.
- How the system is affecting the general environment or "culture" in the company.

States the benefits, supported by clear logic

Whitfield has designed a four-part, multimethod research plan to gather these data. Through our periodic reports, you will be able to see how InfoStream is being integrated into the working life of the company. As a result, you will be able to make, and budget for, such interventions as strategic communications and additional training. You will also find out where employee work processes need to be adjusted to accommodate the new system.

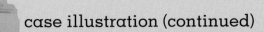

Whitfield Organizational Research   2

Instituting a change of this magnitude *will* generate feedback, whether it is employee grumbling or constructive criticism. Whitfield associates will gather this feedback in a positive, orderly way and compile it into a usable format. The findings will enable RT's management to address initial problems and ward off future problems. The research itself will also contribute to the change management efforts of the company by giving RT's employee stakeholders a voice in the process and allowing their feedback to contribute to the initiative's success.

Deliverables

The information you need will be delivered as shown below. All dates assume a project start date of July 1, 2014.

Approximate Date:	Deliverable:
October 1, 2014	Written report on an **initial** study of 12–14 employees' work processes and attitudes and on a companywide survey.
February 1, 2015	Written report at **midyear** on the same employees' work processes and attitudes and on a second companywide survey.
June 30, 2015	**Year-end** report (written and oral) on employees' work processes and attitudes and on a final companywide survey.

Readers can see the products of the proposed research up front

Anticipated Schedule/Methods

The research will take place from July 1, 2014, the anticipated go-live date for InfoStream at RT, to approximately June 30, 2015, a year later. As shown below, there will be four main components to this research, with Part III forming the major part of the project.

Research Part and Time Frame	Purpose	Methods
Part I (July '14)	Gather background information; recruit research participants	Gather data on RT (history, products/mission, organizational structure/culture, etc.). Interview personnel at RT and at InfoStream about why RT is adopting an ERP system, why RT bought InfoStream, and how employees at RT have been informed about InfoStream. During this period we will also work with the COO's staff to recruit participants for the main part of the study (Part III).

Gives details of the project in a readable format

Whitfield Organizational Research 3

Research Part and Time Frame	Purpose	Methods
Part II (July '14):	Obtain the perspective of the InfoStream launch team	Conduct a focus group with the launch team, with emphasis on their goals for and concerns about the implementation. Anticipated duration of this interview would be one hour, with participants invited to share any additional feedback afterward in person or by email.
Part III (July–Sept. '14; Nov. '14–Jan. '15; Mar.–June '15):	Assess the impact of InfoStream on employee work processes and attitudes	Conduct three rounds of 1–2 hour interviews with approximately 12–14 RT employees to track their use of InfoStream. Ideally, we will have one or two participants from each functional area of the company, with multiple levels of the company represented.
Part IV (September '14, January '15, May '15)	Assess companywide reception of InfoStream	Conduct three Web-based surveys during the year to track general attitudes about the implementation of InfoStream.

This plan yields the following time line:

	7/14	8/14	9/14	10/14	11/14	12/14	1/15	2/15	3/15	4/15	5/15	6/15
Initial research	▓											
Focus group	▓											
1st round of interviews	▓	▓										
1st Web survey			▓									
Initial report				▓								
2nd round of interviews					▓	▓						
2nd Web survey							▓					
Midyear report								▓				
3rd round of interviews									▓	▓		
3rd Web survey											▓	
Year-end report												▓

Timeline makes it easy to see what will happen at each point

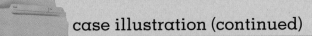

Whitfield Organizational Research   **4**

Interview Structure and Benefits

While Parts I, II, and IV will provide essential information about the project and its reception, the most valuable data will come from Part III, the onsite interviews with selected RT employees. Gathering data in and about the subject's own work context is the only reliable way to learn what is really happening in terms of the employees' daily experience. Following is a description of our methodology for gathering these kinds of data:

Initial interview:

- Gather background information about the participants (how long they have worked at RT, what their jobs consist of, what kind of computer experience they've had, how they were trained on InfoStream).
- Ask them to show us, by walking us through sample tasks, how they use InfoStream.
- Ask them to fill out a questionnaire pertaining to their use of InfoStream.
- Go back over their answers, asking them to explain why they chose the answers they did.
- Ask them either to keep notes on or email us about any notable experiences they have with InfoStream.
- Take notes on any interruption, interactions, and other activities that occur during the interview.

From data gained in these interviews, we will assess how well the participants' current work processes are meshing with InfoStream. We will also document how use of InfoStream is affecting the participants' attitudes and their interactions with other employees and departments. We will check our findings with the participants for accuracy before including these data in the initial report.

Midyear interview:

- Ask the participants if they have any notable experiences to relate about InfoStream and/or if any changes have occurred in the tasks they perform using InfoStream.
- Have the participants fill out the same questionnaire as in the first interviews.
- Discuss with participants the reasons for any changes in their answers since the first questionnaire.
- Observe any interactions or other activities that occur during the interview.
- Check our findings with the participants for accuracy before including these data in the midyear report.

Year-end interviews:

- Will be conducted in the same fashion as the second interviews.
- Will also include questions allowing participants to debrief about the project and about InfoStream in general.

Benefits of this interview method:

- Because researchers will be physically present in the employees' work contexts, they **can gather a great deal of information**, whether observed or reported by the employee, **in a short amount of time**.
- Because employees will be asked to elaborate on their written answers, the researcher **can learn the true meaning of the employee's responses**.

Special section elaborates on the company's unique methodology; helps justify the most expensive part of the plan

Whitfield Organizational Research   5

- Asking employees to verify the researcher's findings **will add another validity check and encourage honest, thorough answers**.

Specific Knowledge Goals

We will design the interviews and the companywide surveys to find out the extent to which

- InfoStream is making participants' jobs easier or harder, or easier in some ways and harder in others.

- InfoStream is making their work more or less efficient.

- InfoStream is making their work more or less effective.

- They believe InfoStream is helping the company overall.

- They are satisfied with the instruction they have received about the system.

- InfoStream is changing their interactions with other employees.

- InfoStream is changing their relations with their supervisors.

- InfoStream is affecting their overall attitude toward their work.

The result will be a detailed, reliable picture of how InfoStream is playing out at multiple levels and in every functional area of RT Industries, enabling timely intervention by RT management.

A tantalizing list of what the readers most want to know whets their desire to hire the proposing company

Cost

Because we are a local firm, no travel or lodging expenses will be involved.

Research Component	Estimated Hours	Cost
Part I (background fact finding)	6 hours	$300
Part II (focus group with launch team)	3 hours (includes preparation and analysis)	$300
Part III (3 rounds of on-site interviews)	474 hours	$18,960
Part IV (3 rounds of Web-based surveys)	48 hours	$1,920
Preparation of Reports	90 hours	$3,600
		Total: $25,080

Cost breakdown justifies the expense but is not so detailed that the readers can nitpick specific items

Credentials

Efficient credentials section focuses only on relevant qualifications

Whitfield Organizational Research has been recognized by the American Society for Training and Development as a regional leader in organizational consulting. We have extensive education and experience in change management, organizational psychology, quantitative and qualitative research methods, and team building. Our familiarity with ERP software, developed through projects with such clients as Orsys and PRX Manufacturing, makes us well suited to serve RT's needs. Résumés and references will be mailed upon request or can be downloaded from www .whitfieldorganizationalresearch.com.

and equipment and supplies needed. Remember that a proposal is essentially a contract. Anticipate and address any issues that may arise, and present your requirements in the most positive light.

7. *Evidence of your ability to deliver.* The proposing organization must sometimes establish its ability to perform. This means presenting information on such matters as the qualifications of personnel, success in similar cases, the adequacy of equipment and facilities, operating procedures, and environmental consciousness. With an external proposal, resist the temptation to include long, generic résumés. The best approach is to select only the most persuasive details about your personnel. If you do include résumés, tailor them to the situation.

8. *Concluding comments.* In most proposals you should urge or suggest the desired action. This statement often occurs in a letter to the readers, but if there is no cover letter or the proposal itself is not a letter, it can form the conclusion of your proposal. You might also include a summary of your proposal's highlights or provide one final persuasive push in a concluding section.

Whatever you're writing—whether a proposal, request, sales message, or some other kind of message—the art of persuasion can be one of your most valuable assets. Adding the tips in this chapter to your general problem-solving approach will help you prepare for all those times in your career when you will need others' cooperation and support.

become a savvy persuader!

- What are six principles for effective persuasion?
- How can you increase the effectiveness of your sales letters?
- How should you format your proposal?

Scan the QR code with your smartphone or use your Web browser to find out at www.mhhe.com/RentzM3e. Choose Chapter > 7 > Bizcom Tools & Tips. While you're there you can view a chapter summary, exercises, PPT slides, and more to become a savvy persuader.

www.mhhe.com/RentzM3e

Researching
and Writing
Reports

chapter eight

How often you write reports in the years ahead will depend on the size and nature of the organization you work for. If you work for an organization with fewer than 10 employees, you will probably write only a few. But if you work for a mid-size or large organization, you are likely to write many. The larger the organization, the greater its complexity; and the greater the complexity, the greater the need for information to manage the organization.

Successful reports and effective research go hand in hand. Caroline Molina-Ray, Executive Director of Research and Publications at Apollo Research Institute, explains why:

"Business leaders must base their decisions on relevant facts—not just on intuition. Effective research provides leaders with facts they can use to plan, evaluate, and improve business performance. To be most useful, research reports must not only include pertinent data but also explain what the data mean and how a decision maker might act on this information."

This chapter will help you gather and prepare information to help solve business problems. ∎

LEARNING OBJECTIVES

LO 8-1 Write clear problem and purpose statements.

LO 8-2 List the likely factors involved in a problem.

LO 8-3 Explain the difference between primary and secondary research.

LO 8-4 Use Internet search engines to gather information.

LO 8-5 Use other Web resources to gather information.

LO 8-6 Evaluate Web sites for reliability.

LO 8-7 Use social networking and social bookmarking sites to gather information.

LO 8-8 Use the library to gather information.

LO 8-9 Use sampling to conduct a survey.

LO 8-10 Construct a questionnaire and conduct a survey.

LO 8-11 Design an observational study for a business problem.

LO 8-12 Conduct an experiment for a business problem.

LO 8-13 Explain the uses of focus groups and personal interviews.

LO 8-14 Discuss important ethical guidelines for research.

LO 8-15 Interpret your findings accurately.

LO 8-16 Organize information in outline form using time, place, quantity, factors, or a combination of these as bases for division.

LO 8-17 Turn an outline into a table of contents whose format and wording are logical and meaningful.

LO 8-18 Write reports that are focused, objective, consistent in time viewpoint, smoothly connected, and interesting.

LO 8-19 Prepare reports collaboratively.

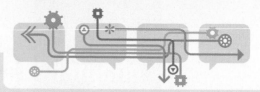
Researching and Writing Reports on the Job

Introduce yourself to the subject of report writing by assuming the role of operations analyst at Technisoft, Inc. Much of your work at this large software company involves getting information for your boss. Yesterday, for example, you looked into the question of excessive time spent by office workers on the Internet. A few days earlier, you worked on an assignment to determine the causes of unrest in one of the branch locations. Before that assignment you investigated a supervisor's recommendation to change an evaluation process. You could continue the list indefinitely because investigating problems is a part of your work.

So is report writing, because you must research and write a report on each of your investigations. A written report has several advantages over other communication forms. Written reports are a good medium for conveying detailed findings. They also make permanent records. Thus, those who need the information contained in these reports can review and study them at their convenience. Plus, written reports are a convenient and efficient means of distributing information because they can be easily routed to a number of readers.

Your report-writing work is not unique to your job. In fact, report writing is common throughout the company. For example, the engineers often report on the technical problems they encounter. The accountants regularly report to management on the company's financial operations. From time to time, production people report on various aspects of operations. The salespeople regularly report on marketing matters. Such reporting is vital to your company's operations—as it is to the operations of all companies.

Writing to external audiences can also be critical to an organization's success. If the organization is a consulting firm, reports to the client may be its primary deliverable. If the company is publicly traded, it is required by law to publish financial reports to the government and to shareholders. Depending on the nature of its business, a company may have to research and write reports for various agencies about its impact on the environment, its hiring practices, or its compliance with quality standards.

Sometimes reports are written by individuals. Increasingly, however, they are prepared in collaboration with others. Even if one person has primary responsibility for a report, he or she will often need contributions from many people. Indeed, report writing draws on a wide variety of communication skills, from getting information to presenting it clearly.

This chapter and the following chapter describe how to prepare this vital form of business communication.

DEFINING REPORTS

You probably have a good idea of what reports are. Even so, you might have a hard time defining them. Some people define reports to include almost any presentation of information, while others use the term to refer only to the most formal presentations. We use this middle-ground definition: A **business report** is an interpret them without personal bias. The word *communication* in our definition is broad in meaning. It covers all ways of transmitting meaning: speaking, writing, using visuals, or a combination of these. The basic ingredient of reports is *factual information*. Factual information is based on events, statistics, and other data. Finally, a business report must *serve a business purpose*. Research scientists, medical doctors, ministers, students, and many

> [A business report is an orderly and objective communication of factual information that serves a business purpose.]

orderly and objective communication of factual information that serves a business purpose.

As an *orderly* communication, a report is prepared carefully. This care in their preparation distinguishes reports from casual exchanges of information. The *objective* quality of a report is its unbiased approach. Good reports present all the relevant facts and

others write reports, but to be classified as a business report, a report must help a business solve its problems or meet its goals.

Business reports can be short or long, formal or informal, electronic or printed, mostly text or mostly visuals. Whatever their specific qualities, though, all reports should help readers make informed business decisions.

communication matters

LO 8-1 Write clear problem and purpose statements.

DETERMINING THE REPORT PROBLEM AND PURPOSE

Your work on a report logically begins with a need, which we refer to in the following discussion as the **problem**. Someone or some group (usually your superiors) needs information for a business purpose. Perhaps the need is for information only; perhaps it is for information and analysis; or perhaps it is for information, analysis, and recommendations (see the Communication Matters box above). Whatever the case, someone with a need will authorize you to do the work. How you define this need (problem) will determine your report's **purpose**.

The Preliminary Investigation

Your first task is to understand the problem. To do this well, you will almost surely have to gather additional information beyond what you've been given. You may need to study the company's files or query its databases, talk over the problem with experts, search through external sources, and/or discuss the problem with those who authorized the report. You should do enough preliminary research to be sure you understand the problem that your report is intended to address.

The Need for Clear Problem and Purpose Statements

Your next task is to clearly state your understanding of the problem and your report's purpose. Clear problem and purpose statements are important for you as you plan and write the report and for those who will read and use the report.

The **problem statement** provides a clear description of the situation that created the need for your report. Problem statements are generally written as declarative statements. For example, a simple one might read "Sales are decreasing at Company X."

You should then write a **purpose statement** (also called the report's objective, aim, or goal). This statement is often written in the form of a question or infinitive phrase. Thus, if your problem is that Company X wants to know why sales are decreasing, your purpose statement may be "to determine the causes of decreasing sales at Company X" or "What are the causes of decreasing sales at Company X?"

Sometimes, as in the preceding example, the purpose will be clearly implied in the problem statement. Other times, the problem will be so complex or general that you will need to put some thought into your report's purpose. For example, the purpose of a report intended to help a company reduce employee turnover could be "to find out why employee turnover is so high," "to find out how other companies have addressed employee turnover," "to find out what makes loyal employees stay," a combination of these, or some other purpose. Consider carefully what approach your report will take to the problem.

These statements will help keep you on track as you continue through the project. In addition, they can be reviewed, approved, and evaluated by people whose assistance may be valuable. Most important, putting the problem and purpose in writing forces you to think them through. Keep in mind, though, that no matter how clearly you try to frame the problem and your research purpose, your conception of them may change as you continue your investigation. As in other types of business writing, report writing often involves revisiting earlier steps (recursivity), as discussed in Chapters 1 and 2.

In your completed report, the problem and purpose statements will be an essential component of the report's introduction and such front matter as the letter of transmittal and executive summary; they will orient your readers and let them know where your report is headed.

LO 8-2   List the likely factors involved in a problem.

DETERMINING THE FACTORS

Once you've defined the problem and identified your purpose, you determine what **factors** you need to investigate. That is, you determine what subject areas you must look into to solve the problem.

3. Financial status

4. Computer systems

5. Product development

6. Human resources

Hypotheses for Problems Requiring Solution

Some problems concern why something bad is happening and perhaps how to correct it. In analyzing problems of this kind, you should seek explanations or solutions. Such explanations or solutions are termed **hypotheses**. Once formulated, hypotheses are tested, and their applicability to the problem is either proved or disproved.

To illustrate, assume that you have the problem of determining why sales at a certain store have declined. In preparing to investigate this problem, you would think of the possible

> **Once you've defined the problem and identified your purpose, you determine what factors you need to investigate.**

What factors a problem involves can vary widely, but we can identify three common types. First, they may be subtopics of the overall topic about which the report is concerned. Second, they may be hypotheses that must be tested. Third, in problems that involve comparisons, they may be the bases on which the comparisons are made.

Use of Subtopics in Information Reports

If the problem is a lack of information, you will need to figure out the areas about which information is needed. Illustrating this type of situation is the problem of preparing a report that reviews Company X's activities during the past quarter. This is an informational report problem—that is, it requires no analysis, no conclusion, no recommendation. It requires only that information be presented. The main effort in this case is to determine which subdivisions of the overall topic should be covered. After thoroughly evaluating the possibilities, you might come up with a plan like this:

Purpose statement: To review operations of Company X from January 1 through March 31.

Subtopics:

1. Production

2. Sales and promotion

explanations (hypotheses) for the decline. You might identify such possible reasons as these:

Purpose statement: To find out why sales at the Springfield store have declined.

Hypotheses:

1. Activities of the competition have caused the decline.

2. Changes in the economy of the area have caused the decline.

3. Merchandising deficiencies have caused the decline.

4. Changes in the environment (population shifts, political actions, etc.) have caused the decline.

You would then conduct the necessary research to test these hypotheses. You might find that one, two, or all apply. Or you might find that none is valid. If so, you would have to generate additional hypotheses for further evaluation.

Bases of Comparison in Evaluation Studies

When the problem concerns evaluating something, either singularly or in comparison with other things, you should look for the bases for the evaluation. That is, you should determine what characteristics you will evaluate and the criteria you will use to evaluate them.

from the tech desk

Report-Writing Tools Help Businesses Succeed

To survive and thrive, businesses must have timely, accurate data about their operations. For many businesses, that means investing in software that will generate the informational reports they need.

The most powerful report-writing tools are those that are integrated with enterprise resource planning (ERP) software, which allows managers real-time access to data about the different facets of the company. These products' report-writing tools make it easy to get a snapshot of any part of business operations, whether it be the current financial picture, the sales history of a certain product, or the status of customers' accounts.

But even small businesses can find electronic assistance for generating reports. Shown here is the title page of a sample home-inspection report created with Horizon software. The software enables home inspectors to create all the necessary components—from transmittal letter to contract to results and recommendations—and then generates a professional-looking report for the customer.

While you may not be able to find software to support your report writing to this extent, you will almost surely use electronically generated reports when preparing your own reports. Be sure to familiarize yourself with any report-writing tools your organization uses so that you do not overlook important data or leave out information that your reader expects to see in your report.

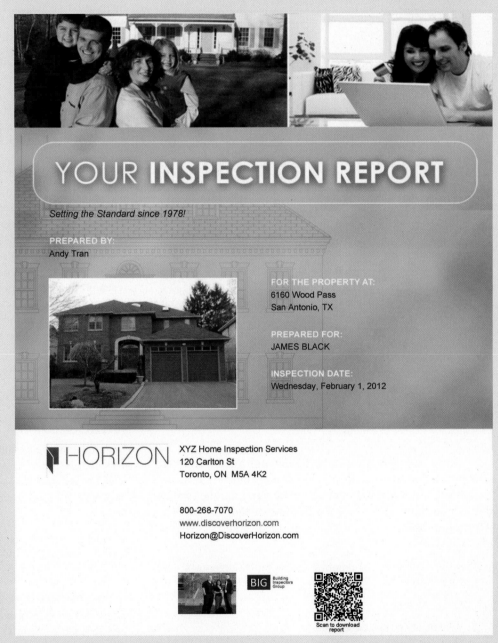

YOUR **INSPECTION REPORT**

Setting the Standard since 1978!

PREPARED BY:
Andy Tran

FOR THE PROPERTY AT:
6160 Wood Pass
San Antonio, TX

PREPARED FOR:
JAMES BLACK

INSPECTION DATE:
Wednesday, February 1, 2012

HORIZON   XYZ Home Inspection Services
120 Carlton St
Toronto, ON M5A 4K2

800-268-7070
www.discoverhorizon.com
Horizon@DiscoverHorizon.com

BIG Building Inspectors Group

Scan to download report

Source: "Professional Reports," *CarsonDunlop*, Carson, Dunlop & Associates Ltd., 2012, Web, 2 June 2013. Reprinted with permission.

Report writing requires hard work and clear thinking in every stage of the process. To understand the problem, identify your report's purpose, and prepare the report that will solve the problem, you may need to consult many sources of information.

Illustrating this technique is the problem of a company that seeks to determine which of three cities would be best for expansion. The bases for comparing the cities are the factors that would likely determine the success of the new branch. After considering such factors, you might come up with a plan like this:

Purpose statement: To determine whether Y Company's new location should be built in City A, City B, or City C.

Comparison bases:

1. Availability of skilled workers

2. Tax structure

3. Community attitude

4. Transportation facilities

5. Nearness to markets

Each of the factors selected for investigation may have factors of its own. In this illustration, for example, the comparison of

transportation in the three cities may well include such subdivisions as water, rail, truck, and air. Workers may be compared by using such categories as skilled workers and unskilled workers. Subdivisions of this kind may go still further. Skilled workers may be broken down by specific positions: engineers, programmers, technical writers, graphic designers. Make as many subdivisions as you need in order to provide a thorough, useful comparison.

LO 8-3   Explain the difference between primary and secondary research.

GATHERING THE INFORMATION NEEDED

You can collect information you need for a project by using two basic forms of research: primary and secondary. **Secondary research** uses material that someone else has published in resources such as periodicals, brochures, books, digital publications, and Web sites. This research is typically conducted before you engage in primary research. **Primary research** is research that uncovers information firsthand. It produces new information through the use of experiments, surveys, interviews, and other methods of direct observation. To be an effective report writer, you should be familiar with the techniques of both primary and secondary research. The following pages describe these techniques.

Conducting Secondary Research on the Internet

One of the most accessible research tools we have is the Internet. That makes the Web a good place to start a research project. Using search engines, other Web-based tools, and online social networks, we can often find all the secondary information we need.

LO 8-4   Use Internet search engines to gather information.

---

using search engines Internet search engines compile indexes of information about Web sites, such as the meta tags (hidden keywords) they use, how often they're visited, and other sites they link to. When you use a **search engine**, you are actually searching its index, not the Web itself. According to Experian Hitwise,[1] the top five search engines are Google, Bing, Yahoo! Search, Ask, and AOL Search, with Google being the most popular of the five. Google, whose simple, clean screens you see in Exhibits 8-1 and 8-2, provides the ability to do a simple search or a more advanced search. As you can see in Exhibit 8-1, even a simple search includes ways to filter the

▼EXHIBIT 8-1 A Simple Search Using Google

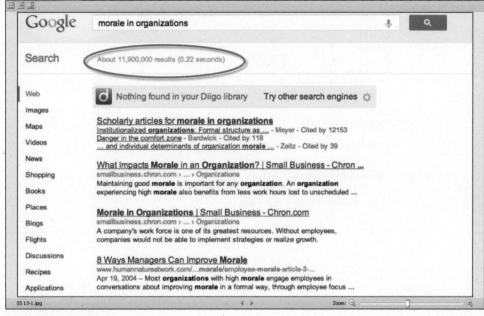

Source: www.google.com

▼EXHIBIT 8-2 A Google Search Using a Filter to Narrow Results

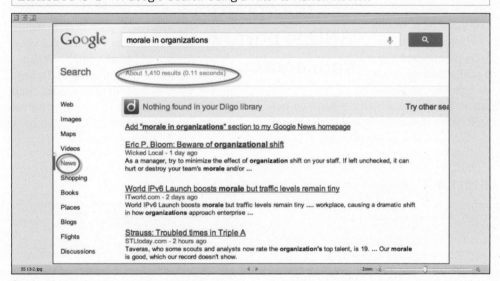

Source: www.google.com

information you are searching. In this case the search phrase "morale in organizations" pulls up 11,900,000 results. If you use the categories on the left side of the screen (such as Videos, News, or Discussions), you can limit your results to those sources, as shown in Exhibit 8-2. In this case, when you filter for News, you receive 1,410 results.

You can use another Google tool, Google Scholar, to search scholarly literature, which includes journals from academic publishers, conference papers, dissertations, academic books, and technical reports. You can perform a simple search in Scholar much as you would in Google's regular search. For example, you could search with the phrase "conducting surveys

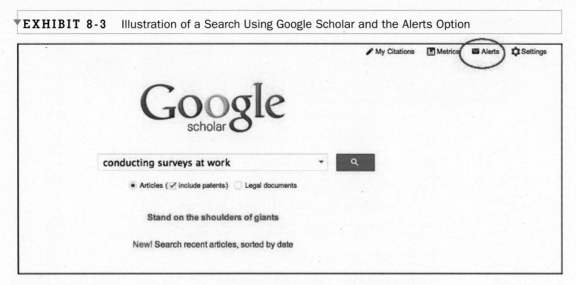

Source: http://scholar.google.com

at work," as in Exhibit 8-3. Google Scholar also has a feature called Alerts, as shown in Exhibit 8-3, which allows you to create an alert for a topic. You will then receive email notification of any new sources on the topic you are researching.

As you can see from the results in Exhibit 8-4, Google Scholar also provides filters on the left side of the screen. In this case the search tool can filter by date and whether or not to include patents and citations. You could also refine your research by using the Google Scholar Advanced Search option (Exhibits 8-5 and 8-6). Finally, when you pull up the results of a search, you can use the Related articles link to find more information on your research topic (Exhibit 8-7).

> ❝ **If your search yields too many citations, you can use the operator AND to narrow your search.** ❞

Whatever search engine you are using, a good command of **Boolean logic** will help you extract the information you need quickly and accurately. Boolean logic uses three primary operators: AND, OR, and NOT.

If your search yields too many citations, you can use the operator AND to narrow your search. When you link two search terms with AND, the search engine will retrieve only those citations that contain both terms. The operator NOT is another narrowing term, instructing the search engine to eliminate citations with a particular term.

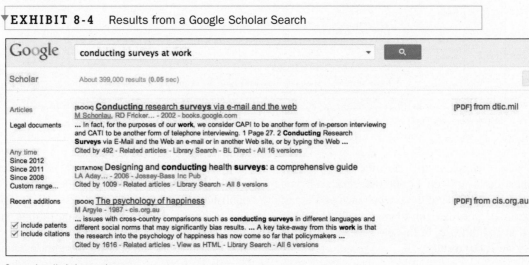

Source: http://scholar.google.com

EXHIBIT 8-5   How to Access Google Scholar's Advanced
             Search

Source: http://scholar.google.com

EXHIBIT 8-6   Illustration of Google Scholar's
             Advanced Search Features

Source: http://scholar.google.com

For example, if you were searching for articles on conducting surveys at work, you could search with the phrase "surveys at work NOT healthcare" to eliminate healthcare-related surveys.

The OR operator can be used to expand the search by adding variations or synonyms to the basic search term. For example, to expand a search for articles on "surveys AND morale," you might add "productivity OR enthusiasm OR confidence." If you have difficulty thinking of terms to broaden your search, look at the keywords or descriptors of the items that have already been identified. Often these will give you ideas for additional terms

to use. If the search still comes up short, you should check for spelling errors or variations. Becoming skilled at using Boolean logic will help you get the Internet-based information you need, and it will also help you search online databases (discussed in a later section) more efficiently.

As search engines evolve to meet the changing needs of the Internet's content and its users, new forms of these tools have emerged as well. **Metasearch tools** allow you to enter the search terms once, run the search simultaneously with several search engines, and view a combined results page. Examples

---

EXHIBIT 8-7   The Related Article Link in a Google Scholar Search

Google	employee morale

Scholar   About 81,800 results (0.05 sec)   ✎ My Citation

Articles

Legal documents

[CITATION] Preserving **employee morale** during downsizing
KE Mishra, GM Spreitzer... - Sloan Management ..., 1998 - Sloan Management Review
Cited by 142 - Related articles - All 4 versions

Any time
Since 2012
Since 2011
Since 2008
Custom range...

A little larceny can do a lot for **employee morale**
LR Zeitlin - Psychology Today, 1971 - ncjrs.gov
Abstract: THE AUTHOR'S THESIS IS THAT THEFT SERVES AS A SAFETY VALVE FOR **EMPLOYEE** FRUSTRATION. IT PERMITS MANAGEMENT TO AVOID THE RESPONSIBILITY AND THE COST OF JOB ENRICHMENT OR SALARY INCREASES AT ...
Cited by 71 - Related articles - Cached - All 3 versions

Recent additions

✓ include patents
✓ include citations

[PDF] The competitive advantage of corporate philanthropy
ME Porter... - Harvard business review, 2002 - expert2business.com
... in 2001. While these campaigns do provide much- needed support to worthy causes, they are intended as much to in- crease company visibility and im- prove **employee morale** as to cre- ate social impact. Tobacco giant Philip ...
Cited by 1296 - Related articles - BL Direct - All 26 versions

Source: http://scholar.google.com

EXHIBIT 8-8   Illustration of the Metasearch Tool Dogpile

"employee morale" and then combines the results and presents them in an easy-to-view form.

Another type of search tool that has emerged is the **specialized search engine**. Four popular examples are Yahoo!: People Search for finding people, Edgar for finding corporate information, FindLaw for gathering legal information, and Mediafinder for finding print items. In 2010 Mediafinder also launched an app for iPhone and iPad that provides access to data on more than 16,000 publications from the United States and Canada. The app allows users to search by title, keyword, and subject.

While these tools help you find relevant Web documents, it is crucial to remember that the tools are limited. You must

> ## You must recognize that not all of the documents published on the Web are indexed and that no search tool covers the entire Web.

of such tools are Dogpile, Kartoo, Mamma, Metacrawler, and Search.com. You will find links to these and other search tools on the textbook Web site. Exhibits 8-8 and 8-9 illustrate how Dogpile searches various search engines for the phrase

evaluate the source of the information critically (see "Evaluating Web sites," pages 209 and 211). Also, you must recognize that not all of the documents published on the Web are indexed and that no search tool covers the entire Web. Skill in using the tools plays a role in finding good Web information, but judgment in evaluating the accuracy and completeness of the search plays just as significant a role.

LO 8-5   Use other Web resources to gather information.

using other web-based resources   There are numerous other Web-based research sources in addition to the ones already mentioned. As technology changes, the list will continue to grow and change, but this section will introduce you to a number of current Web-based resources for research.

You have probably been advised by many of your college instructors not to use **Wikipedia** as a reference

EXHIBIT 8-9   Results of a Search Using Dogpile

when you write essays or other papers. The reason is that Wikipedia is written and maintained by volunteers, and virtually anyone can post and edit articles. Since its launch in 2001, though, Wikipedia has become a more credible and useful resource. And while we do not advise you to use it as your main resource, it can be a useful place to start to learn about a subject that is new to you.

WorldCat (Exhibit 8-10) is an online network of library content and services. You can use it to search the collections of local libraries and libraries around the world for books, CDs, videos, and digital content such as ebooks. You also can find article citations with links to their full text and historical documents and photos. If you have an active membership to the library that owns an item, you can check items out. You may also be able to access electronic databases if you have a valid login for the library that has access to these databases.

RSS (Really Simple Syndication) feeds on Web sites and blogs can be useful research tools as well. News outlets such as *The New York Times* and *CNN* offer these feeds on their Web sites. Scholarly journals also offer RSS feeds for their tables of contents. When you subscribe to an RSS feed, the site pushes its new content to your chosen RSS reader (such as Digg Reader, NewsBlur, or Microsoft Outlook). You can then browse the news you care about all in one place and easily stay current on certain topics.

LO 8-6 Evaluate Web sites for reliability.

---

evaluating Web sites Web sites can be an invaluable source of useful information. But as you know, all are not equally credible. Some may be biased, while others may be inaccurate. So it is important to know how to evaluate Web sites for completeness, accuracy, and reliability.

Although most print sources include items such as author, title of publication, facts of publication, and date, Web sites do not have an established format that helps ensure their credibility. Most users of search engines also do not understand the extent or type of bias involved when search engines present and order their results. And even the best search engines index only a small fraction of the Internet content. So all Web findings must

▼**EXHIBIT 8-10**   Home Screen of the Online Library Collection Worldcat

Source: www.worldcat.org
Screenshot used with OCLC's permission; and WorldCat® is a registered trademark of OCLC Online Computer Library Center, Inc.

be carefully scrutinized, and many should be checked against other sources.

One experimental study found that users of Web site information were particularly susceptible to four types of misinformation: advertising claims, government misinformation, propaganda, and scam sites. Furthermore, the study found that users' confidence in their ability to gather reliable information was not related to their actual ability to judge the information appropriately. The results also revealed that level of education was not related to one's ability to evaluate Web site information accurately.[2]

One solution might be to use only those links posted on trustworthy sites (e.g., Web sites of professional organizations or of government agencies). However, these sites are not comprehensive and are often late in providing links to new sources. Therefore, developing the skill and habit of evaluating Web sites critically is probably a better choice. This skill can be honed by getting into the habit of looking at the purpose, qualifications, validity, and structure of the Web sites you use.

- *Purpose.* Why was the information provided? To explain? To inform? To persuade? To sell? To share? What are the provider's biases? Who is the intended audience? What point of view does the site take? Could it possibly be ironic, a satire, or a parody?

- *Qualifications.* What are the credentials of the information provider? What is the nature of any sponsorship? Is contact information provided? Is it accurate? Is it complete—name, email address, street address, and phone number? Is the information well written, clear, and organized?

- *Validity.* Where else can the information provided be found? Is the information from the original source? Has the information been synthesized or abstracted accurately and in the correct context? Is the information timely?

Managing Citations with Zotero

Zotero is a free citation manager tool you can use to help you collect and organize your research sources. It is an extension for the Firefox Web browser.

The left column of the Zotero screen includes My Library, which contains all the items you save. You can click the button above My Library to create a new collection, which is a folder you can use to help organize the information you add to Zotero.

There are three main ways to add data to Zotero: attaching a Web page, capturing an item, and manually adding an item.

You can attach a Web page as a snapshot by clicking the Create New Item from Current Page button. A snapshot keeps a locally stored copy of a Web page as it was when it was saved and makes it available without an Internet connection.

You can also add information from other sources such as books you find online through the Capture icon (the folder) that appears in the browser address bar once you have downloaded Zotero. This feature lets you automatically create an item. If a full-text PDF is available, it will be automatically attached to the item.

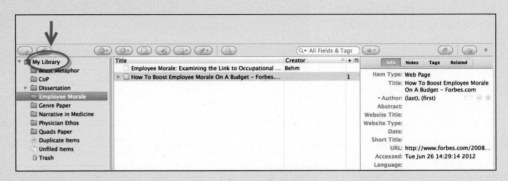

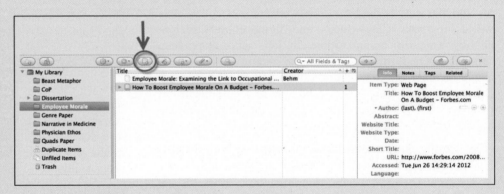

Source: www.zotero.org

Finally, you can manually add resources to Zotero by clicking on the New Item button in the toolbar, selecting the appropriate item, and manually adding in information to the fields.

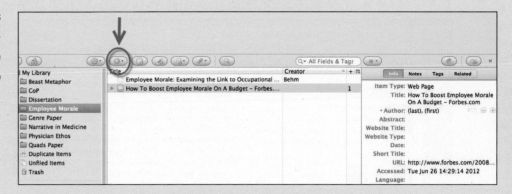

> By critically evaluating the Web sites you use, you will be developing a skill that will help you effectively filter the vast amount of data you encounter.

When was it created? When was it posted? Who links to it? (On Google, you can enter the term "link:" before a Web site's URL in the address bar to find links. If you wanted to find out who links to Toyota's Web site, for example, you would type the following in the search field: link: www.toyota.com.) How long has the site existed? Is it updated regularly? Do the links work? Do they represent other views? Are they well organized? Are they annotated? Has the site received any ratings or reviews? Is the cited information authentic?

• *Structure.* How is the site organized, designed, and formatted? Does its structure provide a particular emphasis? Does it appeal to its intended audience?

By critically evaluating the Web sites you use, you will be developing a skill that will help you effectively filter the vast amount of data you encounter.

LO 8-7 Use social networking and social bookmarking sites to gather information.

taking advantage of social networks Today's businesses take advantage of social media like Facebook and Twitter for marketing purposes, but such networks can also be useful for researching a business problem.

Facebook was launched as a personal social networking service in 2004, but businesses large and small have adopted it as a key marketing venue. In fact, 70 percent of retail merchants now use Facebook to market their products.[3]

Because of its pervasiveness, Facebook also makes an excellent research tool. It is especially useful for surveys (which we discuss later in this chapter). You can also go to company Facebook pages like Ace Hardware's, shown in Exhibit 8-11, to research company-specific information.

▼ **EXHIBIT 8-11**   Illustration of a Company's Facebook Page

Source: http://www.facebook.com/#!/acehardware.

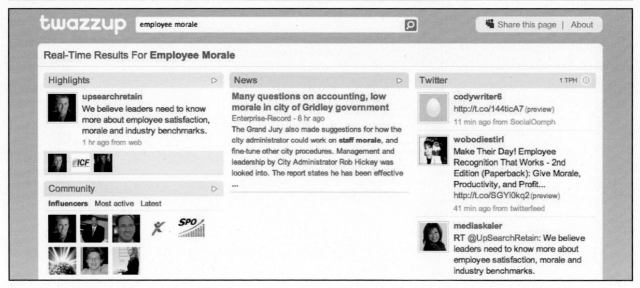

Source: www.twazzup.com

As Chapter 2 explains (and you probably already know), **Twitter** is a microblogging service that lets you send and read messages of up to 140 characters in length. Like blogs, Twitter started out as a personal communication tool, but organizations and companies now liberally populate the "Twitterverse." In fact, according to one study, 51 percent of active Twitter users follow companies, brands, or products on social networks.[4] While most tools that search "tweets" (the name for Twitter messages) are limited in terms of how far back in time they can search, two good real-time search services are Twazzup and Twinitor (Exhibits 8-12 and 8-13).

You can also follow people or topics on Twitter if you have a Twitter account. You can do a simple search for a topic with Twitter's search function, as shown in Exhibit 8-14. To get more information relevant to your particular business

▼ **EXHIBIT 8-13**   Illustration of a Real-Time Twitter Search Using the Tool Twinitor

Source: www.twinitor.com

▼ **EXHIBIT 8-14** Illustration of a Search Using Twitter's Search Function

Source: https://twitter.com

▼ **EXHIBIT 8-15** Illustration of a Twitter Hashtag in a Tweet

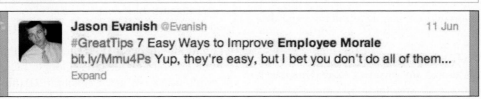

Jason Evanish @Evanish   11 Jun
#GreatTips 7 Easy Ways to Improve **Employee Morale**
bit.ly/Mmu4Ps Yup, they're easy, but I bet you don't do all of them...
Expand

Source: https://twitter.com

problem, you should consider following topics by following specific **hashtags**. Hashtags are created by using the symbol # to mark keywords or topics. These marked topics are then easier to find in Twitter Search. If you find a hashtag you want to follow to learn more about a certain topic, you can click on that link and be taken to all the other tweets in that category. In the example provided in Exhibit 8-15, @Evanish uses the hashtag #GreatTips. If you click on the link for the hashtag, you can see all the results for the discussion, as shown in Exhibit 8-16.

LinkedIn is similar to Facebook in that it connects people, but whereas Facebook stresses social connections, LinkedIn focuses on professional contacts. People generally join LinkedIn to make these connections, but like Facebook, this network can be useful in researching a business problem.

To take full advantage of LinkedIn as a research tool, use the various Search options, particularly People, Companies, and Groups (Exhibit 8-17). If, for example, you want to see what types of employee morale issues other companies are having, you can search with the Companies search. You could use the

▼ **EXHIBIT 8-16** Illustration of the Results of Following a Hashtag

Source: https://twitter.com

Source: www.linkedin.com

Groups search function to research groups that focus on human resources, training, or motivation.

Virtual contacts through LinkedIn, Facebook, and Twitter also have the advantage of giving you access to international data for your research.

Wikis are basically collaborative collections of knowledge. You can find wikis on almost any business topic. One excellent business wiki resource is Smallbusiness.com, shown in Exhibit 8-18, which offers information on everything from tax preparation to time management. Along with using wikis as a research tool, you can create your own wiki for collaborative projects, such as team research projects. There are many free wiki-hosting sites to choose from, including Google Sites, PBworks, PmWiki, and Wikispaces.

Blogs (short for "Web logs") started out as personal diaries or pages in 1994, but they soon became a journalism tool as well.[5] Today many companies maintain blogs, too. Like Facebook and LinkedIn, blogs can be a useful tool for finding information on a business problem.

Often, the challenge is to find blogs that are pertinent to the research you are pursuing. Google blog search (www.google.com/blogsearch) and Technorati (Exhibit 8-19) are two useful blog search engines. Google's blog search works just like a typical Google search except that it limits the results to items posted on blogs. Technorati indexes over 1.3 million blogs, many of them authored by corporations and small businesses.[6] Technorati also helps you determine a blog's standing and influence with its feature Technorati Authority. Authority is calculated on the basis of a site's linking behavior, categorization, and other data over a short period of time. Therefore, a blog's rating will rise and fall rapidly depending on what is being discussed in cyberspace at the moment. Levels of authority range from 0–1,000, with 1,000 being the highest possible authority.[7]

listservs and professional organizations

Professional organizations are another good and sometimes overlooked research tool for business problems. Members of most professional organizations have benefits that

EXHIBIT 8-18   Illustration of a Business-Related Wiki

Source: http://smallbusiness.com/wiki/Main_Page

Top Finance blogs   ⊙ View all Finance blogs

	Top 5 Finance blogs			Top 5 movers	
1	Boomer & Echo — Finance Authority: 957		57	Cash Money Life — Finance Authority: 701	⬆ 38
2	Good Financial Cents -Jeff ... — Finance Authority: 950		28	MintLife Blog | Personal ... — Finance Authority: 774	⬆ 15
3	20's Finances — Finance Authority: 924		50	Best Rates In — Finance Authority: 710	⬆ 14
4	Free From Broke — Finance Authority: 911		41	Canadian Finance Blog — Finance Authority: 734	⬆ 14
5	My University Money — Finance Authority: 875	⬆ 3	46	Novel Investor — Finance Authority: 731	⬆ 12

Source: http://technorati.com/business/finance/

often include access to a member directory, salary surveys, conferences, and educational opportunities. Most organizations today also have Web sites and **listservs** (electronic mailing lists). Because many of these listservs are very active, they can be useful for investigating business problems. You can send a question out to the membership and have responses the same day. You can also use the listserv to send out surveys.

There are any number of organizations to consider joining, including the Association for Business Communication, the Association for Financial Professionals, the Sales and Marketing Professional Association, the American Institute of CPAs, the Society for Human Resource Management, and the American Management Association.

social bookmarking Web sites Social bookmarking is a way for people to organize, store, manage, search, and share their favorite Web resources. Many online bookmark management services have been launched since 1996, including Delicious and Digg (Exhibits 8-20 and 8-21).

A major component of these sites is tagging, which lets users organize their bookmarks in flexible ways and develop shared vocabularies known as *folksonomies*, or *collaborative tagging*. Tagging

Welcome to Delicious!
Delicious helps you find cool stuff and collect it for easy sharing. Dig into stacks created by the community, and then build your own!

👤 Featured Users

netstrider

Fantastic in-depth stacking about science, technology, politics, and culture.

idesign4u

Design, Typography, Business, and the Web.

paolojcruz

▌ Staff Stack Picks

Paper Crafts

Corazones de papel · Pinterest / Home · How About Orange: Paper 3-D holiday ornament instructions · Holiday Card Ornaments · Martha Stewart Holidays · With A Grateful...

Delicious Recipes

Grilled Zucchini and Grape Tomato Salad · Shrimp Tacos I The Pioneer Woman Cooks I Ree Drummond · Easy Green Chile Enchiladas I The Pioneer Woman Cooks I Ree...

The Life and Death of Yolo

YOLO: you only live once. The irony? It feels like this expression is going to live forever.

Source: http://delicious.com

EXHIBIT 8-21 Illustration of the Social Bookmarking Site Digg

Source: http://digg.com

works much like a keyword search. Let's say you are looking for information on companies' use of social media. You can search Delicious with this topic to see if anyone has tagged resources with related terms or phrases and then investigate those sources. This type of search can save you much of the time you would have spent conducting a search from scratch.

LO 8-8 Use the library to gather information.

Conducting Secondary Research in a Library

With so much information available on the Web, it is tempting to think that libraries have become obsolete. But libraries contain a wealth of information that is unavailable anywhere else or available elsewhere only for a fee. You will often find your best information in a library—and probably save money in the process.

General libraries are the best known and the most accessible. General libraries, which include college, university, and most public libraries, are called *general* to the extent that they contain all kinds of materials. Many general libraries, however, have substantial collections in certain specialized areas.

> "You will often find your best information in a library—and probably save money in the process."

Libraries that limit their collections to one type or just a few types of material are considered **special libraries**. Many such libraries are private and do not invite routine public use of their materials. Still, they will frequently permit access for research projects that they consider relevant and worthwhile.

Among the special libraries are those libraries of private businesses. As a rule, such libraries are designed to serve the sponsoring company and provide excellent information in the specialized areas of its operations. Special libraries are also maintained by various types of associations—for example, trade organizations, professional and technical groups, and labor unions. Like company libraries, association libraries may provide excellent coverage of highly specialized areas. A number of public and private research organizations maintain libraries. The research divisions of big-city chambers of commerce and the bureaus of research of major universities, for example, keep extensive collections of material containing statistical and general information on certain geographical areas. State agencies collect similar data.

No matter what type of library you use, you'll want to be familiar with how to consult such resources as online catalogs, databases, and reference materials.

searching the catalog Today most libraries use online catalogs to list their holdings. You can locate sources in these catalogs by using the standard Keyword, Title, Author, and Subject options as well as a few other options. Becoming familiar with such catalogs is essential, especially for the libraries you use frequently. Effective and efficient searching techniques can yield excellent information.

Two options you need to understand clearly are **Keyword** and **Subject**. When you select the Keyword option (Exhibit 8-22), the system will ask you to enter search terms and phrases. It will then search for only those exact words in several of each record's fields, missing all those records using slightly different wording. When you select Subject, the system will scan the Library of Congress subject heading for your search term (Exhibit 8-23). This means that for the most part you need to know the exact heading that the Library of Congress uses.

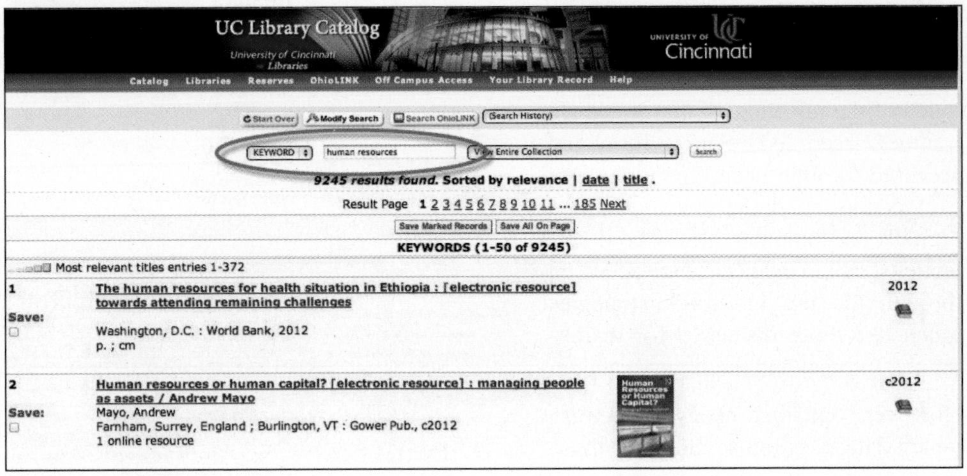

To find possible Library of Congress subject headings for your topic, visit the Library of Congress Authorities Web page at http://authorities.loc.gov and click "Search Authorities." Exhibit 8-24 shows possible headings for a search on "intercultural communication." Sometimes the search engine will cross-reference headings, such as suggesting that you "*See* Intercultural Communication" when you enter "cross-cultural

communication." A Subject search will find all those holdings on the subject, including those with different wording such as "intercultural communication," "international communication," "global communication," and "diversity." If you were to run multiple searches under the Subject option using these terms, you would have more complete information, though you would still miss some titles you would find using Keywords.

searching databases The online catalog helps you identify books and other holdings in your library, and it may help you find some articles. But to do a good job of searching the periodical literature—that is, articles published in newspapers,

▼**EXHIBIT 8-22** Illustration of the Results from a Keyword Search in a Library Catalog

Source: http://uclid.uc.edu

▼**EXHIBIT 8-23** Illustration of the Results from a Subject Search in a Library Catalog

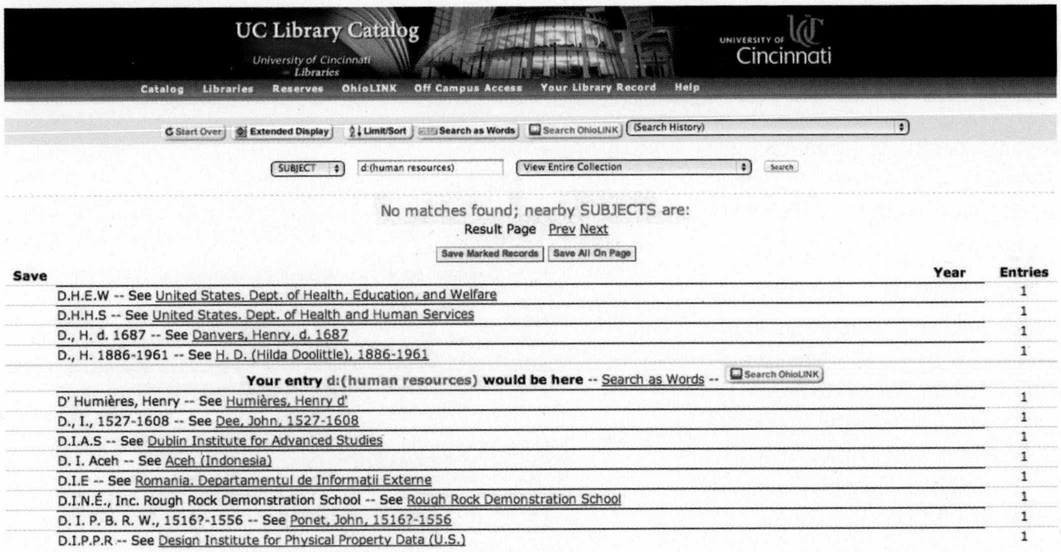

Source: http://uclid.uc.edu

magazines, and journals—you will need to use an online **database**, such as *ABI/Inform* (Exhibit 8-25). As the sophistication and capacity of computer technology have improved, much of the information that was once routinely recorded in print form and accessed through print directories, encyclopedias, and indexes is now stored digitally in computer files. These collections, called databases, are accessed through the use of search strategies much like those discussed for searching the Internet and the library catalog.

However, one first needs to identify which databases to use. Some of those most useful to business researchers are *ABI/Inform, Business Source Premiere, Factiva*, and *LexisNexis Academic. ABI/Inform* and *Business Source Premiere* are two of the most complete databases, providing access to hundreds of business research journals as well as important industry and trade publications. Most of the articles are included in full-text form or with lengthy summaries.

Factiva provides access to current business, general, and international news, including access to various editions of *The Wall Street Journal*. It also includes current information on U.S. public companies and industries. Similarly, *LexisNexis* offers access to current business and international articles, providing them in full text. Additionally, it includes legal and reference information.

If you need information on a particular company, you could use *LexisNexis® Company Dossier*. This database provides complete pictures of companies' financial health, brands, and competitors for both U.S. and international companies. *Hoover's Online* is also an excellent resource for company-specific information, and others include *Business & Company Resource Center, Business Source Complete*, and *D&B's Million Dollar Database*.

consulting reference materials

Along with database sources, you may want to investigate other print and Web-based reference materials for information (Exhibit 8-26, next page). To gather research on a particular industry, for example, you could use *BizMiner*, which offers industry statistical reports and industry financial analysis benchmarks for over 5,000 lines of business and industries. Other industry-specific sources include *Plunkett Research Online* and *Standard & Poor's*

Industry Surveys. To find out about international trade you can use the *(CIA) World Factbook*, which offers information on different countries' histories, governments, economies, geographical traits, and interactions with other countries. *Country Reports* from the Department of State also provides general information, region-specific information, and travel information.

The library materials you choose will be determined by your research question. Exhibit 8-27 lists helpful resources for

---

EXHIBIT 8-24   Illustration of the Library of Congress Authorities Subject Search Results

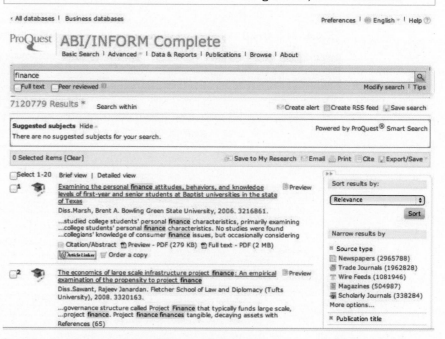

Source: http://catalog.loc.gov

EXHIBIT 8-25   Illustration of a Search Using the *ABI/Inform* Database

Source: http://search.proquest.com.proxy.libraries.uc.edu/abicomplete

Type of Source	Description	Examples
Encyclopedias	Offer background material and other general information. Individual articles or sections of articles are written by experts in the field and frequently include a short bibliography.	*Encyclopedia Americana* *Encyclopaedia Britannica* *World Book* *Encyclopedia of Banking and Finance* *Encyclopedia of Business and Finance* *Encyclopedia of Small Business* *Encyclopedia of Advertising* *Encyclopedia of Emerging Industries*
Biographical Directories	Supply biographical information about leading figures of today or of the past.	*Who's Who in America* *Who's Who in the World* *Who's Who in the East* *Who's Who in the South and Southwest*
Almanacs	Offer factual and statistical information.	*The World Almanac and Book of Facts* *The Time Almanac* *The New York Times Almanac*
Trade Directories	Compile details in specific areas of interest. Variously referred to as *catalogs, listings, registers,* or *source books.*	*The Million Dollar Directory* *Thomas Register of American Manufacturers* *The Datapro Directory* *America's Corporate Families* *Who Owns Whom* *Directory of Corporate Affiliations* *Directories in Print*
Government Publications	Include surveys, catalogs, pamphlets, and periodicals from various governmental bureaus, departments, and agencies.	*Annual & Quarterly Services* (service-industry data) *Census of Wholesale Trade* *Census of Mineral Industries* *Statistical Abstract of the United States* *Survey of Current Business* *Monthly Labor Review* *Occupational Outlook Quarterly* *Federal Reserve Bulletin*
Dictionaries	Provide definitions, spellings, and pronunciations of words or phrases. Electronic dictionaries add other options such as pronunciation in audio files.	*American Heritage Dictionary* *Funk & Wagnalls Standard Dictionary* *Random House Webster's College Dictionary* *Merriam-Webster's Collegiate Dictionary*
Additional Statistical Sources	Provide statistical data.	*Statistical Abstract of the United States* *Standard & Poor's Statistical Service* *Statistical Reference Index*
Business Information Services	Supply a variety of information to business practitioners.	*Corporation Records* *Moody's Investors' Advisory Service* *Value Line Investment Survey* *Gale Business Insights: Global* *Hoover's Online* *Factiva*
International Sources	Supply international corporate information.	*Principal International Businesses* *Major Companies of Europe* *Japan Company Handbook* *International Encyclopedia of the Social Sciences* *International Business Dictionary and References* *International Brands and Their Companies* *Foreign Commerce Handbook* *Index to International Statistics Statistical Yearbook*

How do I find business news and trends?

ABI Inform Complete on ProQuest

Business Source Complete

Factiva (includes Dow Jones, Reuters Newswires and *The Wall Street Journal*, plus more than 8,000 other sources from around the world)

LexisNexis Academic, News and Business sections

Proquest Business Insights

Wilson OmniFile Full Text Mega

How do I find information about companies?

Business & Company Resource Center

Business Source Complete

Companies' own Web sites

Company Dossier (on Lexis/Nexis)

D&B's (Dunn & Bradstreet's) *Million Dollar Database*

Datamonitor 360

Factiva

Hoover's Online

Mergent Online

ORBIS

SEC Filings (on Edgar) at www.sec.gov/edgar/

Standard & Poor's NetAdvantage

Thomson One Banker

Value Line Research Center

How do I find information about particular industries?

ABI/INFORM Complete

BizMiner

Datamonitor 360

Freedonia Focus Market Research

Decision Support Database

Global Market Information Database

IBISWorld

ICON Group International

MarketLine

MarketResearch.com Academic

Mergent Industry Reports

Mintel Market Research Reports

Plunkett Research Online

Standard & Poor's Industry Surveys

How do I find biographical and contact information for businesspeople?

Biographical Dictionary of American Business Leaders

Biography in Context (Galegroup)

Biography Reference Bank (Wilson)

D&B's Million Dollar Database

LexisNexis Academic, Reference/Biographical Information section

Standard & Poor's NetAdvantage (Register of Executives)

How do I find information provided by the U.S. government?

American Community Survey (U.S. Census Bureau) at www.census.gov/acs/www/

American FactFinder at http://factfinder2.census.gov/

Business USA at http://business.usa.gov/

Fedstats at www.fedstats.gov/

FRED (Federal Reserve Economic Data) at http://research.stlouisfed.org/fred2/

Statistical Abstract of the United States at www.census.gov/compendia/statab/

Bureau of Labor Statistics Data at www.bls.gov/home.htm

How do I find out about other countries and international trade?

Country Studies at http://lcweb2.loc.gov/frd/cs/cshome.html

(CIA) World Factbook at www.cia.gov/library/publications/the-world-factbook/

Country Commercial Guides at www.buyusainfo.net/

Country Reports (From the Department of State) at www.state.gov/countries/

Europa World Yearbooks

Global Market Information Database

SourceOECD

WDI Online (World Bank's World Development Indicators)

Yahoo Country Links at http://dir.yahoo.com/Regional/Countries/

How do I find information about cities?

American FactFinder at http://factfinder2.census.gov/

Cities' own Web sites

Complete Economic and Demographic Data Source: CEDDS (Woods & Poole Economics)

County and City Data Book at www.census.gov/statab/www/ccdb.html

SimplyMap

Source: Compiled with the assistance of Senior Business Librarian Wahib Nasrallah, University of Cincinnati.

common research tasks in business. A reference librarian can recommend additional resources to help you with your research task.

Conducting Primary Research with Surveys

When you cannot find the information you need in secondary sources, you must get it firsthand through primary research. One of the most popular primary research tools in business is the **survey**.

The premise of the survey as a method of primary research is simple: You can best acquire certain types of information by asking questions. Such information includes personal data, opinions, behaviors, attitudes, and beliefs. It also includes information necessary to plan an experiment or an observation or to supplement or interpret the data that result.

Once you have decided to conduct a survey, you'll need to make a number of decisions, including what questions to ask and how to ask them. But none of these decisions will be more important than whom to survey. Except for situations in which a small number of people are involved in the problem under study, you won't be able to reach all the people involved. Thus, you'll need to select a **sample** of respondents who represent the group as a whole as accurately as possible. You can select that sample in several ways.

LO 8-9   Use sampling to conduct a survey.

choosing your sampling technique

The type of sampling technique you use will be determined by the purpose of your research. While all samples have some degree of sampling error, you can reduce the error through techniques used to construct representative samples. These techniques fall into two groups: **probability** and **nonprobability sampling**.

Probability samples are based on chance selection procedures. Every element in the population has the same probability of being selected. These techniques include **simple random sampling, stratified random sampling, systematic sampling**, and **area** or **cluster sampling**.

- *Simple random sampling.* By definition, this sampling technique gives every member of the group under study an equal chance of being included. To ensure equal chances, you must identify every member of the group and then, using a list or some other convenient format, record all the identifications. Next, through some chance method, you select the members of your sample.

For example, if you are studying the job attitudes of 200 employees and determine that 25 interviews will give you the information you need, you might put the names of all 200 workers in a container, mix them thoroughly, and draw out 25. Since each of the 200 workers has an equal chance of being selected, your sample will be random and can be presumed to be representative.

- *Stratified random sampling.* Stratified random sampling subdivides the group under study and makes random selections within each subgroup. The distribution of a particular group in the sample should closely replicate the distribution of that group in the entire population.

Assume, for example, that you are attempting to determine the curriculum needs of 5,000 undergraduates at a certain college and that you have decided to survey 20 percent of the enrollment, or 1,000 students. To construct a sample for this problem, first divide the enrollment list by academic concentration: business, liberal arts, nursing, engineering, and so forth. Then draw a random sample from each of these groups, making sure that the number you select is proportionate to that group's percentage of the total undergraduate enrollment. Thus, if 30 percent of the students are majoring in business, you will randomly select 300 business majors for your sample; if 40 percent of the students are liberal arts majors, you will randomly select 400 liberal arts majors for your sample; and so on.

- *Systematic sampling.* In systematic sampling you decide what percentage of a population you are interested in

Researchers frequently survey a sample of the group that is being studied.

sampling, such as 10 percent of 10,000. Then, going down a list of the population's members, you select your participants at regular intervals (e.g, every 9th person).

If you use this method, your sample will not really be random because by virtue of their designated place on the original list, items do not have an equal chance of being selected. Therefore, it is important to make sure your source list for the sample is not organized in a way that would create a biased sample.

- *Area or cluster sampling.* Researchers use area sampling when no master source list of a population is available. For example, if you want to survey employees in a given industry, it is unlikely there is a list of all these employees. An approach you may take in this situation is to randomly select a given number of companies from a list of all the companies in the industry. Then, using organization units and selecting randomly at each level, you break down each of these companies into divisions, departments, sections, and so on until you finally identify the workers you will survey.

to locate members when the population is small or hard to reach. For example, you might want to survey Six Sigma Black Belt certification holders. To get a sample large enough to make the study worthwhile, you could ask those from your town to give you the names of other Black Belt holders. Or perhaps you are trying to survey the users of a project management application. You could survey a user group and ask those members for names of other users. You might even post an announcement on a blog or online forum asking for names.

LO 8-10 Construct a questionnaire and conduct a survey.

constructing the questionnaire Once you have determined whom you will survey, you will need to construct a survey instrument, a **questionnaire**. A questionnaire is simply an orderly arrangement of the survey questions with appropriate spaces provided for the answers. But simple as the finished questionnaire may appear, it is the result of careful planning. You must word your questions so that the results will

[Simple as the finished questionnaire may appear, it is the result of careful planning.]

Nonprobability samples are based on an unknown probability of any one member of a population being chosen. These techniques include **convenience sampling**, **purposeful sampling**, and **referral sampling**.

- *Convenience sampling.* A convenience sample is one whose members are convenient and economical to reach. When professors use their students as subjects for their research, they are using a convenience sample. Researchers generally use this sample to reach a large number quickly and economically. This kind of sampling is best used for exploratory research. A form of convenience sampling is *judgment* or *expert* sampling. This technique relies on the judgment of the researcher to identify appropriate members of the sample. Illustrating this technique is the common practice of predicting the outcome of an election based on the results in a bellwether district.

- *Purposeful sampling.* With purposeful sampling you look for a sample that has certain characteristics. Let's say you want to find out students' attitudes about a new tool to search the university's online library collections. It would be more logical to draw your sample from the students who use the system rather than from all the students at the university.

- *Referral sampling.* Referral samples are those whose members are identified by others. This technique is used

be **reliable**; a test of a questionnaire's reliability is its ability to generate similar results when used in similar circumstances. You also want your questionnaire to be **valid**, measuring what it is supposed to measure.

Keeping these guidelines in mind will help you achieve reliable, valid results:

- *Avoid leading questions.* A **leading question** is one that in some way influences the answer. For example, the question "Is Dove your favorite bath soap?" may lead the respondent to favor Dove. Some people who would say "yes" would name another brand if they were asked, "What is your favorite brand of bath soap?"

- *Avoid absolute terms.* Try not to include words like *always* and *never* in your questions. Using these terms may make respondents unlikely to choose these answers, and the wording of the question could skew your data toward middle selections like *sometimes* or *frequently*.

- *Focus on one concept per question.* **Double-barreled questions** combine multiple questions and lead to inaccurate answers. An example of such a question is "To what extent do managers and co-workers affect your perception of the company?" This question asks a respondent two questions. If the respondent feels that managers do impact their perception but co-workers do not, the answer the respondent provides will not be an accurate reflection of his or her beliefs, and the question will not lead you to reliable data.

- *Make the questions easy to understand.* Questions that not all respondents will clearly understand will generate faulty data. Unfortunately, it is difficult to determine in advance just what respondents will not understand. As will be discussed later, the best means of detecting such questions in advance is to test the questions before using them, but you can be on the alert for a few common sources of confusion.

One source of confusion is vagueness of expression, which is illustrated by the question, "How do you bank?" Who other than its author knows what the question means? Another source is using words respondents do not understand, as in the question, "Do you read your house organ regularly?" The words *house organ* have a specialized, not widely known meaning, and *regularly* means different things to different people.

- *Avoid questions that touch on personal prejudices or pride.* For reasons of pride or prejudice, people cannot be expected to answer accurately questions about certain areas of information. These include age, income status, morals, and some personal habits. How many people, for example, would answer "no" to the question "Do you brush your teeth daily?" How many people would give their ages correctly? How many citizens would admit to fudging a bit on their tax returns?

If such information is essential to the solution of the research problem, use a less direct means of inquiry. To ascertain age, for example, you could ask for dates of high school graduation. From this information, you could approximate age. Or you could provide an age range for a respondent to choose from, such as 20–24, 25–34, 35–44, 45–54, and 54 and older. This technique works well with income questions, too. People are generally more willing to answer questions that provide ranges instead of asking for specifics.

- *Ask only for information that can be remembered.* Since the memory of all human beings is limited, you should design your questionnaire to ask only for information that the respondents can be expected to remember. To be able to do this, you need to know certain fundamentals of memory. *Recency* is the most important principle of memory. People remember insignificant events that occurred within the past few hours. By the next day, however, they will forget some. A month later they may not remember any. You might well remember, for example, what you ate for lunch on the day of the survey, and

perhaps you might remember what you ate for lunch a day, two days, or three days earlier. But you would be unlikely to remember what you ate for lunch a year earlier.

The second principle regarding memory is *significance.* You may long remember minor details about the first day of school, your wedding, or an automobile accident. People readily remember events such as these because in each event there was an intense stimulus—a requisite for retention in memory.

A third principle of memory is that fairly insignificant facts may be remembered over long time periods through *association* with something significant. Although you would not normally remember what you ate for lunch a year earlier, for example, you might remember if the day in question happened to be Christmas Day or your first day at college. Obviously, the memory is stimulated not by the meal itself but by the association of the meal with something more significant.

designing the questionnaire and planning its delivery Overall, the questionnaire should be designed to gather useful information that can be easily tabulated and meaningfully analyzed.

Be sure to enable the respondents to provide the demographic information you need. In some instances, such information as the age, sex, and income bracket of the respondent is vital to the analysis of the problem.

When conducting a survey, ask only for information that is likely to be remembered accurately.

When practical, enable the respondents to check an answer. Easy-to-answer questions will encourage participation, and providing choices will make numerical analysis easier, too. Such questions must always provide for all possible answers, including conditional answers. For example, a direct question may provide for three possible answers: Yes ____, No ____, and Don't know ____ .

Consider using **scaling** when appropriate. It is sometimes desirable to measure the intensity of the respondents' feelings about a given topic, such as a product or company process. In such cases, some form of scaling is useful. The most common forms are ranking and rating.

The **ranking** technique consists simply of asking the respondent to rank a number of alternative answers to a question in order of preference (1, 2, 3, and so on). For example, in a survey to determine consumer preferences for toothpaste, the respondent might be asked to rank toothpastes A, B, C, D, and E in order of preference. The **rating** technique provides a scale showing the complete range of possible attitudes on a topic and assigns number values to the positions on the scale. The respondent must then indicate the position on the scale that corresponds to his or her attitude. Typically, the numeral positions are described by words, as the example in Exhibit 8-28 (next page) illustrates. Because the rating technique deals with the subjective rather than the factual, it is sometimes desirable to use more than one question to cover the attitude being measured. Logically, the average of a person's answers to such questions gives a more reliable answer than does any single answer.

from the tech desk

Web-Based Survey Tools Help Writers Design, Analyze, and Report Results of Questionnaires

Web-based survey tools can help you design professional-looking questionnaires as well as compile and analyze the results. Some tools, such as those offered at Qualtrics.com and SurveyMonkey.com, are available in both free and for-purchase forms.

When preparing your questions, you can choose from several question types, and you can also select your preferred design (e.g., color and layout). You can move the questions to change the order, and you can enable respondents to skip parts of the survey based on their answers to certain questions. You can also design questions that enable respondents to enter comments. All these questions can be saved in a library for reuse. Some of the tools even include libraries of surveys that can be adapted for your particular use.

As shown below, these tools can provide helpful summary reports, even when the survey is still in progress. They also permit you to view the detailed raw data in various forms.

Businesses can use these tools in a variety of applications, including training program evaluations, employee feedback on policies and procedures, longitudinal studies of ongoing practices such as network advertising revenues, opinion surveys of customers and potential customers, and assessments of customer satisfaction.

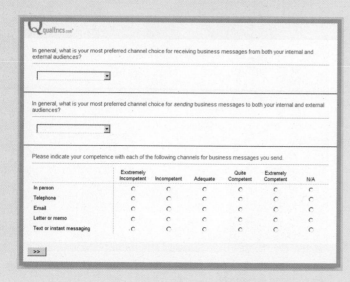

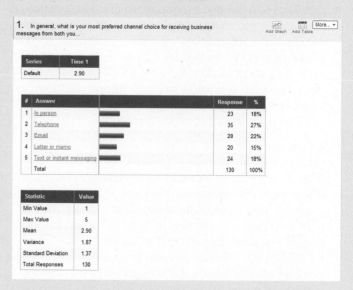

Use the best possible sequence of questions. In some instances, starting with a question of high interest may have psychological advantages. In other instances, it may be best to follow some other order of progression. Frequently, some questions must precede others because they help explain the others. Whatever the requirements of the individual case may be, you'll need to put careful thought into determining the sequence of questions.

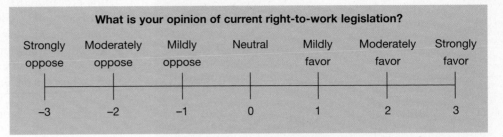

▼**EXHIBIT 8-28** Illustration of a Rating Question

Fairly early in the planning process, you should choose your **survey channel**. You can get responses to your questions in four primary ways: by personal (face-to-face) contact, by phone, by mail (print or digital), or through Web sites (e.g., Facebook). You should select the way that in your unique case yields the best sample and the best results at the lowest cost.

All these decisions should be recorded in a **survey plan**. Your plan should include such logistics as when and where you'll conduct your survey, how many times it will be sent out, and when it will close. It should also include any additional materials you'll need. If you are conducting a mail or Web survey, for example, you'll need to develop an explanatory message that motivates the subjects to respond, tells them what to do,

and answers all the questions they are likely to ask (see Exhibit 8-29). If you are conducting a personal or phone survey, you'll need to develop a script and/or instructions for the surveyors.

conducting a pilot study Before conducting the actual survey, it is advisable to conduct a **pilot study** on your questionnaire and survey plan. A pilot study is a small-scale version of your survey; in essence, it is a form of user testing (described on the next page). You select a few people to use as testers and have them take your survey to identify unclear questions, technological glitches, or other problems. Based on the results, you modify your questionnaire and working plan. Including this step in your survey planning will help you avoid the disappointment (and cost) that results from administering a flawed survey.

▼**EXHIBIT 8-29** Illustration of a Cover Message for an Online Survey

University of Wisconsin
Eau Claire

Distribute Survey
Send Survey

| My Surveys | Create Survey | Edit Survey | Distribute Survey | View Results | Polls | Library | Panels |

Current Survey: Grammar, Mechanics, and Us;

Get Help ▼ Give us your feedback

Survey Link | Send Survey | Mail History | Popup Survey | Embedded Survey | Website Feedback Link | Preview Survey

Grammar, Mechanics, and Usage Survey

| Quick Send | Survey Mailer |

Font   Size   **B** *I* U ...

Dear Business Student:

Thank you for participating in this survey to help us better understand the choices student business writers make regarding grammar, mechanics, and usage in their writing. Your responses will enable us to develop our business writing courses to ensure that we are meeting your needs as well as the needs of audiences who read your work. The survey should take approximately 15 minutes.

Your responses will be anonymous; your responses will not be associated with your name in any way. If you wish to see the results of the survey, you may contact me after November 15. If you have any questions regarding your participation in the survey, please contact me at (715) 836-3640 or ginderpj@uwec.edu. To take the survey, follow this link: http://uweauclaire.qualtrics.com/SESID=SV.

Email Message

From Name: Paula Lentz

From Email Address: ginderpj@uwec.edu

Subject: Business Writing Survey

Systematically observing what happens can be a useful form of primary research.

Conducting Observations and Experiments

Two types of primary research that involve watching and/or recording what happens are **observations** and **experiments**.

LO 8-11 Design an observational study for a business problem.

observations Simply stated, observation is seeing with a purpose. It consists of watching the events involved in a problem and systematically recording what you see. In observation, you do not manipulate the details of what you observe; you take note of situations exactly as you find them.

To see how observation works as a business technique, consider this situation. You work for a fast-food chain, such as McDonald's, that wants to check the quality and consistency of some menu items throughout the chain. By hiring observers, sometimes called mystery shoppers, you can gather information on the temperature, freshness, and speed of delivery of various menu items. This method may reveal important information that other data collection methods cannot.

The observation procedure can be any system that ensures the collection of complete and representative information. But every effective observation procedure includes a clear

> **Researchers conducting experiments are interested in testing the effects of a particular variable on some existing situation or activity.**

focus, well-defined steps, and provisions for ensuring the quality of the information collected. For example, an observation procedure for determining the courtesy of employees toward customers when answering the phone might include counting the number of times each employee used certain polite expressions, checking for the use of other courtesy techniques (e.g., offering further help), and recording how long it took for the employee to solve the customer's problem. In other words, you would have to identify observable courteous behaviors to be able to record them.

One particular observation technique that can be used in business research is **user testing.** User testing, also called **usability testing**, measures a person's experience when interacting with a product such as a document, a mobile device, a Web site, a piece of software, or any number of other consumer products.

In general, user testing measures how well users can learn and use a product and how satisfied they are with that process. When engaging in user testing, a researcher will measure the following factors:[8]

- Ease of learning.
- Efficiency of use.
- Memorability.
- Error frequency and severity.
- Subjective satisfaction.

There are many ways to perform a user test. The most common method is to select a small number of testers (5–15) representative of the people who would be using the product and have them perform a task. To set up a user test of a new tablet your company is developing, for example, you would create a situation (a scenario) in which a person performs a task using the product while observers watch and take notes. While the tester worked on this task, you would watch what he or she does to determine if the tester is having difficulty accomplishing the task or seems to like the device. After the testing session, you would administer a questionnaire to get feedback from the person as well.

LO 8-12 Conduct an experiment for a business problem.

experiments Conducting an experiment can be a useful technique for researching a business problem. Originally developed in the sciences, the experiment is an orderly form of testing. Researchers conducting experiments are interested in testing the effects of a particular variable on some existing situation or activity. Therefore, to conduct an experiment,

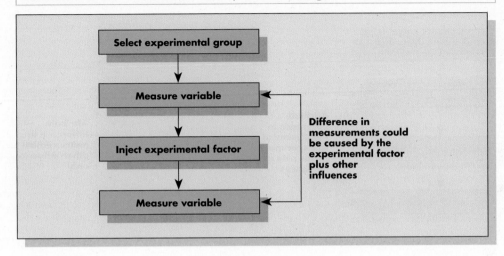

you systematically manipulate one factor of a problem while holding all the others constant. You then measure any changes resulting from your manipulations.

As an example, suppose you are conducting research to determine whether a new package design will lead to more sales. You might start by selecting two test cities, taking care that they are as alike as possible on all the characteristics that might affect the experiment. Then you would secure information on sales in the two cities for a specified time period before the study. Next, for a second specified time period, you would use the new package design in one of the cities and continue to use the old package in the other. During that period, you would keep careful sales records and check to make sure that advertising, economic conditions, competition, and other factors that might have some effect on the experiment remain unchanged. At the end of the study period, you could be relatively confident that any differences you found between the sales in the two cities were caused by the difference in package design.

The simplest experimental design is the **before–after design**. In this design, illustrated in Exhibit 8-30, you select a test group of subjects, measure the variable in which you are interested, and then introduce the experimental factor. After a specified time period, during which the experimental factor has presumably had its effect, you remeasure the variable in which you are interested. If there are any differences between the first and second measurements, you may assume that the experimental factor, plus any uncontrollable factors, is the cause.

You can probably recognize the major shortcoming of this research method: The experimental factor may not explain the change; your results could have been caused by other factors (e.g., changes in the weather, holiday or other seasonal influences on business activity, or advertising for other products).

At best, you have determined only that the variable you were testing *could* have had an effect.

To account for influences other than the experimental factors, you may use designs more complex than the before–after design. These designs attempt to measure the other influences by including some means of control. The simplest of these designs is the **controlled before–after design**.

With this method, you select not one group, but two: the experimental group and the control group. Before introducing the experimental factor, you measure the variable to be tested in each group. Then you introduce the experimental factor into the experimental group only. When the period allotted for the experiment is over, you again measure in each group the variable being tested. Any difference between the first and second measurements in the experimental group can be explained by two causes: the experimental factor and other influences. But the difference between the first and second measurements in the control group can be explained only by other influences because this group was not subjected to the experimental factor. Thus, comparing the "afters" of the two groups will give you a measure of the influence of the experimental factor (Exhibit 8-31).

In a controlled before–after experiment designed to test point-of-sale advertising, you might select Gillette razor blades and Schick razor blades and record the sales of both brands for one week. Next you introduce point-of-sale displays for Gillette only and you record sales for both Gillette and Schick for a second week. At the end of the second week, you compare the results for the two brands. Whatever difference you find in Gillette sales and Schick sales will be a fair measure of the experimental factor, independent of the changes that other influences may have brought about.

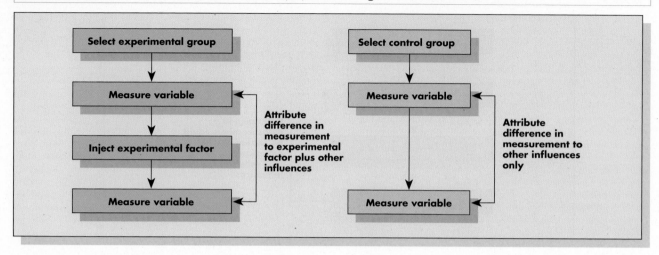

LO 8-13   Explain the uses of focus groups and personal interviews.

Conducting Qualitative Primary Research

The techniques for primary research that we've discussed thus far are (largely) quantitative. That is, they are designed to gather useful numbers. But some primary research methods are qualitative. Qualitative researchers take a more interpretive approach to research. They begin with a more general question about what they want to learn and then study natural phenomena to gather insights into the phenomena or even to learn to ask different questions. Accordingly, they are likely to use such research tools as focus groups and personal interviews, and they will collect mostly verbal data. **Qualitative research** does not enable statistical analysis or the application of the findings to larger populations; rather, it enables you to interpret what the data mean at a more localized level.

Whether you conduct a personal interview or convene a **focus group**, you need to decide how you will record the interactions. You cannot rely on your memory. Sometimes, simply taking notes is sufficient. Other times, you may want to record the session so that you can note nonverbal behaviors (e.g., tone, facial expressions, gestures) that influence the interpretation of a participant's response. Then you would transcribe the notes using a system for coding these nonverbal behaviors in the text of the transcript. You should always ask the participants for their permission to record focus groups or interviews.

conducting focus groups

The purpose of a focus group is to bring together a group of people to find out their beliefs or attitudes about the topic of a research project. For instance, if you want to learn how one of your company's products could be improved, you might gather a group of people who currently use your product and have them discuss what they like or don't like about it.

As the moderator of the discussion, you can structure the conversation and ask questions that will elicit useful data from the participants, or you can

Focus groups can help you learn not only what the target population prefers but why.

simply allow participants to voice their ideas. As you may have experienced, when people discuss a topic in a group, they often generate more or better ideas than they would have working alone. The focus group thus becomes a sort of brainstorming session, which can yield rich data. Of course, as the moderator you also have to make sure that all participants can freely share their ideas. Some of the tips discussed later in this chapter and in Chapter 10 for encouraging participation in group projects and meetings may also help you facilitate focus groups. Because of advances in technology, focus groups can be conducted face to face, online with technologies such as Skype, or even over the phone.

conducting personal interviews If you decide that talking with people one-on-one is the best way to gather data to answer your research question, you will likely conduct face-to-face interviews or phone interviews. People may be willing to share stories and opinions in a personal interview that they might not be comfortable sharing in a larger group.

Preparing for a personal interview is much like preparing for a survey. First, you need to decide whom to interview (your sample). Then you need to construct questions, as you would for a survey. However, the nature of the questions for

treat research participants ethically Many companies, academic institutions, and medical facilities have guidelines for conducting research with human subjects and have institutional review boards (IRBs) to ensure that employees comply with the laws and policies that govern research. Be sure that you are familiar with these policies before conducting research.

The main principle behind such policies is that participants in a research study have the right to informed consent. That is, they have the right to know the nature of their participation in the study and any associated risks. In addition, their participation must be voluntary, and people have the right to discontinue their participation at any time during the study. Just because they agreed to participate at one point does not mean they are obligated to finish the project. Furthermore, participants need to know whether their participation and the data associated with them in the study will be **confidential** (known only to the researcher and participant) or **anonymous** (known only to the participant). If protecting participants' rights will require you to develop a proposal to an IRB, an informed consent letter to the participants, and an informed consent form, be sure you build this process into the planning stage of your project.

[When researching and writing business documents, do not ever lose sight of your main goal: to provide decision makers with reliable information.]

a face-to-face interview will be a bit different. Researchers conducting surveys prefer to use **closed-ended questions** because these force the participants to give only one possible response (e.g., answering a yes/no question, choosing an age range from a list provided by the researcher, or selecting a rating on a scale) and allow for quick data analysis. However, when conducting interviews, many researchers favor **open-ended questions** because the conversational nature of the interview setting enables participants to provide detailed, rich, and varied responses. Furthermore, open-ended questions in personal interviews give researchers the opportunity to ask follow-up questions that they would not be able to ask participants taking a written survey.

LO 8-14 Discuss important ethical guidelines for research.

Conducting Ethical Business Research

Throughout the research process, you need to be sure you are conducting research in an ethical manner. In particular, you should adhere to guidelines for treating research participants ethically, and you should report your research accurately and honestly.

report information accurately and honestly When researching and writing business documents, do not ever lose sight of your main goal: to provide decision makers with reliable information. You will defeat this purpose if you misrepresent your findings.

As you interpret and present secondary information, assess its quality. Does the author draw conclusions that can be supported by the data presented? Are any sources used reliable? Are the data or interpretations biased in any way? Are there any gaps or holes in the data or interpretation? You need to be a good judge of the material, and if it has limitations, you should note them in your document.

Also, be sure to cite your sources. The whole point of including citations is to allow your readers to check your sources for themselves. Any mistakes in your citations may not only frustrate your readers but also make you look inept or dishonest. Be particularly careful to give credit where credit is due. **Plagiarism,** which is submitting another person's published work as your own without properly crediting it, is especially damaging to your credibility. Be sure to follow the guidelines in Bonus Chapter E for correctly citing what you need to cite.

As for primary research, once you have good data to work with, you must interpret them accurately and clearly for your reader.

Here, too, you should acknowledge any limitations of your research. Be careful as well to avoid misleading visuals (see Chapter 3).

LO 8-15 Interpret your findings accurately.

INTERPRETING THE FINDINGS

The next major stage of the report-writing process is to interpret the information you've gathered.

Actually, you will have done a good bit of interpreting already by the time you reach this stage. You had to interpret the elements of the situation to understand the problem and determine your research purpose. You also had to interpret your data as you were gathering them to make sure that you were getting appropriate and sufficient information. But when your research is finished, you will need to formulate the interpretations that will guide the shape and contents of your report.

To do this, keep both your problem and your readers in mind. Your findings will need to apply clearly to the given problem in

Interpreting facts requires not only analytical skills and objective judgment but consideration for ethical issues as well.

order to be viewed as logical solutions. But they will also need to meet the readers' needs in order to be viewed as relevant and helpful. If you have kept your reader-based problem and purpose statements in mind while doing your research, making logical, reader-based analyses of your data should follow naturally.

How you interpret your data will vary from case to case, but the following general advice can help you with this process.

Avoiding Errors in Interpretation

Certain human tendencies lead to error in interpretation. The following list explains how to minimize them:

1. *Report the facts as they are.* Do nothing to make them more or less exciting. Adding color to interpretations to make the report more interesting compromises objectivity.

2. *Do not think that conclusions are always necessary.* When the facts do not support a conclusion, you should just summarize your findings and conclude that there is no conclusion. All too often, report writers think that if they do not conclude, they have failed in their investigation.

3. *Do not interpret a lack of evidence as proof to the contrary.* The fact that you cannot prove something is true does not mean that it is false.

4. *Do not compare noncomparable data.* When you look for relationships between sets of data, make sure they have enough similarities to be comparable. For example, you might be able to draw conclusions about how two groups of employees differ at Company X, but you probably would not be justified in comparing Group A from Company X to Group B from Company Y.

5. *Do not draw illogical cause–effect conclusions.* The fact that two sets of data appear to affect each other does not mean they actually do. They may be only **correlated** (strongly associated for an undetermined reason). Use research and good logic to determine whether a cause–effect relationship is likely.

6. *Beware of unreliable and unrepresentative data.* Much of the information to be found in secondary sources is incorrect to some extent. The causes are many: collection error, biased research, recording mistakes. Beware especially of data collected by groups that advocate a position (political organizations, groups supporting social issues, and other special interest groups). Make sure your sources are reliable. And remember that the interpretations you make are no better than the data you interpret.

7. *Do not oversimplify.* Most business problems are complex, and it can be tempting to settle for easy answers. Avoid conclusions and recommendations that do not do justice to the problem.

8. *Tailor your claims to your data.* There's a tendency among inexperienced report writers to use too few facts to generalize

You're right. This report does make you look like a fool.

Source: © 1985 Dean Vietor. Used with permission.

far too much. If you have learned about a certain phenomenon, do not assume that your interpretations can automatically be applied to similar phenomena. Or if your research has revealed the source of a problem, do not assume that you can also propose solutions; finding solutions can be a separate research project altogether. Make only those claims that are well supported by your evidence, and when you are not sure how strong to make them, use such qualified language as "may be," "could be," and "suggest."

Using Statistical Tools and Visuals to Interpret Data

In many cases, the information you gather is quantitative—that is, expressed in numbers. "You can't manage what you can't measure" is a common business expression, and while nonnumerical data, such as descriptions of customers' experiences or comments by employees, are also extremely valuable, the popularity of this expression rightly suggests that businesses need accurate numbers in order to succeed. As Chapter 1 points out, barcode systems and other "smart machines," which store statistics about their use, are generating huge amounts of numerical information. To use such data intelligently, you must find ways of simplifying them so that your reader can grasp their general meaning.

Statistical techniques provide many methods for analyzing data. By knowing them, you can improve your ability to interpret. Although a thorough review of statistical techniques is beyond the scope of this book, you should know the more commonly used methods, described in the following paragraphs.

Possibly of greatest use to you in writing reports are **descriptive statistics**—measures of central tendency, dispersion, ratios, and probability. Measures of central tendency—the mean, median, and mode—will help you find a value that roughly represents the whole. The measures of dispersion—ranges, variances, and standard deviations—help you describe how spread out the data are. Ratios (which express proportionate relationships) and probabilities (which determine how many times something will likely occur out of the total number of

possibilities) can also help you give meaning to data. **Inferential statistics**, which enable you to generalize about a whole population based on the study of a sample, are also useful but go beyond these basic elements. You will find descriptions of these and other useful techniques in the help documentation of your spreadsheet and statistics software as well as in any standard statistics textbook.

As Chapter 3 points out, visuals are a powerful way to communicate detailed information clearly. But they can also greatly aid your own interpretation of data. Such programs as SPSS and Microsoft Excel make it easy for you to translate different combinations of numbers into visual form so you can actually see patterns and comparisons. Run as many statistical tests and create as many visuals as you need to be able to analyze your findings thoroughly.

When presenting your interpretations, explain the statistical methods you used, and make clear what your tables and charts mean. Remember that statistics and visuals are not an end in themselves: Their ultimate purpose is to help you give readers the findings they need in a form they can understand.

LO 8-16 Organize information in outline form, using time, place, quantity, factor, or a combination of these as bases for division.

ORGANIZING THE REPORT INFORMATION

When you have interpreted your information, you will know your report's main points. Now you are ready to organize this content for presentation. Your goal here is to arrange the information in a logical order that meets your reader's needs.

The Nature and Benefits of Outlining

An invaluable aid at this stage of the process is an **outline**. A good one will show what things go together **(grouping)**, what order they should be in **(ordering)**, and how the ideas relate in terms of levels of generality **(hierarchy)**. Although you can outline mentally, a written plan is advisable for all but the shortest reports. Time spent on outlining at this stage is well spent because it will make your drafting process more efficient and orderly. For longer reports, your outline will also form the basis for the table of contents.

If you have proceeded methodically thus far, you probably already have a rough outline. It is the list of topics that you drew up when planning how to research your problem. You may also have added to this list the findings that you developed when interpreting your data. But when it's time to turn your research plan into a report plan, you need to outline more deliberately. Your goal is to create the most logical, helpful pattern of organization for your readers.

In constructing your outline, you can use any system of numbering or formatting that will help you see the logical structure of your planned contents. If it will help, you can use the conventional or the decimal symbol system to mark the levels. The **conventional outlining system** uses Roman numerals to show the major headings, and letters of the alphabet and Arabic numbers to show the lesser headings, as illustrated here:

Conventional System
   I. First-level heading
      A. Second level, first part
      B. Second level, second part
        1. Third level, first part
        2. Third level, second part
          a. Fourth level, first part
            (1) Fifth level, first part
               (a) Sixth level, first part
  II. First-level heading
      A. Second level, first part
      B. Second level, second part
      etc.

The **decimal outlining system** uses whole numbers to show the major sections, with decimals and additional numbers added to show subsections. That is, the digits to the right of the decimal show each successive level in the outline, as shown here:

Decimal System
  1.0 First-level heading
      1.1 Second level, first part
      1.2 Second level, second part
          1.2.1 Third level, first part
          1.2.2 Third level, second part
              1.2.2.1 Fourth level, first part
                  1.2.2.1.1 Fifth level, first part
                      1.2.2.1.1.1 Sixth level, first part

  2.0 First-level heading
      2.1 Second level, first part
      2.2 Second level, second part
      etc.

Bear in mind that the outline is a tool for you, even though it is based on your readers' needs. Unless others will want to see an updated outline as you work, spend minimal time on its appearance. Allow yourself to change it, scribble on it, depart from

from the tech desk

Brainstorm and Outline with Visualization Tools

Inspiration is a concept mapping tool aimed at helping writers generate ideas and outline their documents. The example shown here demonstrates how individuals or groups can brainstorm the factors of a report that investigates which color laser printer a product design department should purchase. Using either the diagram or outline view (or both), a report writer would list as many ideas as possible. Later the items and relationships can be rearranged by dragging and moving pointers.

The software will update the outline symbols as changes are made. Users can toggle between the different views to work with the mode that works best for them. When ready to write, users can export the outline or diagram to Word or Google Drive.

You can download a free 30-day trial version from www.inspiration.com/freetrial.

Or try the online version, WebspirationPro, available at www.mywebspiration.com/. Both forms are relatively inexpensive, and the

Web-based version is particularly good for collaborative planning and report writing.

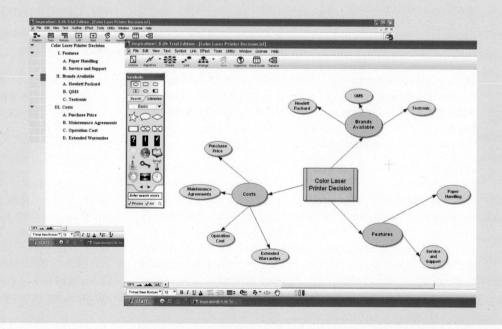

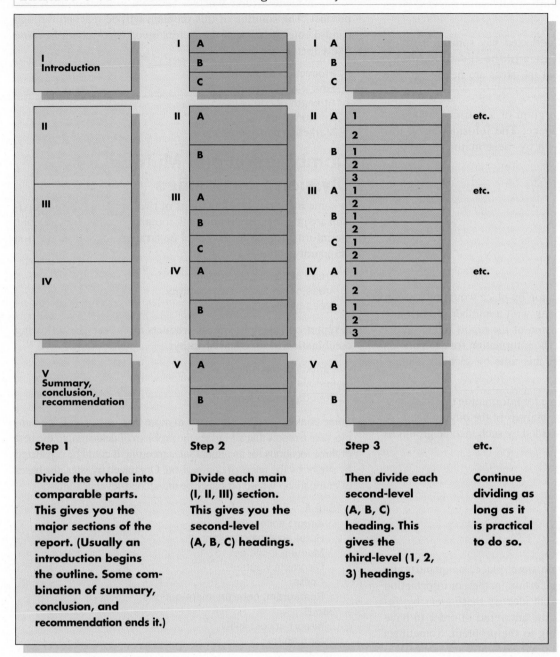

Step 1

Divide the whole into comparable parts. This gives you the major sections of the report. (Usually an introduction begins the outline. Some combination of summary, conclusion, and recommendation ends it.)

Step 2

Divide each main (I, II, III) section. This gives you the second-level (A, B, C) headings.

Step 3

Then divide each second-level (A, B, C) heading. This gives the third-level (1, 2, 3) headings.

etc.

Continue dividing as long as it is practical to do so.

it—whatever seems appropriate as your report develops. For example, you might want to note on your outline which sections will contain visuals, or jot down a particularly good transition between sections that comes to mind. The time to labor over the outline's format and exact wording will be when you use it to create the headings and the table of contents for your finished report.

Organization by Division

One methodical way to create an outline is to use the process of dividing the contents into smaller and smaller sections. With this method, you begin by looking over all your information. You then identify its major parts. This first level of division gives you the major outline parts, indicated in Exhibit 8-32 by the Roman numerals (I, II, III, and so on).

Next, you find ways to subdivide the contents in each major section, yielding the second-level information (indicated by A, B, C). If practical, you keep dividing the contents, generating more levels. This method helps you divide your report into manageable chunks while also creating a logical and clear structural hierarchy.

Division by Conventional Relationships

In dividing your information into subparts, you have to find a way of dividing that will produce approximately equal parts. Time, place, quantity, and factor are the general bases for these divisions.

Whenever the information you have to present has some time aspect, consider organizing it by **time division**. In such an organization, the divisions are periods of time. These time periods usually follow a logical sequence, such as past to present or present to past. The periods you select need not be equal in duration, but they should be about equal in importance.

A report on the progress of a research committee illustrates this possibility. The period covered by this report might be divided into the following comparable subperiods:

Orientation, May–July

Project planning, August

Implementation, September–November

The happenings within each period might next be arranged in order of occurrence, and additional subdivisions might even be possible.

If the information you have collected has some relation to geographic location, you may use a **place division**. Ideally, this division would be such that the areas are nearly equal in importance.

A report on the U.S. sales program of a national manufacturer illustrates division by place. The information in this problem might be broken down by these major geographic areas:

New England
Atlantic Seaboard
South
Southwest
Midwest
Rocky Mountains
Pacific Coast

Another illustration of organization by place would be a report on the productivity of a company with a number of customer service branches. A major division of the report might be devoted to each of the branches. The information for each branch might be broken down further, this time by sections, departments, or divisions.

Quantity divisions are possible for information that has quantitative values. To illustrate, an analysis of the buying habits of potential customers could be divided by such income groups as the following:

Under $30,000
$30,000 to under $45,000
$45,000 to under $60,000
$60,000 to under $85,000
$85,000 to under $100,000
$100,000 and over

Problems often have few or no time, place, or quantity aspects. Instead, they require that certain factors, or information areas, be investigated. You might identify these areas by figuring out what questions must be answered in order to have complete information pertaining to the problem. Sometimes the problem you're investigating will naturally suggest certain subtopics.

An example of **division by factors** is a report that seeks to determine which of three locations is the best for a new office for property management. In arriving at this decision, one would need to compare the three locations based on the factors affecting the office location. Thus, the following organization of this problem would be a possibility:

Location accessibility
Rent
Parking
Convenience to current and new customers
Facilities

Another illustration of organization by factors is a report advising a manufacturer whether to begin production of a new product. The solution of this problem will be reached through careful consideration of the factors involved. Among the more likely factors are these:

Production feasibility
Financial considerations
Strength of competition
Consumer demand
Marketing considerations

Combination and Multiple Division Possibilities

In some instances, combinations of two or more bases of division are possible. In a report on a company's sales, for example, the information collected could be arranged by a combination of quantity and place:

Areas of high sales activity
Areas of moderate sales activity
Areas of low sales activity

A report on sales of cyclical products might use the following combination of time and quantity:

Periods of low sales
Periods of moderate sales
Periods of high sales

Some contents can be organized in more than one way. For example, take a report that addresses the problem of determining the best of three locations for an annual sales meeting. It could be organized by site or by the bases of comparison. Organized by sites, the bases of comparison would probably be the second-level headings:

Site A
    Airport accessibility
    Hotel accommodations
    Meeting facilities
    Favorable weather
    Costs
    Restaurant/entertainment options
Site B
    Airport accessibility
    [and so on]
Site C
    Airport accessibility
    [and so on]

Organized by bases of comparison, cities would probably be the second-level headings:

Airport accessibility
    Site A
    Site B
    Site C
Hotel accommodations
    Site A
    Site B
    Site C

Meeting facilities
    Site A
    Site B
    Site C
    [and so on]

Both plans would be logical. However, the organization by cities separates information that has to be compared, thus making it difficult to see which city is the best on each criterion. In the second outline, the information that has to be compared is close together. You can determine which city has the best hotel accommodations after reading only one section of the report. In this example, then, the second way would be preferable.

Nevertheless, the two plans show that some problems can be organized in more than one way. In such cases, you must compare the possibilities carefully to find the one that most helpfully presents the report information.

LO 8-17 Turn an outline into a table of contents whose format and wording are logical and meaningful.

From Outline to Table of Contents

When you are ready to prepare the table of contents for your report, you will be, in essence, turning the outline that helped you write into an aid for the reader. Because it will be your public outline, the table of contents needs to be carefully formatted and worded.

True, you will probably design the table of contents late in the report-writing process. We discuss it here as a logical conclusion to our discussion of outlining. But if others involved in the project want to see a well-prepared outline before your report is done, you can use the following advice to prepare that outline.

Note also that what we say about preparing the headings for the table of contents also applies to writing the headings for the report sections. The two sets of headings, those in the table of contents and those in the report itself, should match exactly. Using Word's Styles to format your headings and its Table of Contents generator to create your table of contents will ensure this consistency.

formatting decisions Whatever format you used for your personal outline, you now need to choose one that your reader will find *instructive, readable,* and *appropriate.* You create an instructive format by clearly indicating the hierarchy of the information. You should use form (font selection, size, style, and color) and placement (location and indentation) to distinguish among the levels of your contents, as illustrated by the table of contents of the sample long report in Bonus Chapter D. You make the format readable by using ample vertical white space between topics and enabling readers to see at a glance how the report is organized. Using leaders (dots with intervening spaces) between your topics and your page numbers can also enhance readability.

communication matters

Contrasting Headings from a Sample Report

Talking Headings	Topic Headings
Introduction to the Problem	Introduction
Authorization by Board Action	Authorization
Selection of the Potential Sites	Purpose
Reliance on Government Data	Sources
Factors to Be Discussed	Preview
Community Attitudes toward a New Plant	Community Attitudes
Favorable Reaction of All Towns to a New Employer	New Plant
Mixed Attitudes of All Towns toward Our Labor Policies	Labor Policies
Labor Supply and Prevailing Wage Rates	Labor Factors
Prevalence of Unskilled Labor in San Marcos	Unskilled Workers
Concentration of Skilled Workers in San Marcos	Skilled Workers
Mixed Patterns of Wage Rates	Wage Rates
Nearness to Suppliers	Available Suppliers
Location of Ballinger, Coleman, and San Marcos in Farming Areas	Adequate Areas
Relatively Low Production Near Big Spring and Littlefield	Inadequate Areas
Availability of Utilities	Utilities
Inadequate Water Supply for All Towns but San Marcos	Water
Unlimited Supply of Natural Gas for All Towns	Natural Gas
Electric Rate Advantage of San Marcos and Coleman	Electricity
General Adequacy of All Towns for Waste Disposal	Waste Disposal
Adequacy of Existing Transportation Systems	Transportation
Surface Transportation Advantages of San Marcos and Ballinger	Surface
General Equality of Airway Connections	Air
A Final Weighting of the Factors	Conclusions
Selection of San Marcos as First Choice	First Choice
Recommendation of Ballinger as Second Choice	Second Choice
Lack of Advantages in Big Spring, Coleman, and Littlefield	Weaker Choices

An appropriate format is one that your reader expects. Some business readers view the conventional outlining system (Roman numerals, letters, and Arabic numbers) and the decimal system (as in 1.2.1) as adding unnecessary clutter to the table of contents. Instead, they prefer the use of form and placement to show them how the parts relate to each other. However, in the military and some technical environments, the decimal system is expected, and in other contexts, your readers may want the full numerals and letters of the conventional system. In our examples, we use format rather than numbering to indicate levels of information, but be sure to use whatever format your readers will prefer.

topic or talking headings

In selecting the wording for your table of contents headings, you have a choice of two general forms: topic headings and talking headings. **Topic headings** are short constructions, frequently consisting of one or two words. They merely identify the topic of discussion, as in "Cost" or "Space Requirements." **Talking headings** also identify the subject matter to be covered, but they go a step further: They also indicate what is said about the subject. In other words, talking headings summarize the material they cover, as in "Increase in Cost of Operation" or "Less Space Required." See the Communication Matters feature on the previous page for an extended example.

> As a general rule, you should write headings at each level of the table of contents in the same grammatical form.

Which of these forms is better? The answer depends on the situation. Talking heads would be appropriate if your readers are extremely busy, trust your judgment, and are likely to skim the supporting facts. Topic headings, because they do not announce the point of the section, are better for readers who want to see the facts before being told what to think about them.

parallel construction

As a general rule, you should write headings at each level of the table of contents in the same grammatical form. In other words, equal-level headings should be **parallel** in structure. For example, if the first major heading is a noun phrase, the rest of the major heads should be noun phrases. If the first second-level heading under a major head is an *-ing* phrase, all second-level headings in the section should be *-ing* phrases.

This rule is not just an exercise in grammar; its purpose is to show similarity. As you will recall from Chapter 4, parallelism helps your readers understand which topics are alike and go together. If you state similar topics in different forms, your logic will become blurry, and your reader will have trouble following you. It is usually considered permissible to vary the form from one section and level to another; that is, the second-level heads in one section need to match, but they do not need to match the second-level heads in the other sections, and the third-level heads do not need to match the second-level heads. Just be sure that the headings on each level of each section are parallel.

The following headings illustrate violations of parallelism:

Programmer Output Is Lagging (sentence).

Increase in Cost of Labor (noun phrase)

Unable to Deliver Necessary Results (adjective phrase)

Making the headings all noun phrases would fix the problem:

Lag in Programmer Output

Increase in Cost of Labor

Inability to Deliver Necessary Results

Or you could make all the headings sentences, like this:

Programmer Output Is Lagging.

Cost of Labor Is Increasing.

Information Systems Cannot Deliver Necessary Results.

Here's a different kind of faulty parallelism:

Managers Prefer an Intranet

U.S. Employees Prefer a Social Media Site

A Newsletter Is Preferred by Overseas Employees

The third heading is "off." Can you see why? If you answered that it switches from active to passive voice, you're right.

concise wording

Your headings should be as concise as possible while still being clear and informative. Although the following headings are informative, their excessive length obviously hinders their communication effectiveness:

Personal Appearance Enhancement Is the Most Desirable Feature of Contact Lenses That Wearers Report.

The Drawback of Contacts Mentioned by Most People Who Can't Wear Them Is That They Are Difficult to Put in.

More Comfort Is the Most Desired Improvement Suggested by Wearers and Nonwearers of Contact Lenses.

Obviously, the headings contain too much information. Just what should be left out depends on your judgment. Here is one possible revision:

Most Desirable Feature: Personal Appearance

Prime Criticism: Difficulty of Insertion

Most Desired Improvement: Comfort

In your effort to be concise, should your headings omit *a*, *an*, and *the*, as some of the examples above do? Authorities on readability recommend including these words in body text, but there appears to be no consensus on whether to use or omit them in headings and titles. See what your teacher or boss prefers, and whichever way you choose, be consistent throughout your report.

variety of expression In the wording of headings, as in all other forms of writing, you should use some variety of expression. Repeating words too frequently makes for monotonous writing. The following outline excerpt illustrates this point:

> Oil Production in Texas
>
> Oil Production in California
>
> Oil Production in Louisiana

As a rule, if you make the headings talk well, there is little chance of monotonous repetition. The headings in the preceding example can be improved simply by making them talk:

> Texas Leads in Oil Production.
>
> California Holds the Runner-up Position.
>
> Rapidly Gaining Louisiana Ranks Third.

will be much more difficult. Allow yourself to move along, stitching together the pieces. Once you have a draft to work with, you can perfect it.

When revising, let the advice in the previous chapters be your guide. As with all the business messages previously discussed, reports should communicate as clearly and quickly as possible. Your readers' time is valuable, and you risk having your report misread or even ignored if you do not keep this fact in mind. Use both words and formatting to get your contents across efficiently.

You can help your reader receive the report's message clearly by giving your report some specific qualities of well-written reports. Two critical ingredients are a reader-centered beginning and ending. Such characteristics as objectivity, consistency in time viewpoint, coherence, and interest can also enhance the reception of your report. We review these topics next.

Beginning and Ending

Arguably the most critical parts of your report will be the beginning and ending. In fact, researchers agree that these are the most frequently read parts of a report.

> "Arguably the most critical parts of your report will be the beginning and ending."

The table of contents is an important preview of your report. Your goal is to use headings that will make it interesting, precise, and logically structured.

LO 8-18 Write reports that are focused, objective, consistent in time viewpoint, smoothly connected, and interesting.

WRITING THE REPORT

By the time you write your report, you will have already done a good deal of writing. You will have written—and probably rewritten—problem and purpose statements to guide you through your research. You will have collected written data or recorded your findings in notes, and you will have organized your interpretations of the data into a logical, reader-centered structure. Now it is time to flesh out your outline with clearly expressed facts and observations.

When you draft your report, your first priority is to get the right things said in the right order. As Chapter 2 advises, you should not strive for a perfect draft the first time around. Understand that some pieces will seem to write themselves, while others

Whatever other goals it may achieve, the opening of your report should convey what problem you studied, how you studied it, and (at least generally) what you found out. Why? Because these are the facts that the reader most wants to know when he or she first looks at your report.

Here is a simple introduction that follows this pattern:

> In order to find out why sales were down at the Salisbury store, I interviewed the manager, observed the operations, and assessed the environment. A high rate of employee turnover appears to have resulted in a loss of customers, though the deteriorating neighborhood also seems to be a contributing factor.

In a formal report, some brief sections may precede this statement of purpose (for example, facts about the authorization of the study), and there might be extensive front matter (for example, a title page, letter of transmittal, table of contents, and executive summary). What follows the purpose statement can also vary depending on the size and complexity of the report (for example, it may or may not be appropriate to go into more detail about the research methods and limitations or to announce specifically how the following sections will be organized). But whatever kind of report you are writing, make sure that the beginning gets across the subject of the report, what kind of data it is based upon, and its likely significance to the reader.

Your ending will provide a concise statement of the report's main payoff—whether facts, interpretations, or recommendations. In a short report, you may simply summarize your findings with a brief paragraph, since the specific findings will be easy to see in the body of the report. In a longer report, you should make this section a more thorough restatement of your main findings, formatted in an easy-to-read way. Both the gist ("So what did you find out?") and the significance ("Why should I care?") of your report should be clear.

Being Objective

As we have said, a good report is objective; it presents all relevant facts and interprets them logically, without bias. Your objectivity should be evident in both your content and your writing style.

objectivity as a basis for believability
An objective report has an ingredient that is essential to good report writing—**believability**. Powerful assertions made in emotionally charged language may at first glance appear to strengthen your report. But if bias is evident at any point in a report, the reader will question the credibility of the entire report. Maintaining objectivity is, therefore, the only sure way to make report writing believable.

the question of impersonal versus personal writing
Recognizing the need for objectivity, early report writers worked to develop an objective style of writing. Since the source of bias in reports was people, they reasoned that objectivity was best attained by emphasizing facts rather than the people involved in writing and reading reports. So they tried to take the human beings out of their reports. The result was **impersonal writing**—that is, writing in the third person, without *I*, *we*, or *you* perspectives.

In recent years, some writers have opposed this approach. They argue that **personal writing** is more forceful and direct than impersonal writing. They point out that writing is more conversational and therefore more interesting if it brings both the reader and the writer into the picture. They contend that objectivity is an attitude—not a matter of pronoun use—and that a report written in the personal style can be just as objective as a report written in the impersonal style. These writers argue that impersonal writing frequently leads to an overuse of the passive voice and a dull writing style. (While this last claim may be true, impersonal writing need not be boring. Newspaper articles use the impersonal style while still maintaining interest.)

As with most controversies, the arguments on both sides have merit. In some situations, personal writing is better. In other situations, impersonal writing is better. And in still other situations, either type of writing is good.

Your decision should be based on the facts of each report situation. First, you should consider the expectations of those for whom you are preparing the report. If your readers prefer an impersonal style, use it—and vice versa. Then you should consider the formality of the situation. In general, personal writing is appropriate for informal situations and impersonal writing for most formal situations.

Here are contrasting examples of the personal and impersonal style:

Personal	Impersonal
Having studied the advantages and disadvantages of using coupons, I recommend that your company not adopt this practice. If you used coupons, you would have to absorb their cost. You would also have to hire additional employees to take care of the increase in sales volume.	A study of the advantages and disadvantages of using coupons supports the conclusion that the Mills Company should not adopt this practice. The coupons themselves would cost extra money. Also, use of coupons would require additional personnel to take care of the increase in sales volume.

Notice that both versions are active, clear, and interesting. Strive for these effects no matter which style you choose.

Being Consistent with Time

A report that has illogical time shifts—for example, one that says "The managers responded . . ." (past tense) in one place but "The employees say . . ." (present tense) in another place—confuses the reader. Thus, it is important that you maintain a **consistent time viewpoint**.

You have two main choices of time viewpoint: past or present. Although some authorities favor one or the other, either viewpoint can produce a good report. The important thing is to be consistent—to select one time viewpoint and stay with it. In other words, you should view all similar information in the report from the same position in time.

If you adopt the **past-time viewpoint**, you treat the research, the findings, and the writing of the report as past. Thus, you would report the results of a recent survey in past tense: "Twenty-two percent of the managers *favored* a change." You would write a reference to another part of the report this way: "As Part II *indicated*, . . ." Your use of the past-time viewpoint would have no effect on references to current and future happenings. It would still be proper to write a sentence like this: "If the current trend *continues*, 30 percent *will favor* a change by 2015." Prevailing concepts and proven conclusions are also exceptions. You would present them in present tense. For example, you would write "Solar energy *is* a major potential source of energy" and "The findings *indicate* that managers are not adequately trained."

Writing in the **present-time viewpoint** presents as current all information that can logically be assumed to be current at the time of writing. All other information is presented in its proper place in the past or future. Thus, you would report the results of a recent survey in these words: "Twenty-two percent of the managers *favor* a change." You would refer to another part of the text like this: As Part II *indicates*, . . ." But in referring to an earlier survey, you would write: "In 2009 only 12 percent *held* this opinion." And in making a future reference, you would write: "If this trend continues, 30 percent *will hold* this opinion by 2015."

Including Transitions

A well-written report reads as one continuous story, with the parts smoothly connected. Much of this flow is the result of good, logical organization. But more than logical order is needed in long reports. As Bonus Chapter D points out, a special coherence plan may be needed as well. In all reports, however, lesser transitional techniques are useful to connect information.

As Chapter 4 explains, transitions are words or sentences that show the relationships between parts of a sentence, paragraph, or document. In reports, they may appear at the beginning of a part as a way of relating this part to the preceding part. They may appear at the end of a part as a forward look. Or they may appear within a part as words or phrases that help move the flow of information.

sentence transitions Throughout the report you can improve the connecting network of thought by using sentence transitions. They are especially helpful between major parts of the report.

In the following example, the sentences first summarize what has been said so far. The last sentence then transitions to the next topic.

> These data show clearly that alternative fuel cars are the most economical. Unquestionably, their operation by gas and hydrogen and their record for low-cost maintenance give them a decided edge over gas-fueled cars. Before a definite conclusion about their merit can be reached, however, one more vital comparison should be made.

Clearly the next part of the report will discuss one more comparison.

Here is another example of a forecasting sentence:

> At first glance the data appear convincing, but a closer observation reveals a number of discrepancies.

The reader knows to expect a discussion of the discrepancies next.

Topic sentences (discussed in Chapter 4) can also be used to link the various parts of the report. Note in the following example how the topic sentences (in italics) maintain the flow of thought by emphasizing key information.

> *The Acura accelerates faster than the other two brands, both on a level road and on a 9 percent grade.* According to a test conducted by *Consumer Reports*, Acura reaches a speed of 60 miles per hour in 13.2 seconds. To reach the same speed, Toyota requires 13.6 seconds, and Volkswagen requires 14.4 seconds. On a 9 percent grade, Acura reaches

Transitions provide important bridges from one report section to the next.

the 60-miles-per-hour speed in 29.4 seconds, and Toyota reaches it in 43.3 seconds. Volkswagen is unable to reach this speed.

Because it carries more weight on its rear wheels than the others, Acura has the best traction of the three. Traction, which means a minimum of sliding on wet or icy roads, is important to safe driving, particularly during the cold, wet winter months. Since traction is directly related to the weight carried by the rear wheels, a comparison of these weights should give some measure of the safety of the three cars. According to data released by the Automobile Bureau of Standards, Acura carries 47 percent of its weight on its rear wheels. Nissan and Toyota carry 44 and 42 percent, respectively.

transitional words Although the most important function of transitions is to connect the major parts of the report, transitions are also needed between the lesser parts. If the writing is to flow smoothly, you will need to connect clause to clause, sentence to sentence, and paragraph to paragraph, as Chapter 4 advises.

Numerous transitional words are available. The following list shows such words and how you can use them.

Relationship	Word Examples
Listing or enumeration of subjects	In addition
	First, second, . . .
	Besides
	Moreover
Contrast	On the contrary
	In spite of
	On the other hand
	In contrast
	However
Likeness	Also
	Likewise
	Similarly
Cause–effect	Thus
	Because of
	Therefore
	Consequently
	For this reason
Explanation or elaboration	For example
	To illustrate
	For instance
	Also
	Too

Helpful as transitions are, you should use them only when they are needed—when including them would provide a useful preview or leaving them out would produce abruptness. For example, avoid such boring, unnecessary transitions as "This concludes the discussion of Topic X. In the next section, Y will be analyzed."

Maintaining Interest

Like any other form of writing, report writing should be interesting. Actually, interest is as important as the facts of the report because communication is not likely to occur without it. Readers cannot help missing parts of the message if their attention is allowed to stray. (If you have ever tried to read dull writing when studying for an exam, you know the truth of this statement.)

To write interestingly, avoid business clichés and unnecessarily abstract language. Remember that behind every fact and figure there is life—people doing things, machines operating, a commodity being marketed. A technique of good report writing is to bring that life to the surface by using concrete words and active-voice verbs as much as possible. Keeping your wording efficient also helps maintain the reader's interest.

But you can overdo efforts to make report writing interesting. Such is the case whenever your reader's attention is attracted to how something has been said rather than to what has been said. Effective report writing simply presents information in a clear, concise, and interesting manner. Report-writing style is at its best when the readers are prompted to say "Here are some interesting facts" rather than "Here is some interesting writing."

LO 8-19 Prepare reports collaboratively.

WRITING REPORTS COLLABORATIVELY

In your business career, you are likely to participate in numerous collaborative writing projects, and the end product of many of them is likely to be a report. Group involvement in report preparation is becoming increasingly significant for a number of reasons. For one, the specialized knowledge of different people can improve the quality of the work. For another, the combined talents of the members are likely to produce a document better than any one of the members could produce alone. A third reason is that dividing the work can reduce the time needed for the project. And fourth, many different software tools allow groups to collaborate easily and well from different places, making group work a viable option.

Determining the Group Makeup

The first step is to decide who will be in the group. The availability and competencies of the people in the work situation involved are likely to be the major factors. At a minimum, the group will consist of two. The maximum will depend on the number actually needed to do the project. As a practical matter, however, a maximum of five is a good rule since larger groups tend to lose efficiency. All major areas of specialization needed to investigate the problem should be represented by the team members.

In most business situations the highest ranking administrator in the group serves as leader. In groups made up of equals, a leader usually is appointed or elected. When no leader is so designated, the group works together informally. In such cases, however, an informal leader usually emerges. Especially with group writing projects, it is a good idea to have one person in charge of overseeing the entire process.

communication matters

Does Your Group Have Emotional Intelligence?

Ever since the publication of Daniel Goleman's *Emotional Intelligence: Why It Can Matter More Than IQ* in 1995, companies have been looking for ways to cultivate the emotional intelligence (EI) of its members.

But groups can enhance their collective EI, too. According to Vanessa Urch Druskat and Steven B. Wolff of the *Harvard Business Review*, "Group EI norms build the foundation for true collaboration and cooperation—helping otherwise skilled teams fulfill their highest potential."

What kinds of things should a group do to channel its members' insights and emotions into positive results? Here's a partial list from Druskat and Wolff:

- Encourage all members to share their perspectives before making key decisions.
- Handle confrontation constructively. If team members fall short, call them on it by letting them know the group needs them.
- Regularly assess the group's strengths, weaknesses, and modes of interaction.
- Create structures that let the group express its emotions.
- Cultivate an affirmative environment.
- Encourage proactive problem solving.

And try to keep things fun. In one company, the industrial-design firm IDEO, participants throw stuffed toys at anyone who prematurely judges ideas during brainstorming sessions.

Source: "Building the Emotional Intelligence of Groups," *Harvard Business Review* 1 Mar. 2001, *HarvardBusiness.org*, Harvard Business Publishing, Web, 3 June 2013.

Creating the Ground Rules

In organizations where teamwork is common, the ground rules for participation in a group may be understood. But students and working professionals alike may find it helpful to establish explicit guidelines for the participants.

Some rules may govern the members' interactions. For example, a rule might be "Listen respectfully and actively to what others are saying, without interrupting." Or it might instruct members to use "I" language ("I think . . .") rather than "you" language ("The problem with your idea is . . .") when disagreeing. Others might cover more logistical issues, such as conscientiously doing one's share of the work, keeping the group informed if problems arise, and being on time with one's contributions.

Ideally, the group will generate its own ground rules to which all members will agree. Some instructors find that actually drawing up a contract and having each member sign it is a good way to get group work off to a good start and prevent problems down the line.

Choosing the Means of Collaboration

Not that many years ago, groups needed numerous face-to-face meetings in order to get their work done. Today there are many other venues for group interactions. Your group should put careful thought into the choice of media that will enable effective collaboration while taking into account members' time constraints, distance from each other, and technological preferences.

If possible, you should have at least two face-to-face meetings—one at the start of the project and another near the end (for example, when doing the final revisions). You can meet either physically or virtually (using an online meeting tool such as Skype). But the bulk of the collaborating may take place by email, by discussion board, or through such online collaborative authoring tools as Google Drive or wikis. Whatever tools you use, it is important to choose them consciously and create any ground rules that will apply to their use.

Making a Project Plan

Especially when the desired outcome is a coherent, effective report, the group should structure its tasks to meet the project's goals. Using the steps discussed in the next section and any additional considerations, the group should prepare a timeline that clearly states or shows the deadline for each task. A Gantt chart can be very useful along these lines (see Chapter 3), but even a simple list or table can suffice. In addition, your plan should make clear who is responsible for what. If your group has taken an inventory of its strengths before its planning, you can match members up with what they do best (for example, doing research or revising a document).

Your plan can also describe in some detail the desired form and style of the final document (such as which template it will use or whether or not it will use "you"). The more the group determines such matters up front, the less scrambling it will need to do at the end to generate a coherent, consistent-looking report.

Many reports written in business are produced in collaboration with others. Although you will do some work individually, you can expect to plan, organize, and revise the report as a group.

Researching and Writing the Report

However the group decides to operate, the following activities typically occur, usually in the sequence shown.

1. *Determine the Problem and Purpose.* As in all report projects, the participants must determine just what the report must do. Thus, the group should follow the preliminary steps of determining the problem and purpose of the project as discussed previously. They also need to develop a coherent, shared sense of the report's intended readers and their needs.

2. *Identify the Factors.* The group next determines what needs to be studied in order to achieve the report's purpose. This step involves determining the factors of the problem, as described earlier in the chapter. An advantage of collaboration is that several minds are available for the critical thinking that is so necessary for identifying the factors of the problem.

3. *Gather the Needed Information.* Before the group can begin writing the report, it must get the information needed. This activity can include any of the types of research discussed in this chapter. In some cases, however, group work begins

after the information has already been assembled, thus eliminating this step.

4. *Interpret the Information.* Determining the meaning of the information gathered is the next logical step for the group. In this step, the participants apply the findings to the problem and select the appropriate information for the report. In doing so, they also give meaning to the facts collected. The facts do not speak for themselves. Rather, group participants must think through the facts, identify their significance, and interpret them from the readers' points of view.

5. *Organize the Material.* Just as in any other report-writing project, the group next organizes the material selected for presentation. They will base the report's structure on time, place, quantity, factor, or other relationships in the data.

from the tech desk

Comment and Review Tools Help Writers Track Changes to Their Documents

The commenting and reviewing tools in most word processors help people work together on documents asynchronously. When others review content and edit your document digitally, the commenting tool allows them to express opinions and concerns while the tracking tool makes their editing changes clearly visible. The tools allow you to accept or reject their suggestions individually or all at once.

In the example shown here, the reviewer clicked Word 2013's Review tab (circled in red) to reveal the commenting and reviewing tools. By clicking "Track Changes" (highlighted in blue on the toolbar), the reviewer had Word keep track of all changes being made to the document as well as any comments being added. Word's tracking system allows reviewers to use a variety of colors so that others can easily determine whom the changes belong to. The commenting tool identifies each reviewer, too. Clicking on "Reviewers" in the dropdown menu will show which people have reviewed the document.

When the writer opens the reviewed document, he or she will see all the comments and edits. The "Accept" and "Reject" options (circled in dark blue), which also appear when you right-click on a given edit, enable the writer to keep the desired changes and remove the rest.

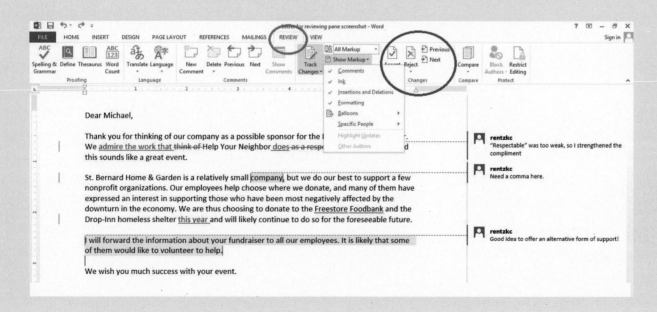

6. *Plan the Report's Components and Style.* A next logical step is planning the makeup of the report. In this step the formality of the situation, the anticipated length of the report, and the intended audience need to be considered (see Chapter 9). In addition, the team needs to agree on such matters as the report's style, the kinds of headings to use, and whether to use the present- or past-time viewpoint.

7. *Assign Parts to Be Written.* After the planning has been done, the group next turns its attention to the writing. The usual practice is to assign each person a part of the report.

8. *Write Parts Assigned.* Following comes a period of individual work. Each participant writes his or her part. Each should apply the ideas in Chapter 4 about word selection, sentence design, paragraph construction, and tone.

9. *Revise Collaboratively.* The group meets and reviews each person's contribution and the full report. This should be a give-and-take session with each person actively participating. It requires courteous but meaningful criticisms. It also requires that the participants be open minded, remembering that the goal is to construct the best possible document.

10. *Edit the Final Draft.* After the group has done its work, one member is usually assigned the task of editing the final draft. This editor gives the document a consistent style and serves as the major proofreader. However, since the document reflects on all members, they should assist with the final proofreading.

If all the work has been done with care and diligence, this final draft should be a report better than anyone in the group could have prepared alone.

In a sense, though, you will never write any report by yourself, even if you are the sole author. You will likely have input from others (including the person who requested the report), and you can consult many different kinds of resources. If you analyze the problem carefully, aggressively seek the information you need, and take pride in preparing a well-structured, well-written product, you can meet any report-writing challenge successfully.

get ready for your report-writing challenges!

- What are five rules for writing better business reports?
- What are some effective headings for different types of reports?
- Where can you find resources for a report on planning an effective email campaign?
- How should you conduct a research interview?

Scan the QR code with your smartphone or use your Web browser to find out at www.mhhe.com/RentzM3e. Choose Chapter 8 > Bizcom Tools & Tips. While you're there, you can view a chapter summary, exercises, PPT slides, and more to get ready for your report-writing challenges.

www.mhhe.com/RentzM3e

Writing **Short Reports**

chapter nine

With a general understanding of what reports do, how to research them, and how to write them, you are ready to consider the many varieties of reports. We focus in this chapter on the shorter forms—the reports that enable much of any organization's work.

While we describe several common categories of short reports, we do not cover every conceivable type. This would not be possible given the hundreds or even thousands of different kinds of reports that companies produce. The specific requirements for any report you will write will come from the specific situation in which you'll be writing it.

But the guidelines in this and the preceding chapter can help you meet any report-writing challenge. Just remember to learn all you can about your readers, their needs, and their expectations. If possible, find a report like the one you're about to write. Then use your good business-writing judgment to prepare an attractive, easy-to-read report that delivers exactly what your readers need. ■

LEARNING OBJECTIVES

LO 9-1 Explain the makeup of reports relative to length and formality.

LO 9-2 Discuss the four main ways that the writing in short reports differs from the writing in long reports.

LO 9-3 Choose an appropriate form for short reports.

LO 9-4 Adapt the procedures for writing short reports to routine operational reports, progress reports, and problem-solving reports, as well as to minutes of meetings.

workplace scenario

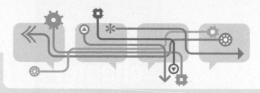

Preparing Different Types of Business Reports

You are again an operations manager at Technisoft (as introduced in Chapter 8). Writing reports is a significant part of your job. Most of the time, these reports concern routine topics, such as human resource policies, administrative procedures, and work flow. Following established company practice, you write these reports in simple email form.

Occasionally, however, you have a more involved assignment. Last week, for example, you investigated a decline in performance of one of the development teams. Because your report on this investigation was written for the benefit of ranking company administrators, you used a more formal style and format.

Then there was the report you helped prepare for the board of directors last fall. That report summarized the environmental impact of the company's operations. A number of executives contributed to this project, but you were the coordinator. Because the report was important and was written for the board, you made it as formal as possible.

Clearly, reports vary widely. This chapter will help you determine your reports' makeup, style, form, and contents. It will then focus on the types of short reports that are likely to figure in your business-writing future.

["The more complex the problem and the more formal the situation, the more elaborate the report is likely to be."]

LO 9-1 Explain the makeup of reports relative to length and formality.

AN OVERVIEW OF REPORT COMPONENTS

As you prepare to write any report, you will need to decide on its makeup. Will it be a simple email? Will it be a long, complex, and formal report? Or will it fall between these extremes?

To a great extent, your decisions will be based on the report's anticipated length and formality. The more complex the problem and the more formal the situation, the more elaborate the report is likely to be. Conversely, less complex problems and less formal situations will require less elaborate reports. Adjusting your report's form and contents based on its likely length and formality will help you meet the reader's needs in each situation.

In the subsections that follow, we first explain how to decide which components to use for a given report. We then briefly review the purpose and contents of each of these components.

The Report Classification Plan

The diagram in Exhibit 9-1 can help you construct reports that fit your specific need. At the top of the "stairway" are the most **formal reports**. Such reports have a number of pages that come before the report itself, just as this book has pages that come before the first chapter. Typically, these **prefatory pages**, as they are called, are included when the situation is formal and the report is long. The exact makeup of the prefatory pages may vary, but the most common parts, in this order, are title fly, title page, letter of transmittal, table of contents, and executive summary. Flyleaves (blank pages at the beginning and end that protect the report) also may be included.

As the need for formality decreases and the problem becomes smaller, the makeup of the report changes. Although the changes that occur are far from standardized, they follow a general order. First, the title fly drops out. This page contains only the report title, which also appears on the next page. Since the title fly is used primarily for reasons of formality, it is the first component to go.

On the next level of formality, the executive summary and the letter of transmittal are combined. When this stage is reached, the report problem is simple enough to be summarized in a short space. As shown in Exhibit 9-1, the report at this stage has three prefatory parts: title page, table of contents, and combined transmittal letter and executive summary.

At the fourth step, the table of contents drops out. Another step down, as formality and length requirements continue to decrease, the combined letter of transmittal and executive summary drops out. Thus, the report commonly called the **short report** now has only a title page and the report text. The title page remains

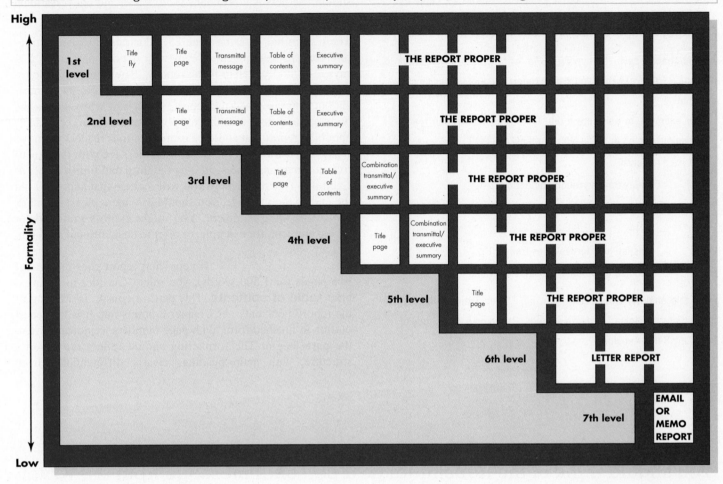

to the last because it serves as a useful cover page. In addition, it contains the most important identifying information.

Below the short-report form is a form that presents the information as a **letter report**. And finally, for short problems of less formality, the **email** or **memo** form is used.

This is a general analysis of how reports are adapted to the problem and situation. While it won't cover every report, it can be relied upon for most reports you will write.

The Report Components

To be able to decide which parts of a long, formal report to include in your reports, you need a basic understanding of each part. This section describes the different report components represented in Exhibit 9-1. The sample long report in Bonus Chapter D contains examples of these.

title pages The first two pages of a long, formal report—the **title fly** and **title page**—contain identification information.

As we have said, the title fly contains only the report title; it is included simply to give a report the most formal appearance. The title page, as illustrated on page 256, is more informative. It typically contains the title, identification of the writer and reader, and the date.

Although constructing title pages is easy, composing the title is not. In fact, on a per-word basis, the title requires more time than any other part of the report. A good title efficiently and precisely covers the contents. Consider building your title around the five Ws: *who, what, where, when,* and *why* (see the Communication Matters feature on the next page). Sometimes *how* may be important as well. You may not need to use all the Ws, but they can help you check the completeness of your title. Remember that a good title is concise as well as complete; be careful not to make your title so long that it is hard to understand or so short that it is not meaningful. A subtitle can help you be both concise and complete, as in this example: "Employee Morale at Florida Human Resource Offices: Results from a 2013 Survey."

communication matters

Creating a Report Title with the 5 Ws and 1 H

As this chapter says, the five Ws (*who, what, where, when, why*) and one H (*how*) can help you craft a report title that is precise and informative.

For example, to generate a title for a recommendation report about sales training at Nokia, you might ask yourself . . .

Who?	Nokia
What?	Sales training recommendations
Where?	Implied (Nokia regional offices)
When?	2014
Why?	Implied (to improve sales training)
How?	Studied the company's sales activities

From this analysis would come the title "Sales Training Recommendations for Nokia," with the subtitle "Based on a 2014 Study of Company Sales Activities."

with the recipient. Except in cases of extreme formality, you should use personal pronouns (*you, I, we*) and conversational language.

The transmittal letter on page 257 illustrates the usual structure for this component. Begin with a brief paragraph that says, essentially, "Here is the report." Briefly identify the report's contents and purpose and, if appropriate, its authorization (who assigned the report, when, and why). Focus the body of the message on the key points of the report or on facts about the report that could be useful for your readers to know. If you are combining the transmittal message with the executive summary, as represented by the third and fourth levels of Exhibit 9-1, here is where you will include that summary. At the end of the message, you should provide a pleasant and/or forward-looking comment. You might express gratitude for the assignment, for example, or offer to do additional research.

table of contents If your short report goes much over five pages (or 1,500 words), you might consider including a brief **table of contents**. This part, of course, is a listing of the report's contents. As Chapter 8 points out, it is the report outline in finished form, with page numbers to indicate where the parts begin. The formatting should reflect the report's structure, with main headings clearly differentiated from

> You should think of the transmittal as a personal message from the writer to the reader, with much the same contents you would use if you were handing the report over in a face-to-face meeting with the recipient.

In addition to displaying the report title, the title page identifies the recipient and the writer (and usually their titles and company names). The title page also contains the date unless it is already in the title of the report. You can find attractive designs for title pages in the report templates available on the Internet (see From the Tech Desk, page 254).

transmittal message As the label implies, the **transmittal message** is a message that transmits the report to the reader. In formal situations, it usually takes letter form. In less formal situations (e.g., when delivering a report to internal readers whom you know fairly well), the report can be transmitted orally or by email. Whatever the case, you should think of the transmittal as a personal message from the writer to the reader, with much the same contents you would use if you were handing the report over in a face-to-face meeting

subheadings. The section titles should state each part's contents clearly and match the report's headings exactly. The table of contents may also include a list of illustrations (or, if long, this list can stand alone). If a separate table of contents would be too formal, you can use the introduction of your report to list the topics the report will cover.

executive summary The **executive summary** is the report in miniature. For some readers it serves as a preview to the report, but for others—such as busy executives who may not have time to read the whole report—it's the only part of the report they will read. Because of this latter group of readers, the summary should be self-explanatory; that is, readers shouldn't have to read other parts of the report in order to make sense of the summary. As pointed out previously, whether the executive summary is one of the prefatory parts, is included in the transmittal message, or is part of the report proper depends on how long and how formal the report is.

You construct the executive summary by summarizing the parts of the report in order and in proportion. You should clearly identify the topic, purpose, and origin of the report; state at least briefly what kind of research was conducted; present the key facts, findings, and analysis; and state the main conclusions and recommendations. If you include these parts in this order, which usually matches the order of the report contents, your summary will be written in the indirect order. But sometimes writers use the direct order by starting with the conclusions and recommendations and then continuing with the other information. Exhibit 9-2 shows the difference between these two structures, and Exhibit 9-3 gives examples. Whichever order you choose, the executive summary will need to be a masterpiece of economical writing.

It may be desirable to include other report components not discussed here—for example, a copy of the message that authorized the report, various appendices containing supplementary material, a glossary, or a bibliography. As with any writing task, you will need to decide what parts to provide given the facts of the situation and your readers' preferences.

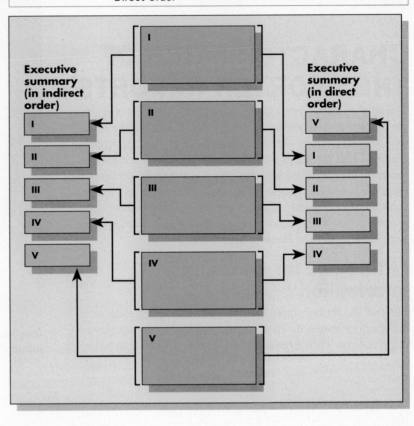

▼EXHIBIT 9-2   Diagram of the Executive Summary in Indirect and Direct Order

▼EXHIBIT 9-3   Example of an Executive Summary in Direct and Indirect Order

[Direct]

EXECUTIVE SUMMARY

To enhance the performance of Nokia's salespeople, this report recommends adding the following topics to Nokia's sales training program:

- Negative effects of idle time
- Projection of integrity
- Use of moderate persuasion
- Value of product knowledge

Supporting these recommendations are the findings and conclusions drawn from a five-day observational study of 20 productive and 20 underperforming salespeople. The study also included an exit interview and a test of the salesperson's product knowledge.

The data show that the productive salespeople used their time more effectively than did the underperforming salespeople. Compared with the latter, the productive salespeople spent less time being idle (28% vs. 53%). They also spent more time in contact with prospects (31.3% vs. 19.8%) and more time developing prospects (10.4% vs. 4.4%).

Observations of sales presentations revealed that productive salespeople displayed higher integrity, used pressure more reasonably, and knew the product better than underperforming salespeople. Of the 20 productive salespeople, 16 displayed images of moderately high integrity (Group II). Underperforming group members ranged widely, with 7 in Group III (questionable) and 5 each in Group II (moderately high integrity) and Group IV (deceitful). Most (15) of the productive salespeople used moderate pressure, whereas the underperforming salespeople tended toward extremes (10 high pressure, 7 low pressure). On the product knowledge test, 17 of the productive salespeople scored excellent and 3 fair. In the other group, 5 scored excellent, 6 fair, and 9 inadequate.

[Indirect]

EXECUTIVE SUMMARY

Midwestern Research associates was contracted to study the performance of Nokia's salespeople. A team of two researchers observed 20 productive and 20 underperforming salespeople over five working days. The study also included an exit interview and a test of the salesperson's product knowledge.

The data show that the productive salespeople used their time more effectively than did the underperforming salespeople. Compared with the latter, the productive salespeople spent less time in idleness (28% vs. 53%). They also spent more time in contact with prospects (31.3% vs. 19.8%) and more time developing prospects (10.4% vs. 4.4%).

Observations of sales presentations revealed that productive salespeople displayed higher integrity, used pressure more reasonably, and knew the product better than underperforming salespeople. Of the 20 productive salespeople, 16 displayed images of moderately high integrity (Group II). Members of the underperforming group ranged widely, with 7 in Group III (questionable) and 5 each in Group II (moderately high integrity) and Group IV (deceitful). Most (15) of the productive salespeople used moderate pressure, whereas the underperforming salespeople tended toward extremes (10 high pressure, 7 low pressure). On the product knowledge test, 17 of the productive salespeople scored excellent and 3 fair. In the other group, 5 scored excellent, 6 fair, and 9 inadequate.

On the basis of these findings, this report recommends adding the following topics to Nokia's sales training program:

- Negative effects of idle time
- Projection of integrity
- Use of moderate persuasion
- Value of product knowledge

CHARACTERISTICS OF THE SHORTER REPORTS

The shorter report forms (those at the bottom of the stairway) are by far the most common in business. These are the every-day working reports—those used for the routine information reporting that is vital to an organization's communication. Because these reports are so common, we devote this chapter to them. This section describes four general differences between short and long reports. You can see these differences yourself by comparing the sample short reports in this chapter to the sample long report in Bonus Chapter D.

Little Need for Introductory Information

Most of the shorter, more informal reports require little or no introductory material. These reports typically concern day-to-day problems. Their lives are short; they are not likely to be kept

Many routine reports, such as inventory reports and expense reports, are submitted on smartphones and other hand-held devices.

> **Because shorter reports usually solve routine problems, they are likely to be written in the direct order.**

on file very long. They are intended for only a few readers, and these readers are likely to understand their context and purpose. If readers do need an introduction, a brief one usually suffices.

Determining what introductory material to provide is simply a matter of answering one question: What does my reader need to know before reading the information in this report? In very short reports, an incidental reference to the problem or to the authorization of the investigation will be sufficient. In some cases, however, you may need a detailed introduction comparable to that of the more formal reports.

Reports need no introductory material if their very nature explains their purpose. This holds true for personnel actions, weekly sales reports, inventory reports, and some progress reports.

Predominance of the Direct Order

Because shorter reports usually solve routine problems, they are likely to be written in the direct order. That is, the report will begin with its most important information—usually the conclusion and perhaps a recommendation. Business writers use this order because they know that busy readers typically want the key point quickly.

The form that the direct order takes in longer reports is somewhat different. The main findings will be somewhere up front—either in the letter of transmittal, executive summary, or both—but the report itself may be organized indirectly. The introduction will present the topic and purpose of the report, but the actual findings will be brought out in the body sections, and their fullest statement will usually appear in the conclusions or recommendations section.

As you move down the structural ladder toward the more informal and shorter reports, however, the need for the direct order in the report itself increases. At the bottom of the ladder, the direct order is more the rule than the exception.

Illustrating the direct arrangement is the following beginning of a report on a personnel issue:

> The hiring committee recommends appointing Sue Breen as our new Corporate Communications Officer.
>
> We interviewed three candidates for the position . . . [*The rest of this paragraph describes the candidates.*]

While all three candidates had strengths, Ms. Breen emerged as the top candidate, for these reasons:

- She was the most experienced of the three candidates, with 27 years' experience in corporate communication.

- She had the most expertise with the widest variety of communication media, from annual reports to blog posts to intranets and social networking, and she understood the advantages and disadvantages of each.

- She has an impressive track record. At Gemini Web Conferencing, she launched a corporate communications program that . . . [*The report continues to make the case and then reiterates its recommendation at the end.*]

As you can see, this report states its main point first and then supplies the supporting information.

In contrast, a report written in the indirect order presents the supporting information before stating its main conclusion or recommendation.

Using the personnel issue from the last example, the indirect arrangement would appear like this:

The hiring committee interviewed three candidates for the position of Corporate Communications Officer: . . . [*The opening paragraph briefly describes the candidates.*]

While all three candidates had strengths, Ms. Breen emerged as the top candidate, for these reasons:

- She was the most experienced of the three candidates, with 27 years' experience in corporate communication.

- She had the most expertise with the widest variety of communication media, from annual reports to blog posts to intranets and social networking, and she understood the advantages and disadvantages of each.

"This has so many different fonts in it, I thought it was a ransom note."

communication matters

Are Tweets, Blog Comments, and Text Messages Undermining Your Report-Writing Skills?

If you are like most college students, you do a lot of short messaging, whether in the form of Tweets, online comments, or text messages. Is this heavy use of brief communications making it more difficult for you to write an effective report?

Researchers disagree on whether brief messaging has undermined students' writing skills. Some argue that it has impaired students' use of correct grammar, while others argue that students have become more comfortable with writing because they're writing more than ever.

One thing is certain, though: The work that goes into an effective report, even a short one, is very different from what's required for an effective short message. If most of what you write takes the form of a brief comment, you will need to make a special effort to be patient and deliberate when tackling a longer project.

Check out the "15 Cures for Bad Writing in the Twitter Age" at Forbes.com, and take the report-writing advice in this book seriously. An employee who can write an on-target, clear report is more valuable, and more promotable, than one who can't.

Source: Brett Nelson, "15 Cures for Bad Writing in the Twitter Age," *Forbes*, Forbes .com LLC, 21 Dec. 2012, Web, 12 June 2013.

- She has an impressive track record. At Gemini Web Conferencing, she launched a corporate communications program that . . . [*The list continues to make the case.*]

In light of these assets, we recommend that Sue Breen be appointed as our new Corporate Communications Officer.

Deciding whether to use the direct order is best based on a consideration of your readers' likely use of the report. If your readers need the report conclusion or recommendation as a basis for an action that they must take, directness will speed their effort by enabling them to quickly receive the most important information. If they have confidence in your work, they may choose to skim or not even read the rest of the report before acting on your information. Should they desire to question any part of the report, however, the material is there for their inspection.

On the other hand, if there is reason to believe that it would be better for your readers to arrive at the conclusion or recommendation only after a logical review of the analysis, you should organize your report in the indirect order. This arrangement is

from the tech desk

Using a Report Template for a Polished Look

When preparing a report, consider using a pre-designed template to give your report a professional, consistent design.

In Word 2013, as in Word 2010, you can access the available report templates by clicking File > New and entering *report* in the "Search for online templates" field. When you find a template you like, such as the one shown here, you can download it to your computer for your current and future use.

If you're not pleased with the color scheme, you can click Design > Themes (in Word 2010, Page Layout > Themes) to select a different palette of colors, as shown. If you like, you can also change the fonts and margins.

For most business reports, you'll want to choose a relatively conservative design like the one shown here. The more visually elaborate designs are better for special publications, such as annual reports and sales proposals.

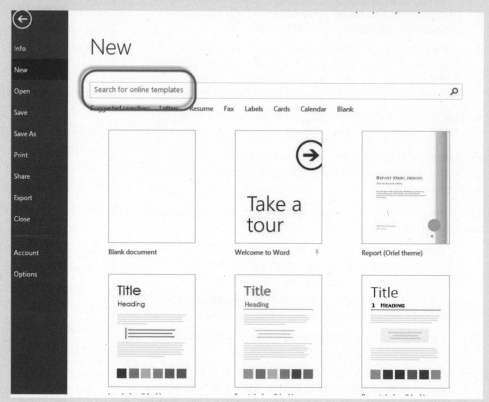

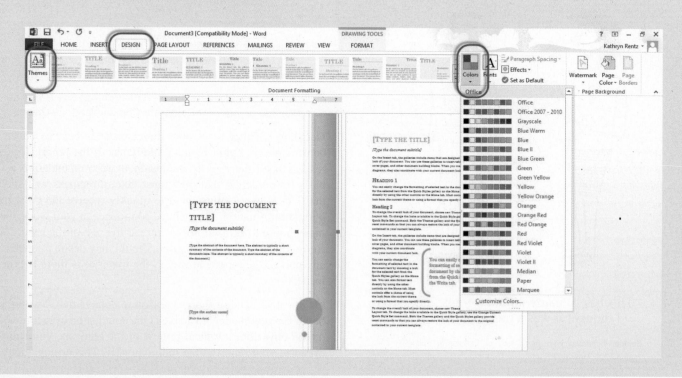

especially preferable when you will be recommending something that you know your readers will not favor or want to hear. For example, in the preceding illustration, if you suspect that one of the executives to whom you're making your hiring recommendation prefers another candidate to Sue Breen, you'll want to make your case before stating the committee's decision. Presenting the supporting data before the recommendation prepares resistant readers to accept your solution to the report problem.

A More Personal Writing Style

The writing in shorter reports tends to be more personal than in long reports. That is, the shorter reports are likely to use the personal pronouns *I, we,* and *you* rather than only the third person.

Several factors account for this tendency toward personal writing in shorter reports. First, short-report situations usually involve personal relationships. Such reports tend to be from and to people who know each other and who normally address each other informally when they meet. In addition, shorter reports are apt to involve personal investigations and to represent the observations, evaluations, and analyses of their writers. Finally, shorter reports tend to deal with day-to-day problems. These problems are informal by their very nature. It is logical to report them informally, and personal writing tends to produce this informal effect.

As explained in Chapter 8, your decision about whether to write a report in personal or impersonal style should be based on the situation. Convention favors using impersonal writing for the most formal situations. For most short reports, personal writing is likely to be preferable because of their relatively routine nature.

Less Need for a Structured Coherence Plan

A long, formal report usually needs what we call a "structured coherence plan"—a network of introductions, conclusions, and transitions that guides the reader through the report (see Exhibit D-1 in Bonus Chapter D). Creating such a plan means giving the report an overview and a conclusion, providing the same for the individual sections, and incorporating transitions that bridge each section to the next. Such devices enable the reader to know at every point where he or she is in the report and how the current section is related to the overall goal of the report.

Short reports, because they are short, generally do not need an elaborate coherence plan. Readers will not need many reminders of what they just read or previews of what they're about to read. The report introduction (which should contain an overview), clear headings, and brief transitional devices (such as "Second," "next," and quick references to previous points) will usually be sufficient to keep readers on track.

LO 9-3 Choose an appropriate form for short reports.

FORMS FOR SHORT TO MID-LENGTH REPORTS

As noted earlier, the shorter report forms are by far the most numerous and important in business. In fact, the three forms represented by the bottom three steps of the stairway in Exhibit 9-1—short reports, letter reports, and email or memo reports—make up the bulk of written reports.

The Short Report

One of the more popular of the less formal report forms is the short report. Representing the fourth and fifth steps in the formality stairway, this report consists of only a title page and text or a title page, combined transmittal message/summary, and text. Its popularity may be explained by the middle-ground impression of formality that it conveys. The short report is ideally suited for the short but somewhat formal problem.

Like most of the less formal report forms, the short report may be organized in either the direct or indirect order. If the report is addressed to an internal audience, it will likely use the direct order unless there is reason to believe that readers will resist the conclusions or recommendations. Reports written to external audiences may or may not state the main conclusions or recommendations in the opening paragraphs, but it is customary to include these in the transmittal message since it doubles as an executive summary.

When you open your short report with the main results of your investigation, the next section usually provides background

Though missing certain components of long, formal reports, short reports require many of the same analytical and organizational skills used to develop longer reports.

This report, with its title page and combined letter of transmittal and executive summary, would fall on the fourth level of Exhibit 9-1. It is organized indirectly in order to prepare the reader for the students' recommendations.

Increasing Student Patronage at Kirby's Grocery

Title makes the topic and purpose of the report clear

May 28, 2014

Prepared for:

Mr. Claude Douglas, Owner
Kirby's Grocery
38 Lance Avenue
Crestview, IN 45771

Page uses an attractive but simple template

Prepared by:

Kirsten Brantley, Business Communication Student
College of Business
P. O. Box 236
Metropolitan University
Crestview, IN 45770-0236

METROPOLITAN UNIVERSITY

College of Business, P. O. Box 236, Crestview, IN 45770-0236
Phone: (421) 555-5555, Fax: (421) 555-5566

May 28, 2014

Mr. Claude Douglas
Kirby's Grocery
38 Lance Avenue
Crestview, IN 45771

Dear Mr. Douglas:

As you requested, our Business Communication class conducted a study to
determine ways to increase Metropolitan University students' awareness of Kirby's
Grocery and to attract more students to your store. This report presents the results.

To gather our information, we first interviewed assistant manager Bradley Vostick,
who leads the store's marketing efforts. To get a veteran customer's perspective, we
also interviewed our professor, Beth Rawson, a long-time Kirby's customer. Next
we researched MU campus events, publications, transportation issues,
demographics, and the Lance Avenue area surrounding Kirby's. Finally, the class did
walkthroughs of Kirby's to gain firsthand reactions to the store as well as quantified
data in the form of an exit survey on overall reactions to Kirby's.

We found that Kirby's is part of a niche market that offers a wide variety in a small
space, much like the surrounding Lance Avenue area. Given these findings, we
recommend the following:

- Targeting health-conscious, older MU students who enjoy shopping.
- Adding Facebook to your current promotional strategies.
- Focusing on your strengths to make these shoppers aware of the experience
  that is Kirby's Grocery.

Thank you for allowing us the opportunity to do this real-world project. We enjoyed
learning about your store and hope that our research will bring many more MU
students to Kirby's Grocery.

Sincerely,

Kirsten Brantley

Kirsten Brantley
For Professor Beth Rawson's Business Communication Class

Identifies the project and delivers the report

Combines letter of transmittal and executive summary

Ends with goodwill comments

Increasing Student Patronage at Kirby's Grocery

Introduction

Kirby's Grocery is a full-service neighborhood grocery store in the vicinity of Metropolitan University. With its appealing range of products and proximity to the university, its potential to attract student customers is great. Yet according to a survey conducted by Professor Beth Rawson's Fall 2013 Business Communication class, only one in three MU students has ever shopped at Kirby's. As a follow-up to this study, our Business Communication class conducted research to determine how Kirby's might attract more MU student shoppers. This report presents our results and recommendations.

Opening gives the context and purpose in a nutshell

Research Methods

The research for this study was conducted in three phases:

- Phase One: As preparation for the observational part of the study, the class gathered supplemental information about a variety of topics related to Kirby's. It was carried out by groups of three to four students, with each group focusing on one of seven particular "beats." These beats included MU campus events and publications, MU's demographic information, and interviews with Kirby's customers and Bradley Vostick of Kirby's.

- Phase Two: This was the main phase of our research. For this phase, 13 pairs of students visited Kirby's during the week of March 7–14, 2014, to gather observational data. Each pair consisted of an observer, who made oral comments about any and all aspects of the store, and a recorder, who recorded these observations. On average, each pair spent approximately 40 minutes in the store, and each pair was required to make a small purchase. At the end of their visits, the observers all completed exit surveys to quantify their overall reactions to Kirby's and to provide some demographic information about themselves.

Detailed description of research builds confidence in the validity of the findings

- Phase Three: Given our findings, we gathered information to develop our recommendations for marketing Kirby's to MU students.

The following sections describe the study participants, present the observational data, and offer our recommendations.

Preview adds to report's coherence

Demographics of the Participants

Paragraph leads into the table

While our observer number of 13 was very small in comparison to the entire MU population of 33,000, it was actually a fairly representative sample in terms of student diversity, gender, and age. The following table shows how these observers compared to the general MU population.

Lone table in nonacademic report does not need to be numbered

Demographics of Store Observers		
	Study Participants	**MU Population**
Diversity	77% (11) European American 23% (2) African American 77% (11) US citizens	71.5% European American 12% African American 83% US citizens
Gender	47% (6) female, 54% (7) male	54.2% female, 45.8% male
Average Age	24	23

Report includes a special section to further support the validity of the findings

Paragraphs interpret and elaborate on the table

Compared to the MU population, our sample of 13 was relatively diverse. In addition to including 2 African American participants, it included 2 non-U.S. participants, one from Russia and one from Sweden.

The gender ratios were also relatively close. The MU population is 54.2% female to 45.8% male. Again, our observers came very close to this ratio, with 47% female and 54% male.

The average age for full-time students at Metropolitan University is 23. The average age of our observers was 24. This included a student who is 41 years old, but even without this outlier, our group closely represented the average MU student in terms of age.

In addition, all students in our class (26) were juniors and seniors at MU, and most were business students with some background in marketing. Through class discussion, we were able to bring our collective perspective as MU students to bear on our observations.

Qualitative Findings

Section preview adds coherence

This section presents the qualitative results from our observational research, broken down into two categories: perceived strengths and perceived weaknesses.

2

Perceived Areas of Strength

Three main positive reactions came out of this research:

- Students were impressed with the wide product variety.
- Students were happy to see organic and health food products.
- Kirby's employees provided excellent service to their customers.

Product variety was seen as the strongest asset of Kirby's. Of the 13 observation surveys, 11 mentioned the large variety of products offered by Kirby's as a positive aspect of the store. Overall, the consensus was that the selection offered by Kirby's in such a small space was impressive. This was especially apparent in the beer aisle, which received the most praise of any section in Kirby's for a selection that rivals that of specialty stores.

Organic and health food products also received a large positive reaction, being mentioned by almost two-thirds of the observers. While health may not be the main concern of the stereotypical college student, it is important to older students. Since the average age of an MU student is 23, it is likely that Kirby's organic and health food selection can also be a strong selling point for getting more students into the store.

Customer service was the third most mentioned positive aspect of Kirby's. During observations, employees at Kirby's were consistently happy and helpful. The long-time customer we interviewed, Professor Rawson, cited Kirby's excellent customer service as one of the reasons she continues to shop there. She commented that the employees of Kirby's care more about their customers' shopping experiences than employees of other stores, and the data compiled by the class supported this claim.

Not only were students asked if they needed help, but they also observed that employees knew their customers, which makes the customer feel more like family than just another person wanting groceries. When observers made their small purchases, they noted that they were treated in the same friendly manner as customers who were checking out with filled carts.

Perceived Areas of Weakness

Two main negative reactions came out of this research:

- Students observed a number of dirty shelves and floors.
- The placement of some products was confusing.

About half of the observers made note of areas of Kirby's that seemed to need a good cleaning, with specific remarks about stained floor tiles, produce on the floor

Section starts with a helpful summary

Paragraphs present and interpret data

Another helpful summary sets up the discussion

3

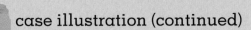

that hadn't been swept up, and dusty shelves. This was a significant negative for several student shoppers, as dirty floors and shelves don't shed the kindest light on the products for sale, especially when compared to the seeming sterility of larger grocery stores like Kroger and Biggs.

But the largest negative reaction to Kirby's concerned the random-seeming placement of a variety of products. Examples included cakes next to whole turkeys, freezers next to greeting cards, and hot peppers next to candies. This confused student shoppers because the placement of some of these products did not fall in line with the aisle signs. While Kirby's is a smaller grocery store and is certainly impacted by the limits of space, the class as a whole felt that more could be done to eliminate this random product layout.

More helpful data and interpretation make the writer's point

Quantitative Findings

The following figure contains the composite results of the 13 exit surveys. Students were asked to rate their Kirby's experience on a variety of topics on a scale of one to five, with one being the worst and five being the best. The observers took the survey immediately after completing their walkthroughs of the store.

Paragraph introduces the figure

Lone figure in a nonacademic report does not need to be numbered

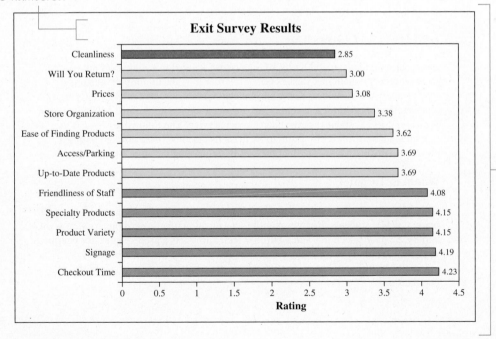

Visual, textual, and numerical elements work together to present specific findings clearly

4

These results simply quantify what was discovered in the qualitative phase of the observation. Checkout time, friendliness, variety, and specialty products were the highest scorers, while cleanliness received the lowest rating. Still, Kirby's scored above average in all categories, which is an important positive result.

Along these lines, the "Will You Return?" score is encouraging. Though it may seem negative when compared to the other reactions, a score of 3.00 is actually above average. This means that with only one trip through Kirby's, there is a better than average chance that students will return to purchase groceries.

Paragraphs help the reader interpret the findings

Prices ranked just above "Will You Return?" at 3.08. Again, its comparative rating is low, but it is still above average. Price is an important consideration for student shoppers, and Kirby's has a tough time competing with prices offered by the nearby Kroger store, its main competitor. Even given these two facts, though, observers felt that Kirby's had slightly better than average prices, which is another important positive result.

Interesting to note are the scores for Signage, Ease of Finding Products, and Store Organization. Signage was rated high at 4.19, while Ease of Finding Products was rated lower (3.62), and Store Organization even lower (3.38). These numbers correlate with the student shopper reactions. While Kirby's has excellent signs, some of the aisles are difficult to find because of displays blocking them, and the placement of certain products adjacent to other unrelated products only increases this confusion.

Recommendations

Paragraph summarizes the main impression and leads into recommendations

From our compiled data on Kirby's Grocery, we learned that Kirby's is not a large, faceless grocery chain that slashes prices in order to make up for poor customer service and variety. Rather, Kirby's is a niche-market store that offers wide variety and excellent service for a decent price. Yet only about a third of the MU population is aware of its excellent variety and customer service. The following recommendations for Kirby's aim to increase its customer base through targeted awareness-raising advertising that focuses on promoting the store's uniqueness.

Whom to Target

Since Kirby's is something of a niche-market store, it will only truly appeal to a niche market of MU students. Kirby's should focus on attracting those shoppers who want variety, healthy selections, and a familiar feel. Kirby's is a grocery store for people who want shopping to be an experience, not an errand. It can therefore be especially appealing to the more mature, somewhat alternative segment of the MU population.

5

How to Reach Them

Recent research supports the claim

While Kirby's has been using the MU Coupon Book for campus ads and participating in campus events such as Welcome Week, the store's marketing efforts do not take advantage of a key fact about modern students: They are heavy users of the social networking site Facebook. According to a poll conducted by researchers at Harvard University, 80% of all 18- to 29-year-olds and 90% of four-year college students now have a Facebook account. Facebook thus represents a huge opportunity for Kirby's to reach its student market.

Sentence shows knowledge of the store's current efforts

Informal citation of sources is acceptable in an informal report

You can take advantage of this site in two main ways: creating a page and advertising.

Creating a Free Online Profile

Link makes it easy to learn more

Kirby's currently has an attractive Web site, but it is unlikely to receive many student visitors. Creating an appealing presence on Facebook (*www.facebook.com*) can help you build a student fan base and generate a buzz about your store. Once you register at the site, Facebook makes it easy to create your "page." Here you can feature major selling points, photos, directions, a link to your Web site, and other static material. The page can also feature such dynamic material as posts to your "wall" (an interactive message board), polls, quizzes, and special promotions.

Once the page is posted, users can learn about your site using Facebook's search engine or any other search engine. If they like what they see and want to be kept updated about special offerings, they can click a link to include Kirby's in their network. This will place your logo and name on their own Facebook pages, where all their friends will see it and perhaps consider becoming "fans" of Kirby's themselves. In this way, your store can take advantage of "viral marketing."

Using Paid Advertising

Paragraphs adapt research to reader's needs

It is also possible to advertise on Facebook. You can go to *www.facebook.com/Ads* to get advice on creating ads and to create your ads. Facebook allows you to target your ads to certain readers, so you could target your ads to MU students with particular interests. You can also track the success of several ads to see which are the most successful.

The cost for these ads varies depending on whether you choose to pay per view (how many times users see your ad) or per click (how many times users click on your Facebook page), as well as on what other advertisers targeting the same market are offering to pay. Using the Facebook help page "Campaign Cost and Pricing," you can indicate how much you are willing to spend each day, and Facebook will calculate a "cost per click" estimate for you. If few other businesses are competing for the same users, the cost may actually be lower than this, whereas it will be higher if many businesses are trying to reach the same demographic.

6

What to Say

Kirby's advertisements need to focus on promoting the store's variety and uniqueness to the MU student body. It has to dispel the idea that it is simply another grocery store and emphasize that it encompasses a wide variety in a small space, just like the eclectic Lance Avenue area in which it is located.

Promotions for students should focus on its wide beer selection (perhaps mentioning the exact number of domestic and imported brands offered), on its organic and health food selection, and on its unique products such as sushi and fresh peanut butter. These advertisements will not appeal to the entire MU student body, but it will make the students who are likely to value Kirby's strengths aware of these and give them a desirable, convenient alternative to shopping at a giant superstore.

Paragraphs use the main findings to suggest a marketing strategy

Conclusion

We found that Kirby's Grocery offers a shopping experience that simply cannot be found at larger grocery stores. The variety and customer service are top notch, and this makes a trip to Kirby's an experience rather than an errand (perhaps this could be a slogan?). By taking advantage of the large-scale marketing opportunities offered by Facebook, Kirby's can raise awareness of its many positive qualities among the MU student body well beyond what it currently achieves with its current outreach efforts. This strategy is also appealing because it can greatly increase Kirby's visibility at no to very little cost.

Ending wraps up the report in a positive way

7

on the report problem and what you did to investigate it. In other words, it provides your problem and purpose statements, as described in Chapter 8. When your report uses the indirect order, a coherent statement of your problem and purpose open the report, as illustrated by the sample long report in Bonus Chapter D.

Even if you have provided your recommendations up front, you should reiterate and perhaps expand on them at the end of your report. Readers do not mind this kind of redundancy as long as the recommendations are helpfully restated, not just copied and pasted from the front of the report. Plus, stopping short of your main points at the end of your report would end it too abruptly.

The mechanics of constructing the short report are much the same as the mechanics of constructing the more formal, longer types. The short report uses the same form of title page and page layout. Like the longer reports, it uses headings, though usually only one or two levels because of its brevity. Like any other report, the short report uses visuals, an appendix, and a bibliography when these are needed.

Letter Reports

The second of the more common shorter report forms is the letter report—that is, a report in letter form. Letter reports are used primarily to present information to people outside the organization. For example, a company's written evaluation of its experience with a particular product may be presented in letter form and sent to the person who requested it. An outside consultant may deliver his or her analyses and recommendations in letter form. Or the officer of an organization may report certain information to the membership in a letter.

Typically, the length of letter reports is three to four pages or less, but they may be longer or shorter.

As a general rule, letter reports are written personally, using *I, you,* and *we* references. Exceptions exist, of course, such as letter reports for very important readers—for example, a company's board of directors. But since letters are traditionally a personal form of communication, letter reports tend to use the personal style.

Letter reports may use either the direct order or the indirect order. Those in the direct order begin with the main finding or recommendation. Sometimes they use a subject line to announce the topic of the report so the first paragraph of the letter can get right to the main point. The subject line usually appears after the salutation. It commonly begins with the word *subject* followed by a description of the report's contents, as the following example illustrates:

> Subject: Review of Travel Expenditures of Association Members, Authorized by Board of Directors, January 2014

Association members spent 11 percent more on travel to asssociation meetings in 2013 than they did in 2012, and they expect another increase in travel expenses this year.

Indirect-order letters tend not to use a subject line, and they open with brief background information, such as who authorized the report and the topic. A letter report written to the executives of an organization, for example, might use the following indirect opening:

> As authorized by your board of directors January 6, this report reviews member expenditures for travel for 2012 and 2013. It is based on . . . [whatever research was performed].

Following the introduction would be a logical presentation and analysis of the information gathered. After this presentation would come the conclusion or recommendation that the facts lead to.

With either the direct or indirect order, a letter report may close with whatever friendly goodwill comment fits the occasion.

Email and Memo Reports

As we noted in Chapter 2, email is heavily used in business. It has largely eclipsed memos, but memos are still written, especially in cases where computers aren't easily accessible or the writer prefers to deliver a print message. Both email and memos can be used for internal reports—that is, for reports written by and to people in an organization.

Because email and memos are primarily communication between people who know each other, they are usually informal.

> " Letter reports are used primarily to present information to people outside the organization. "

Sometimes it is courteous to deliver a report in both hard copy (to keep your reader from having to print it) and electronic form (so he or she can easily distribute it).

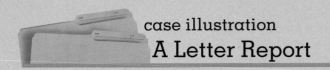

This direct-order letter report compares two hotels for a meeting site. It starts with the recommendation, followed by well-organized supporting facts. The personal style is appropriate in this situation.

To make the letter easy to share with the other board members, the writer might send it to the reader as an email attachment.

INTERNATIONAL COMMUNICATION ASSOCIATION

314 N. Capitol St. NW • Washington, DC 20001 • 202.624.2411

October 26, 2014

Professor Helen Toohey
Board of Directors
International Communication Association
Thunderbird American Graduate School of International Management
15249 N. 59th Ave.
Glendale, AZ 85306-6000

Dear Professor Toohey:

Subject: Recommendation of Convention Hotel for the 2015 Meeting

The Hyatt Hotel is my recommendation for the International Communication Association meeting next October. The Hyatt has significant advantages over the Marriott, the other potential site for the meeting.

One of the Hyatt's main advantages is its downtown location, which will appeal to convention goers and their spouses. The accommodations, including meeting rooms, are adequate in both places, although the Marriott does have a more contemporary decor. But the Hyatt room costs are approximately 15 percent lower than those at the Marriott. Thus, although both hotels are adequate, the Hyatt appears to be the better choice because of its location and cost advantages.

Recommendation and key facts are up front

Bases of comparison (factors) permit hotels (units) to be compared logically

The Hyatt's Favorable Downtown Location

The older of the two hotels, the Hyatt, is located in the heart of the downtown business district, making it convenient to the area's major mall as well as the other downtown shops. The Marriott, on the other hand, is approximately nine blocks from the major shopping area. Located in the periphery of the business and residential area, it provides little location advantage for those wanting to shop. It does, however, have shops within its walls that provide for virtually all of the guests' normal needs. Because many members will bring spouses, however, the downtown location does give the Hyatt an advantage.

Short sentences and transitional words increase readability and move ideas forward

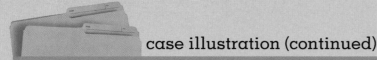

Alternate placement of topic sentences offers variety

Board of Directors   -2-   October 26, 2014

Adequate Accommodations at Both Hotels

Talking headings (all noun phrases) make the key points stand out

Both hotels can guarantee the 600 rooms we will require. Because the Marriott is newer (built in 2006), its furnishings are more modern. The 16-year-old Hyatt, however, is well preserved, and its guest rooms are elegant and comfortable.

The Marriott has 11 small meeting rooms, and the Hyatt has 13. All are adequate for our purposes. Both hotels can provide the nine we need. For our opening session, the Hyatt would make available its Capri Ballroom, which can easily seat our membership. It would also serve as the site of our presidential luncheon. The assembly facilities at the Marriott appear to be somewhat crowded, although the management assures me that their largest meeting room can hold 600. Pillars in the room, however, would make some seats undesirable. In spite of the limitations mentioned, both hotels appear to have adequate facilities for our meeting.

Logical paragraphing helps the different points stand out

Lower Costs at the Hyatt

Both the Hyatt and the Marriott would provide nine rooms for meetings on a complimentary basis. Both would provide complimentary suites for our president and our executive director. The Hyatt, however, would charge $500 for use of the room for the opening session. The Marriott would provide this room without charge.

Analysis is adapted to the readers' needs

Convention rates at the Hyatt are $169 for singles, $179 for double-bedded rooms, and $229 for suites. Comparable rates at the Marriott are $189, $199, and $350. Thus, the savings at the Hyatt would be approximately 15 percent per member.

Cost of the dinner selected would be $35 per person, including gratuities, at the Hyatt. The Marriott would meet this price if we would guarantee 600 plates. Otherwise, they would charge $38. Considering all of these figures, the total cost picture at the Hyatt is the more favorable one.

In conclusion, while both hotels would meet our needs, the significant location and cost advantages of the Hyatt make it the more desirable site for next year's conference.

Repetition of the key point provides a sense of closure

Sincerely,

Willard K. Mitchell

Willard K. Mitchell
Executive Secretary

communication matters

When Is a Report not a Report?

One answer: When it is a *white paper*.

White papers are often categorized as reports because they share many characteristics with reports. They're based on research (many have footnotes or a references page); they present numerical data in tables and graphs; they have a title page and often have a table of contents and/or an executive summary (see the first two pages of the sample below); and they use levels of headings to indicate the document's structure. In other words, they look like reports.

They differ from reports, though, in one crucial way: They are almost always intended to be persuasive rather than objective.

White papers look like problem-solving reports, and in a way they are. They start by identifying a problem that their intended readers—usually other businesses—are likely to have. For example, a white paper may discuss the need for effective waste management or for an effective check-out system. To establish the existence and extent of the problem, the author will present data

gathered from creditable sources. But then the solution the paper proposes will be the authoring company's own products or services (see the third sample page on the next page, promoting Newsweaver software). In this part, little effort is made to objectively or thoroughly compare all possible solutions.

White papers have become a common selling technique. You can find an abundance of white papers by typing "white papers on _____" into your search engine's search field and filling in the blank with virtually any

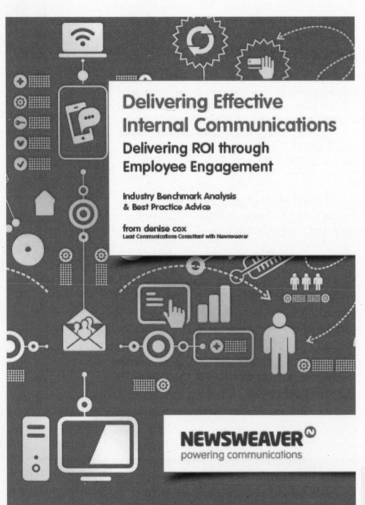

Executive summary

The world of internal communications (IC) is evolving rapidly as organisations realise the positive impact effective communication and engaged employees have on business performance.

In this whitepaper Newsweaver's Lead Communications Consultant, denise cox, presents new research and benchmarks for internal communicators: shining a light on how IC is developing, and offering expert interpretation and insight that helps you apply the results to your organisation.

denise goes on to offer best practice advice and techniques you can use to deliver more engaging and effective communications, boost employee engagement and demonstrate the ROI of your internal communications.

Key internal communication benchmarks

- Roles and titles within the IC function still show significant variation, but standards are slowly emerging
- Employee engagement is the biggest goal for IC professionals today
- The biggest challenge IC teams face is managing information overload
- Intranet and email are the most popular IC channels, and perceived to be most effective
- Demonstrating communication effectiveness and ROI is growing in importance for IC teams

Key best practice takeaways

- Engagement is about being timely targeted and relevant - sending the right information to the right employee at the right time
- Giving employees choice empowers, fuels engagement and promotes buy-in
- Personalisation improves perceived relevance, which boosts engagement
- Social elements entice engagement, enable employee voice, and offer valuable insight into how employees interact with your content
- Measuring and benchmarking are the key informed decisions that deliver sustainable employee engagement
- Measuring engagement helps demonstrate the value of internal communications to the organisation.

Data collated from a Newsweaver-commissioned study conducted across Newsweaver's network of more than 400 Blue Chip and FTSE 100 companies spanning 96 countries. The results have been cross referenced, where relevant, with the Melcrum "Key Benchmarking Data for Communicators" report.

2 Delivering Effective Internal Communications

opens to click-throughs and more - you get a really clear picture of the content employees find most engaging, and that helps you deliver more engaging content in future

• **Engagement across the organisation:** this gives you a high level breakdown of engagement across different parts of the organisation or different regions. It helps you identify where the problems are, letting you pinpoint and eliminate barriers to engagement, and switch to alternative channels (like face-to-face) to make sure important information gets through where necessary

• **What devices are employees using to engage:** how are employees accessing your communications? What proportion are accessing it from their desk? How many are using their mobile device? How does that impact your content and your design?

• **Social engagement:** measuring how employees interact with social elements in your messaging can be a key indicator of enhanced engagement. It also lets you assess which social elements are relevant / work best for your organisation, and lets you experiment with new features to see how they impact engagement

Newsweaver's Internal Connect Reporting Dashboard

18 Delivering Effective Internal Communications

---

business-related issue. White papers are popular because they educate business-people about common problems and helpful products.

It's important to understand the difference between a report and a white paper. Readers bring different expectations to these documents. Failing to make clear which type you're writing is unethical, and it can sway readers to make a decision on insufficient or misinterpreted data.

You can learn more about white papers and their uses by visiting whilepaperguy.com, whitepapersource.com, and many other Web resources.

Source: Denise Cox, *Delivering Effective Internal Communications*, Newsweaver, 2013. Reprinted with permission.

Some, however, are formal, especially reports directed to readers high in the administration of the organization. In fact, some email and memo reports rival the longer forms in formality. Like the longer forms, they may use headings to display content and visuals to support the text. For the longer email reports, writers will often choose to make the report itself an attached document and use the email message as a transmittal message.

Because they are largely internal, email reports tend to be problem-solving reports. They are intended to help improve operations, lay the groundwork for an innovation, solve a problem, or otherwise assist decision makers in the organization.

Written Reports in Other Forms

While most written reports in business will take the form of short reports, letter reports, or email or memo reports, they can take a variety of other forms as well. The report featured in Exhibit 9-4 appeared in an online newsletter by the John Deere company. It used objectively gathered and reported evidence to persuade readers of the value of a John Deere product. Research can also be reported in pamphlets, white papers (see the

Communication Matters feature), and other publications, and many reports are uploaded to the Web as stand-alone documents in PDF format. You can apply your report-writing knowledge to all these forms and more. Just be sure to choose the appropriate form for your readers and purpose.

LO 9-4 Adapt the procedures for writing short reports to routine operational reports, progress reports, and problem-solving reports, as well as to minutes of meetings.

COMMON TYPES OF SHORT REPORTS

Because organizations depend heavily on short reports, there are many varieties, written for many different purposes. We cover some of the most common types here, categorized on the basis of their main purpose, but the form they take will vary from company to company. Also, most companies will have

Guess Row Study

Introduction

Producers are becoming increasingly interested in using GPS-based guidance systems as either an enhancement to or in replacement of mechanical markers in planting operations. Recently, a study was conducted to assess the accuracy of mechanical markers for planting operations, and compare this to manual GPS guidance and GreenStar™ AutoTrac. While the pass-to-pass accuracy of the StarFire receiver has been determined to be +/- 4 inches by the University of Illinois, the measurements taken in this study are more indicative of what producers are achieving in the field.

The guess row is the distance between the outside rows of two side-by-side passes made in the field (Figure 1). If the planter is set up to plant 30-inch rows, the perfect guess row width would be 30-inches, Significant guess row variability causes difficulty in using mismatched row equipment (i.e. 16 row planter harvested with a 12 row corn head). In addition, wide guess rows may allow weed escapes that decrease yields.

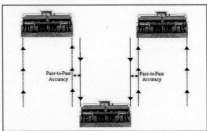

Figure 1: Pass-to-pass accuracy

Methods

Servi-Tech, a crop consulting company offering services primarily in Kansas, Nebraska, and Iowa, measured the guess row width on a number of fields in these three states. Twenty consecutive guess row measurements were taken in each field. The data presented in this discussion are from corn and soybean fields planted in 30-inch rows only. The distribution of fields is shown in Table 1.

Table 1 Distribution of Fields Measured

	Mechanical Markers	Manual GPS Guidance	AutoTrac	All Data
Number of Fields Measured	70	9	19	98

Note: all planting using AutoTrac was done with JD8000T tractors.

In addition to guess row width and guidance type, additional information was recorded for each field, including:

- Crop
- Tillage system
- Field topography
- Tractor and planter model
- Planter width
- Approximate planting speed
- Time of day planted

After preliminary analysis of the data, it was determined that in some fields consistent planter draft to one direction (often caused by improper set up) resulted in guess row variability over and above that attributable to the guidance system used. In order to remove these fields from the analysis, odd and even passes were analyzed and 35 fields were removed.

Results

While the average guess row width for all marker systems was similar (Figure 2), both the manual GPS

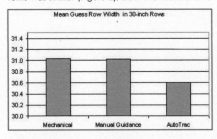

Figure 2: Mean Guess Row Width for all guidance systems

guidance and AutoTrac systems exhibited much less variability than that of mechanical markers (Figure 3; Table 2). Range in guess row width for manual GPS guidance and AutoTrac was nearly half that of mechanical marker systems. The standard deviation for manual GPS guidance and AutoTrac planted guess rows was 2.1 inches compared to 3.3 inches for the mechanical marker planted fields.

Conclusions

Guess row width in fields planted with GPS-based guidance systems was less variable than those planted with mechanical marker systems. On average, GPS-based guidance systems were as accurate as those fields planted with mechanical marker systems. The test

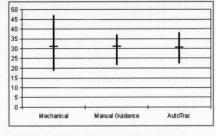

Figure 3: Mean Guess Row Width and Range

data showed many row crop planters were not set up properly (unit spacing or marker arm mis adjustment), causing them to pull one direction consistently or resulting in varied pass-to pass 'guess row' width. Experienced operators may compensate for this automatically when using mechanical marker systems, so proper planter setup is more important with automatic steering systems.

While the guess row widths for both the manual GPS guidance system and AutoTrac were similar, Auto-Trac would be expected to deliver this accuracy over a wide range of conditions consistently. Manual GPS guidance system accuracy depends on how well the operator can follow the steering cues. As time progresses, the operator may become fatigued, resulting in a reduced accuracy. This is not a factor with GreenStar AutoTrac.

Table 2

	All Data	Mechanical	Manual Guidance	AutoTrac
Maximum	50	47	37	38
Minimum	19	19	22	22.5
Mean	31.3	31.0	31.0	30.6
Range	31	28	15	15.5
Standard Deviation	3.4	3.3	2.1	2.1
Coefficient of Variation	0.11	0.11	0.07	0.07

The bottom line is that GreenStar AutoTrac was found to be capable of delivering the same to slightly better accuracy than producers achieve using mechanical markers in the fields used for the study. Coupled with the fact that AutoTrac users will be less fatigued at the end of the day, GreenStar AutoTrac may be able to help your operation during one of the most important times of the growing season, planting. For more information about GreenStar Parallel Tracking or Auto-Trac, visit your local John Deere dealer.

Source: Deere & Company, *Growing Innovations*, Winter 2002: 3, Web, 16 June 2010. Reprinted with permission.

---

developed unique types of reports to accomplish particular goals. Always consider your company's typical ways of reporting when deciding what to report and how.

Routine Operational Reports

The majority of the reports written within companies are **routine operational reports** that keep supervisors, managers, and team members informed about the company's operations. These can be daily, weekly, monthly, or quarterly reports on the work of each department or even each employee. They

can relate production data, information on visits to customers, issues that have arisen, or any kind of information that others in the organization need on a routine basis.

The form and contents of these reports will vary from company to company and manager to manager. Many will be submitted on predesigned forms. Others may not use forms but will follow a prescribed format. Still others will be shaped by the writer's own judgment about what to include and how to present it.

The nature and culture of the organization heavily influence the forms taken by these reports. For example, one innovative format

communication matters

The Monetary Value of a Good Report

A well-researched report can save a business thousands of dollars or generate thousands of dollars in income. That's why companies are willing to pay hundreds of dollars for industry reports like those prepared by such information brokers as ReportLinker. The 32-page sample report shown here carries a price tag of $1,020—but the cost is minimal for a company that can really benefit from the information.

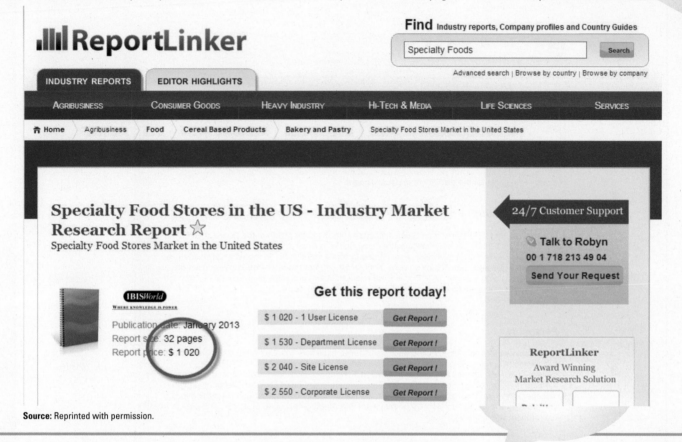

Source: Reprinted with permission.

for weekly reporting is the 5-15 report.[1] The name comes from the fact that it is intended to be read in 5 minutes and written in 15 minutes. Its typical three-part contents are a description of what the employee did that week, a statement about the employee's morale and that of others he or she worked with, and one idea for how to improve operations. Clearly, this format would work best in an organization where employees have nonroutinized jobs and the management values the employees' opinions.

Whatever the form, the routine operational report should convey clearly and quickly what readers most need and want to know about the time period in question. It is also an opportunity for you, the writer, to showcase your ability to provide needed information on deadline.

When using standardized forms for periodic reports, you should consider developing a template macro or merge document with your word-processing software. A macro would fill in all the standard parts for you, pausing to let you fill in the variable information. A template merge document would prompt you for the variables first, merging them with the primary document later. However standardized the process, you will still need to be careful to gather accurate information and state it clearly.

Progress Reports

You can think of an internal **progress report** as a routine operational report except that it tends to be submitted on an as-needed basis, and, as its name implies, it focuses on progress toward a specific goal. If you are working on a project for an external client, you may also need to submit progress reports to show that your work is on track. For example, a fundraising organization might prepare weekly summaries of its efforts to achieve its goal. Or a building contractor might prepare a report on progress toward completing a building for a customer. Typically, the contents of these reports concern progress made, but they also may include such related topics as problems encountered and projections of future progress.

A Progress Report in Email Form

This email report summarizes a sales manager's progress in opening a new district. It begins with the highlights—all a busy reader may need to know. Organized by three categories of activity, the factual information follows. The writer–reader relationship justifies a personal style.

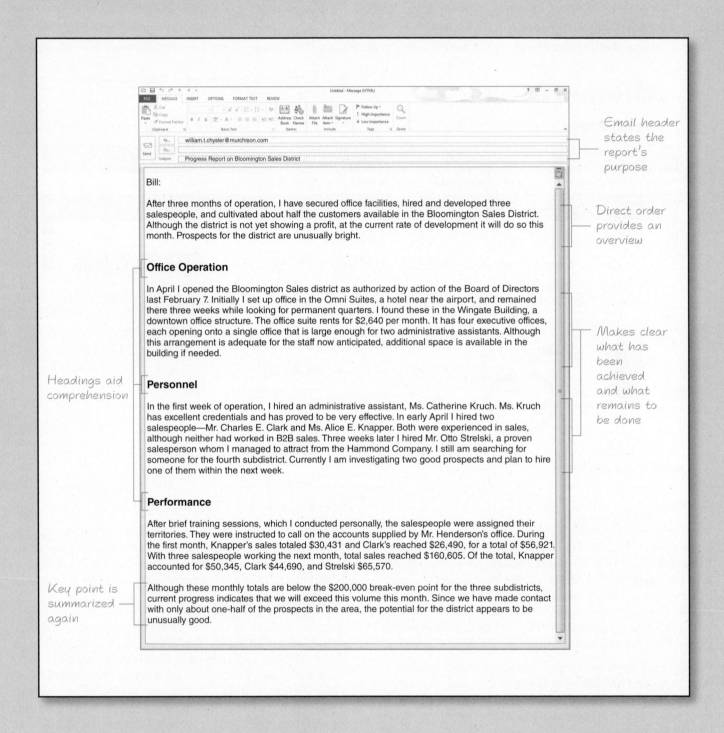

Email header states the report's purpose

Direct order provides an overview

Makes clear what has been achieved and what remains to be done

Headings aid comprehension

Key point is summarized again

Text of the email:

To: william.t.chysler@murchison.com
Cc:
Subject: Progress Report on Bloomington Sales District

Bill:

After three months of operation, I have secured office facilities, hired and developed three salespeople, and cultivated about half the customers available in the Bloomington Sales District. Although the district is not yet showing a profit, at the current rate of development it will do so this month. Prospects for the district are unusually bright.

Office Operation

In April I opened the Bloomington Sales district as authorized by action of the Board of Directors last February 7. Initially I set up office in the Omni Suites, a hotel near the airport, and remained there three weeks while looking for permanent quarters. I found these in the Wingate Building, a downtown office structure. The office suite rents for $2,640 per month. It has four executive offices, each opening onto a single office that is large enough for two administrative assistants. Although this arrangement is adequate for the staff now anticipated, additional space is available in the building if needed.

Personnel

In the first week of operation, I hired an administrative assistant, Ms. Catherine Kruch. Ms. Kruch has excellent credentials and has proved to be very effective. In early April I hired two salespeople—Mr. Charles E. Clark and Ms. Alice E. Knapper. Both were experienced in sales, although neither had worked in B2B sales. Three weeks later I hired Mr. Otto Strelski, a proven salesperson whom I managed to attract from the Hammond Company. I still am searching for someone for the fourth subdistrict. Currently I am investigating two good prospects and plan to hire one of them within the next week.

Performance

After brief training sessions, which I conducted personally, the salespeople were assigned their territories. They were instructed to call on the accounts supplied by Mr. Henderson's office. During the first month, Knapper's sales totaled $30,431 and Clark's reached $26,490, for a total of $56,921. With three salespeople working the next month, total sales reached $160,605. Of the total, Knapper accounted for $50,345, Clark $44,690, and Strelski $65,570.

Although these monthly totals are below the $200,000 break-even point for the three subdistricts, current progress indicates that we will exceed this volume this month. Since we have made contact with only about one-half of the prospects in the area, the potential for the district appears to be unusually good.

A Memo Progress Report on a Class Project

As a student, you will sometimes need to submit a progress report on a complex project, such as a group project. In such cases, the same goals apply here as to workplace progress reports—to show that you understand the project's purpose, that you have made good progress, that you have a good sense of what remains to be done, and that you're headed toward a successful conclusion.

Memo Report

Date: May 12, 2014

To: Professor Rodriguez

From: Sam Ellis SE

Subject: Group Four's Progress on the Report for Ms. Herbert

Our group consists of Bo Riddle, Ina Ward, Tiffany Paine, and me. We have made good progress on gathering information to help Ms. Herbert decide whether to include personality testing in her hiring process for Fashion Sense.

Direct opening orients the reader and announces good progress on the project

Research Topics and Methods

In our first group meeting, on May 3, we decided to investigate the following topics, assigned as shown:

List shows that the group is well organized

- Types of personality testing available (Tiffany)
- Use of personality testing in the retail industry (Ina)
- Cost of personality testing (Bo)
- Possible legal risks of personality testing (me)

To research the types of testing available, we conducted mostly Internet research. We did both Internet and library research to investigate the remaining topics. We also interviewed Amy Loehmann, a professor in the law school, about possible legal issues.

Discussion of research methods builds credibility

Findings So Far

In the first week of work, we have gathered these main findings:

Informative headings and bulleted lists make the contents easy to digest

- Companies use many different kinds of personality tests, such as the Big Five personality test (see queendom.com) and tests based on four personality types (see MaximumAdvantage.com). But the most popular is the Myers-Briggs Type Indicator. Given its proven track record, Ms. Herbert is likely to want to use some version of this test if she adopts personality testing.
- Large retail stores such as Macy's do use personality testing, but we have not yet found much use of such testing for smaller businesses like Ms. Herbert's.
- Personality testing can be extremely expensive if the company hires a trained consultant to conduct the testing. For large-scale testing, these consultants charge thousands of dollars. However, there are many small testing outfits that provide relatively simple, yet valid, tests. For example, for about $500, Proven Results will test a company's high-performing employees and develop a personality test based on those employees' traits (ProvenResults.com). This test can then be given to job applicants to determine their suitability for the work.
- There are definitely legal risks to personality testing. Mainly, one must be sure not to ask questions that discriminate against the test takers on the basis of religion, race, or gender. If a company hires a consultant or firm like ProvenResults to do the testing, that person or firm will be the responsible legal party. If one doesn't hire a third party, it is advisable to have an attorney review the testing procedure before it is used.

Convincing details attest to the group's hard work

Next Steps

A good plan for completing the work builds the instructor's confidence in the group

We will continue to explore the kinds of tests that might be suitable for Ms. Herbert. Specifically, we will try to find out what kinds of personality tests the smaller retail companies use and which of these might be appropriate for Fashion Sense.

We will also try to determine the most cost-effective method for Ms. Herbert to adopt and acquire contact information for any testing services that look promising.

We are on track to have a complete draft of our report prepared by the May 22 deadline. This is an interesting project, and we believe we will be able to help Ms. Herbert make a well-informed decision.

Conclusion shows good awareness of the next major deadline and of the end goal

Progress reports follow no set form. They can be quite formal, as when a contractor building a large manufacturing plant reports to the company for whom the plant is being built. Or they can be very informal, as in the case of a worker reporting by email to his or her supervisor on the progress of a task being performed. Some progress reports are quite routine and structured, sometimes involving filling in blanks on forms devised for the purpose. Most, however, are informal, narrative reports, illustrated by the examples on pages 272 and 273. As these examples show, you should organize and format the "story" of your progress for easy comprehension.

As with most reports, you have some choice about the tone to use when presenting your information. With progress reports, you want to emphasize the positive if possible. The overall message should be "I (or we) have made progress." The best way to convey this message confidently, of course, is to be sure that you or your team has in fact made progress on the task at hand.

Problem-Solving Reports

Many short reports are **problem-solving reports**. These reports help decision makers figure out what to do any time a problem arises within an organization—which is often. For example, a piece of equipment may have broken down, causing mayhem on the production line. Or employees may have gotten hurt on the job. Or, less dramatically, a company procedure may have become outdated, or a client company may want to know why it's losing money. If we define *problem* as an issue facing the company, we could include many other scenarios as well—for example, whether or not a company should adopt flextime scheduling or what location it should choose for a new store. Whatever the context, the writer of a problem-solving report needs to gather facts about the problem or issue, define it clearly, research solutions, and recommend a course of action.

Like progress reports, problem-solving reports can be internal or external. Internal problem-solving reports are usually assigned, but sometimes employees may need to write unsolicited problem-solving reports—for example, if they must recommend that a subordinate be fired or if they feel that a change in procedure is necessary. External problem-solving reports are most often written by consulting companies for their clients. In these cases, the report is the main product that the client is paying for.

A type of problem-solving report that deserves special attention is a **feasibility study**. For these reports, writers study several courses of action and then propose the most feasible, desirable one. For instance, you might be asked to compare Internet service providers and recommend the one that suits the company's needs and budget best. Or you might investigate what type of onsite childcare center, if any, is feasible for your organization. Sometimes feasibility studies are not full-blown problem-solving reports. They may offer detailed analysis but stop short of making a recommendation. The analysis they provide nevertheless helps decision makers decide what to do.

In fact, many short reports that help solve company problems may not be complete problem-solving reports. As explained in the previous chapter, decision makers who assign research reports may not want recommendations. They may want only good data and careful analysis so that they can formulate a course of action themselves. Whether you are preparing an internal or external report, it is important to understand how far your readers want you to go toward proposing solutions.

You have some latitude when deciding how direct to make your opening in a problem-solving report. If you believe that your readers will be open to any reasonable findings or recommendations, you should state those up front. If you think your conclusions will be unexpected or your readers will be skeptical, you should still state your report's purpose and topic clearly at the beginning but save the conclusions and recommendations until the end, after leading your readers through the details. Exhibit 9-5, a pattern for a problem-solving report used by the U.S. military, follows this indirect plan. As always, try to find out which method of organization your readers prefer.

© 2009 Ted Goff

"I suggest we take a look at the floating, glowing incident report first."

EXHIBIT 9-5 · Military Form for Problem-Solving Report

DEPARTMENT OF THE AIR FORCE
HEADQUARTERS UNITED STATES AIR FORCE
WASHINGTON, DC 20330

REPLY TO
ATTN OF   AFODC/Colonel Jones

SUBJECT   Staff Study Report

TO:

PROBLEM

1. --
---.

FACTORS BEARING ON THE PROBLEM

2. Facts.

   a.--
-----------------------------------.
   b--.

3. Assumptions.

4. Criteria.

5. Definitions.

DISCUSSION

6. --.

7. --.

8. --.

CONCLUSION

9. --.

ACTION RECOMMENDED

10. ---.

11. ---.

JOHN J. JONES, Colonel, USAF   2 Atch
Deputy Chief of Staff, Operations   1. ----------------
   2. ----------------

While they usually propose action, problem-solving reports are not true persuasive messages. Because they have either been assigned or fall within an employee's assigned duties, the writer already has a willing reader. Furthermore, the writer has no obvious personal stake in the outcome the way he or she does with a persuasive message. However, when writing a problem-solving report, especially one that makes recommendations, you do need to show that your study was thorough and your reasoning sound. The decision makers may not choose to follow your advice, but your work, if it is carefully performed, still helps them decide what to do and reflects positively on you.

Meeting Minutes

Many short reports in business, especially internal ones, do not recommend or even analyze. Instead, they describe. Trip reports, incident reports, and other such reports are meant to provide a written record of something that happened. Whatever their type and specific purpose, they all share the need to be well organized, easy to read, and factual. Perhaps the most common of these reports is **minutes** for meetings. We thus single them out for special emphasis.

Minutes provide a written record of a group's activities, which can include announcements, reports, significant discussions, and decisions. They include important details, but they are primarily a summary that reports the gist of what happened, not a verbatim transcript. Minutes include only objective data; their writer carefully avoids using such judgmental words as *excellent* or *impractical* or such descriptive words as *angrily* or *calmly*. However, if the group passes a resolution that specific wording be officially recorded, a writer should include it. Accurate minutes are important because they can sometimes have legal significance, such as when shareholders or boards vote to approve certain corporate activities.

The expected format varies across organizations, but any format you use should enable the reader to easily review what happened and retrieve particular information. Headings in bold or italics are usually appropriate, and some writers find that numbering items in the minutes to agree with the numbering of a meeting's agenda is helpful. Additional advice on writing minutes can be found on pages 290–291 in Chapter 10.

The sample on the next page illustrates typical minutes. The following preliminary, body, and closing items may be included.

Preliminary Items

- Name of the group.
- Name of the document.

It takes both skill and judgment to prepare accurate, neutral minutes for an important meeting.

Minutes of the Policy Committee
Semiannual Meeting
November 21, 2014, 9:30–11:30 A.M., Conference Room A

Present: Elaine Horn (chair), D'Marie Simon, DeAnne Overholt, Michelle Lum, Joel Zwanziger, Rebecca Shuster, Jeff Merrill, Donna Wingler, Chris Woods, Tim Lebold (corporate attorney, guest).

Absent: Joan Marian, Jeff Horen (excused), Leonna Plummer (excused)

Complete preliminary information provides a good record

Approval of Minutes

Minutes from the May 5, 2014, meeting were read and approved.

Announcements

Headings help readers retrieve information

Chris Woods invited the committee to a reception for Milton Chen, director in our Asia region. It will be held in the executive dining room at 3:00 P.M. tomorrow. Chris reminded us that Asia is ahead of the United States in its use of wireless technology. He suggested that perhaps we can get an idea of good policies to implement now.

Old Business—Email Policy

Joel Zwanziger reported the results of his survey on the proposed email policy. While 16 percent of the employees were against the policy, 84 percent favored it. A January 1, 2015, implementation is planned, subject to its distribution to all employees before the Christmas break.

Web-Surfing Policy

D'Marie Simon reported on her study of similar companies' Web-surfing policies. Most have informal guides but no official policies. The guidelines generally are that all surfing must be related to the job and that personal surfing should be done on breaks. The committee discussed the issue at length. It approved a policy that reflects the current general guidelines.

Discussions are summarized and actions taken are included

Temp Policy

Tim Lebold presented the legal steps we need to take to get our old and new temporary employees to sign a nondisclosure agreement prior to working here, as we've been discussing in relation to a new temp policy. The committee directed Tim to begin the process so that the policy could be put in force as soon as possible.

New Business—Resolution

Michelle Lum proposed that a resolution of thanks be added to the record recognizing Megan for her terrific attention to detail as well her clear focus on keeping the committee abreast of policy issues. It was unanimously approved.

Resolutions often include descriptive language

Next Meeting

Closing gives reader complete needed facts

The next meeting of the committee will be May 3, 2014, from 9:30–11:30 A.M. in Conference Room A.

Adjournment

The meeting was adjourned at 11:25 A.M.

Respectfully submitted,

Elaine Horn

Elaine Horn

Signing signifies the minutes are an official record

- Type of meeting (monthly, emergency, special).
- Place, date, and time called to order.
- Names of those attending including guests (used to determine if a quorum is present).
- Names of those absent and the reasons for absence.

Body Items

- Approval of the minutes of previous meeting.
- Meeting announcements.
- Old business—Reports on matters previously presented.
- New business—Reports on matters presented to the group.

Closing Items

- Place and time of the next meeting.
- Notation of the meeting's ending time.
- Name and signature of the person responsible for preparing the minutes.

When you are responsible for preparing the minutes of a meeting, you can take several steps to make the task easier. First, get an agenda in advance. Use it to complete as much of the preliminary information as possible, including the names of those expected to attend. If someone is not present, you can easily move that person's name to the absentee list. You might even set up a table in advance with the following column headings to facilitate your note taking:

Topic	Summary of Discussion	Action/Resolution

Bear in mind that meeting minutes, while they look objective, almost always have political implications. Because minutes are the only tangible record of what happened, meeting participants will want their contributions included and cast in a positive light. Since you will not be able to record every comment made, you will need to decide which ones to include, whether or not to credit a particular speaker, how to capture the group's reaction, and so forth. Use your good judgment when translating a rich oral event into a written summary.

getting it done with reports!

- What are the steps for writing an executive summary?
- What's a feasibility study and how do you write one?
- How should you tell your organization's "story" in an annual report?

Scan the QR code with your smartphone or use your Web browser to find out at www.mhhe.com/RentzM3e. Choose Chapter 9 > Bizcom Tools & Tips. While you're there, you can view exercises, PPT slides, and more to see how to get it done with reports.

www.mhhe.com/RentzM3e

Communicating **Orally**

COMPANY
SEMINAR

ten

Although you may find written communication more challenging and may spend a lot of time planning, drafting, and revising, you are likely to spend more time communicating orally than you are in any other work activity. Much of the oral communication that goes on in business is the informal, person-to-person communication that occurs whenever people get together in meetings, over the phone, or in casual conversations. However, businesspeople also deliver formal presentations and speeches such as sales pitches, keynote speeches, or presentations to stakeholders or boards of directors. In fact, researchers consistently note the growing importance of oral communication skills even as technologies such as email, blogs, and other social networking opportunities demand more of our writing skills. According to one estimate, "speakers address audiences an astonishing 33 million times each day . . . and businesspeople give an average of 26 presentations a year."[1] ■

LEARNING OBJECTIVES

LO 10-1 Discuss talking and its key elements.

LO 10-2 Explain the use of physical aspects such as posture, walking, facial expression, and gestures in effective oral communication.

LO 10-3 Explain the challenges of listening and how to overcome them.

LO 10-4 Explain the techniques for conducting and participating in meetings.

LO 10-5 Describe good phone and voice mail techniques.

LO 10-6 Determine an appropriate topic, purpose, and structure for a speech or presentation.

LO 10-7 Identify and select appropriate presentation methods.

LO 10-8 Choose the means of audience feedback.

LO 10-9 Plan visuals to support oral communication.

LO 10-10 Plan and deliver effective Web-based presentations.

LO 10-11 Work effectively with a group to prepare and deliver a team presentation.

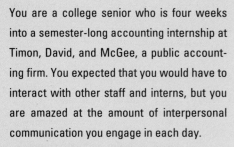

workplace scenario

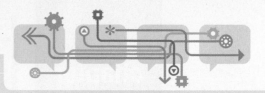

Speaking and Listening Like a Professional Businessperson

You are a college senior who is four weeks into a semester-long accounting internship at Timon, David, and McGee, a public accounting firm. You expected that you would have to interact with other staff and interns, but you are amazed at the amount of interpersonal communication you engage in each day.

Take today, for example. This morning you had to give a presentation to the firm's partners and managers regarding the research you did on a newly passed U.S. tax law. You thought you were thorough and quite clear in answering the questions your supervisor asked you to research, but the looks on the partners' faces and their numerous questions indicated they either didn't understand you or hadn't been listening to you. Regardless, you felt your report would have been received better had your oral communication skills been more polished.

Then, you returned from lunch to find seven voice mails. You needed to return five of the seven calls because the caller left incomplete or vague messages. Not having to return those phone calls would have saved you 45 minutes. You feel pretty comfortable communicating over the phone, but you do wish your co-workers had better phone skills.

In the evening you returned to campus to lead a meeting of the Student Accounting Association. It was a disaster—everyone talking at once, arguing, interrupting. Everyone wanted to talk; no one wanted to listen. Fortunately, you were able to accomplish most of your agenda items, but you know your group needs to improve its dynamic.

You are fast realizing just how much you will rely on good oral communication in your professional career and vow to learn strategies and techniques for improving your skill. The following review of oral communication will help you.

> As a first step in improving your talking ability, think for a moment about the qualities you like in a good talker—one with whom you would enjoy talking in ordinary conversation.

LO 10-1   Discuss talking and its key elements.

CONVERSING INFORMALLY

Most of us do a reasonably good job of informal talking. In fact, we do such a good job that we often take talking for granted and overlook the need to improve our talking ability. To improve our talking, we need to be aware of its nature and qualities.

As a first step in improving your talking ability, think for a moment about the qualities you like in a good talker—one with whom you would enjoy talking in ordinary conversation. Then think about the opposite—the worst conversationalist you can imagine. With these two images in mind, you can form a good picture of the characteristics of good talking. The following section covers the most important of these.

Elements of Professional Talking

The techniques of good talking use four basic elements: (1) voice quality, (2) style, (3) word choice, and (4) adaptation to your audience.

voice quality   Voice quality primarily refers to the pitch and resonance of the sounds a speaker makes and the sounds an audience hears, but it also includes speed and volume. Speaking in a monotone is not likely to hold your listeners' interest for long. In addition, you should present the easy parts of your message at a fairly fast rate and the hard parts and the parts you want to emphasize at a slower rate. A slow presentation of easy information is irritating, while quick presentation of difficult information may be hard to understand. And not varying the speaking speed can make a presentation boring.

A common problem related to the pace of speaking is the incorrect use of pauses. Properly used, pauses emphasize upcoming subject matter and are an effective means of gaining attention. But frequent pauses for no reason are irritating and break the

communication matters

Finding Your Professional Voice

This chapter recommends that you improve your talking voice with self-analysis and practice. The experts at the *Excellent Work* blog provide several tips for cultivating a talking voice that serves you well in any professional setting. Here is a summary of those tips:

- *Breathing:* Breathe from your diaphragm. Doing so will give you better control over your talking because at the end of the exhale, you create pauses that give your listener a chance to process what you say.

  To breathe from your diaphragm, take a deep breath. As you exhale, talk. *Excellent Work* recommends counting to 10 or reciting the months of the year or days of the week. As you're talking, increase the volume by using your abdominal muscles rather than your throat for volume. How do you know if you're using your diaphragm? This comes from us, not from the folks at *Excellent Work:* If you're breathing from the diaphragm, your abdomen should extend when you inhale; if you're breathing from your throat, your

chest will extend, your abdomen will cave in, and your shoulders will rise.

- *Pitch:* High-pitched or monotone voices can be annoying. Using a lower pitch and varying your pitch holds listeners' attention. One way to practice lower and varied pitch is by humming.

- *Volume:* When you speak, ask your audience if your volume is too loud or too soft and regulate it appropriately.

- *Pace:* Speaking either too quickly or too slowly will cause your audience to tune out. Record yourself to assess your pace or get feedback from your audience.

- *Articulation:* Have you been told that you mumble? Exaggerating lip movement, practicing difficult tongue twisters until you can say them as quickly and clearly as possible, and exaggerating vowel sounds are all ways to improve articulation.

- *Timing:* Marking points in your speech where you want to take a break lets you plan your pauses. To add emphasis to your point, take longer or more breaths after you make the point.

- *Anxiety:* Are there people you talk with who make you nervous or situations where you find yourself so nervous that you cannot speak well? Before engaging in conversation or speaking before a group, do some simple exercises to release tension: Turn side to side, roll your head in half circles, roll your shoulders, shift your rib cage from side to side, yawn, or stretch. You can also touch your toes and then slowly roll up, raising your head last.

- *Posture:* Standing up straight (not slouching) will help you breathe better, and better breathing leads to a better talking voice.

- *Self-assessment:* Many people do not like to listen to their own voice, but even if it's painful to do so, record your voice using different pitches, pacing, articulations, etc. When you find a voice you like, practice that voice until it becomes your habit.

Source: "10 Tips to Improve Your Speaking Voice," *Excellent Work*, Excellent Work, 9 July 2012, Web, 22 July 2013.

---

listeners' concentration. Pauses become even more irritating when the speaker uses fillers such as *uh, like, you know,* and *OK.*

Furthermore, regardless of audience size, variety in volume is good for interest and is one way of emphasizing the more important points. Of course, your volume should be greater for a large audience than for a small audience, unless you are using a microphone.

style   Style refers to a set of voice behaviors that makes your voice unique. It is the way that pitch, speed, and volume combine to give personality to your oral expression. What is the image your talking projects? Does it project sincerity? Is it polished? Smooth? Rough? Dull? After your honest assessment, you should be able to improve your style. Another good way to improve your presentation skills is to watch

others. Watch your instructors, your peers, television personnel, professional speakers, and anyone else whom you might learn from. You can also watch top corporate executives on webcasts and video presentations. Analyze these speakers to determine what works for them and what does not, and then imitate those techniques that you think would help you. The Communication Matters tips above will also help you develop a professional voice and style.

word choice   A third quality of talking is word choice. Of course, word choice is related to one's vocabulary. The larger your vocabulary, the more choices you have. Even so, you should keep in mind the need for the listener to understand the words you choose. In addition, the words you choose should convey courtesy and respect for the listener's knowledge of the subject matter—that is, they should not talk

communication matters

The Art of Negotiation

One common type of interpersonal communication you'll engage in as a professional is that of negotiation. You might negotiate a raise, a contract with a client, or even a deal with co-workers over who gets the office with a window. Whatever you're negotiating, you'll present your best professional image, achieve your communication goals, and preserve your listener's goodwill by heeding the following advice from author and entrepreneur Kevin O'Leary:

1. Know What You Want. Be sure you know, exactly, the goals of the negotiation and how to make your point in a way that makes sense to your audience. Also anticipate your audience's questions and have answers ready to show you've thought through your points.

2. Know That Your Audience Is Judging You. Your audiences are judging you on everything from how you're dressed to your body language to how well you're using their time. If you do not look the part, are not confident, or don't get to your point immediately, you lose your credibility.

3. Help Others Help You. Focus on benefits to your audience. Your audiences are not likely to negotiate if they cannot see any outcomes that benefit them.

4. Know Your Facts. Know everything you can about the audience or audiences you're negotiating with (and about the issue you're negotiating). If you look like you haven't done your homework, you lose your advantage or give your audience little reason to continue negotiating.

5. Know When to Walk Away. The goal of negotiation should be that everyone leaves with something he or she is satisfied with. When this doesn't happen (or looks like it's not going to happen), it's all right to walk away.

6. Remember the Bottom Line. In business, the success of a negotiation may be the extent to which you show the audience how your idea helps the bottom line (making money). Depending on what you're negotiating, if the negotiation does not look financially advantageous, your audience may lose interest.

7. Don't Get Greedy. Once you've gotten what you want, be done. If you press the issue, you may quickly lose what you've just gotten.

Source: Kevin O'Leary, "How to Win in Business Negotiations: Never Forget, It's All About the Money," *Huffington Post—Business: Canada,* The Huffington Post, Inc., 26 July 2012, Web, 26 July 2013.

---

down to or above the listener. Consider, too, that the vocabulary you use with your friends outside the workplace may not be professionally appropriate for use in the workplace—no matter how friendly you are with your co-workers, superiors, or subordinates.

adaptation Adaptation is the fourth quality of good talking. As you learned in earlier chapters, adaptation means fitting the message to the intended listener. To illustrate, the voice, style, and words in an oral message aimed at superiors or professionals you don't know well would be different from the same message aimed at subordinates. Similarly, these qualities might vary in messages delivered in different cultures as well as different social situations, work situations, and classrooms.

Courtesy in Talking

Our review of talking would not be complete without a comment about the need for courtesy. We have all been frustrated by talkers who drown out others with their loud voices, who interrupt while others are talking, who attempt to dominate others in conversation. Good talkers encourage others to make their voices heard.

This emphasis on courtesy does not suggest that you should be submissive in your conversations—that you should not be assertive in pressing your points. It means that you should accord others the courtesy that you expect of them.

LO 10-2 Explain the use of physical aspects such as posture, walking, facial expression, and gestures in effective oral communication.

PREPARING YOURSELF TO SPEAK

While much of your success as a speaker depends on the quality of your voice, your success also depends on the traits and behaviors that accompany your oral communication. Of course, the content of a message must be solid, but if your delivery is poor, people may ignore or simply miss your most important points. The following sections provide strategies for ensuring that your audience will understand and respond positively to your message.

Appealing Personal Traits

An important preliminary to good oral reports or speeches is to analyze yourself as a speaker. In oral communication you, the speaker, are a very real part of the message. The members of your audience not only take in the words you communicate but also form an impression of you. And how they perceive you can significantly affect how they respond. You should thus do your best to project personal traits that will appeal to your audience. Those that follow are particularly important.

confidence A primary characteristic of effective oral reporting is confidence—your confidence in yourself and the confidence of your audience in you. The two are complementary: Your confidence in yourself tends to produce an image that gives your audience confidence in you, and your audience's confidence in you can give you a sense of security that increases your confidence in yourself.

Typically, you earn your audience's confidence through repeated contact with them. When you speak to a room full of strangers, you must find other ways to gain their confidence. Preparing your presentation diligently and practicing it thoroughly will give you confidence and help you project it. Another confidence-building technique is an appropriate physical appearance. Looking like those your audience respects gives you credibility and helps you get into your role as presenter. Yet another confidence-building technique is simply to talk in strong, clear tones. Speaking as though you are relaxed and self-assured will actually help you feel this way.

from the tech desk

Presentation Delivery Tools Help You Convey Your Message Effectively

Have you ever used PowerPoint's Presenter View? As you can see in the screenshot to the right, its tools can enhance the smoothness of a presentation. While your slides are being displayed to the audience in Slide Show view, Presenter View lets you view not only the slides but also your notes. Additionally, you see the upcoming slide as well as the elapsed time since the beginning of the presentation. Furthermore, a menu under the current slide allows you to start or end the show on one click, black out that screen to bring the attention back to you, zoom in on one section of your slide, and use a laser pointer.

Using one monitor or two, you can access presenter view in PowerPoint 2013 by going to Slide Show > Monitor > Presenter View (make sure the box is checked). Then open your presentation in Slide Show view > click the "..." at the bottom of the slide show > click "Show Presenter View."

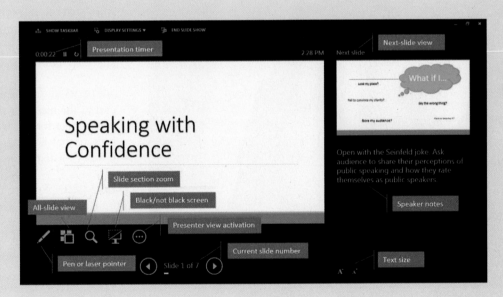

competence Audiences expect speakers to be knowledgeable on the topic they're discussing. Do your homework so that you'll have the knowledge your audience will require of you. Anticipate the listeners' likely questions—and if you're asked something you don't know, don't try to fake an answer. Instead, offer to find the information and share it with the audience later (e.g., by email). Besides gathering the necessary knowledge, spare no effort in making the presentation itself well designed and well written. Grammatical errors or poorly designed slides can sabotage an otherwise great speech.

sincerity Your listeners are quick to detect insincerity. And if they detect it, they are likely to give little weight to what you say. On the other hand, sincerity—when combined with competence—generates trust and conviction. The way to project an image of sincerity is clear and simple: You must *be* sincere. Be yourself, and acknowledge your limits. If you can't answer a question, say so and invite others in attendance to offer their ideas. Graciously compliment those ideas and add what you do know. Show that your topic is of genuine interest to you and that it is more important than your ego.

friendliness A speaker who projects an image of friendliness has a significant advantage. Audiences like friendly people, and they are generally receptive to what friendly people say. To project friendliness, smile, make eye contact, and learn and use audience members' names. You can also watch yourself in a mirror as you practice speaking to improve your projection of friendliness.

Appropriate Appearance and Physical Actions

In all types of formal and informal oral communication, whether face-to-face, online, or video, your listeners will be looking at you and your surroundings. What they will see is a part of the message and can affect the success of your communication. Do your best to make these visual elements contribute to, not detract from, your talking.

the communication environment Whenever you'll be visible, a background will be visible as well. For face-to-face communication, try to visit the room ahead of

> Anticipate the listeners' likely questions—and if you're asked something you don't know, . . . offer to find the information and share it with the audience later (e.g., by email).

Good presenters project confidence, competence, sincerity, and friendliness.

time to see what adjustments to the setting you might want to request. For example, you might want the podium moved or the lighting managed in a certain way or perhaps the tables and seating for a meeting arranged in a particular order. For virtual presentations, be sure to test how you'll look on camera so you can check the backdrop, the lighting, and how much of you will show. In both environments, ensure that your voice will be appropriately amplified and that extraneous noise will be kept to a minimum.

personal appearance Your personal appearance is a part of the message your audience receives. Dress in a manner appropriate for the audience and the occasion. You should also be sure that nothing about your appearance (e.g., hairstyle or jewelry) is distracting.

posture Posture is likely to be the most obvious physical trait of yours that your audience will see. Even listeners not close enough to detect facial expressions and eye movements can see the general form of the body. Stand or sit straight instead of slouching. Try to appear naturally poised and alert.

walking Your audience also forms an impression from the way you walk. A strong, sure walk to the speaker's position

Effective presenters use friendly facial expressions and make direct eye contact with members of their audience.

The appropriateness of physical movements is related to personality, physical makeup, and the size and nature of the audience. Also, a speaker appearing before a formal group tends to use more conservative gestures (e.g., slight head and hand movements) than one before an informal group. Which physical movements you should use is a matter for your best judgment.

LO 10-3 Explain the challenges of listening and how to overcome them.

LISTENING

Up to this point, our review of oral communication has been about sending information (talking). However, good business communicators must also ensure that they are good listeners.

conveys an impression of confidence. Hesitant, awkward steps convey the opposite impression.

Walking during the presentation can be good or bad, depending on how you do it. Some speakers use steps forward and to the side to emphasize points. Too much walking, however, detracts from the message. You would be wise to walk only when you are reasonably sure that this will have the effect you want. And be sure not to walk away from the microphone.

facial expression When speaking, be sure to avoid those facial expressions that convey unintended meanings. A smile, a grimace, and a puzzled frown may be appropriate or inappropriate, depending on the context. Choose those expressions that best convey your intended meaning.

Eye contact is important. The eyes, which have long been considered "mirrors of the soul," provide most listeners with information about the your sincerity, goodwill, and flexibility. Making eye contact tends to show that you have a genuine interest in your audience.

gestures Like facial expressions, gestures are strong, natural aids to speaking. It is natural, for example, to emphasize a plea with palms up and to show disagreement by shaking one's head. Raising first one hand and then the other usually indicates contrasting points, while a shrug might mean "either way is fine" or "who knows?"

> "A smile, a grimace, and a puzzled frown may be appropriate or inappropriate, depending on the context."

The Nature of Listening

When listening is mentioned, we think primarily of the act of sensing sounds. Viewed from a communication standpoint, however, the listening process involves the addition of filtering and remembering.

sensing How well we sense the words around us is determined by two factors. One factor is our ability to sense sounds—how well our ears can pick them up.

The other factor is our attentiveness or concentration. Our concentration on the communication varies from moment to moment. It can range from almost totally blocking out sounds to concentrating on them very intensely. From your own experience, you can recall moments when you were oblivious to the words spoken around you and moments when you listened with intensity. Most of the time, your listening falls somewhere between these extremes.

filtering From your study of the communication process in Chapter 1, you know that interpretation enables you to give meanings to the sounds you hear. In this process, your personal context affects how you filter these sounds and give meaning to incoming messages. This filter is formed by the unique contents of your mind: your knowledge, emotions, beliefs, biases, experiences, and expectations. In addition, larger social or workplace cultures will affect how you filter sounds. Thus, you sometimes give messages meanings that are different from the meanings that others give them.

communication matters

What's in a Handshake?

A handshake is a common part of the greeting in U.S. business culture, especially in formal situations, in situations where individuals have not seen each other in a long time, or when people are meeting for the first time.

In fact, your handshake may be among the most critical nonverbal behaviors in the initial stage of any business relationship, such as the job interview or a meeting with a potential client. The professional handshake requires more than simply extending your hand. It requires that you practice so you are prepared for any situation. Here are the rules according to Innovative Training and Communication Solutions:

- Know when to shake hands. Shake hands when (1) you are introduced to someone, (2) someone introduces himself or herself, (3) you introduce yourself to someone, (4) the conversation is over.
- Shake for no more than three "pumps." Another rule is to shake for about three seconds. Any longer than that and you make the situation awkward for everyone.
- Shake from the elbow, not your shoulder. This allows you to have a smooth handshake that is not rough or too forceful for the other person.
- Don't be a "dead fish" or "wet fish." Just as you do not want to shake too roughly, you do not want a limp handshake either. If your hands are sweaty, have some way to wipe them off (on your slacks or in the bathroom) before meeting people.
- Don't be a "bone crusher." Don't grip the other person's hand too hard. Not only do you not want to inflict pain, you don't want to appear too aggressive. Use the same force you would use to turn a door handle.
- Don't give the "little lady" handshake. Extend your whole hand, not just your fingertips. Extending just your fingertips signals that you are weak (especially if you are a woman).
- Shake with one hand, not two. Using both hands is too personal for a business setting. It also makes you look like you're trying too hard.

Source: Amy Castro, "7 Handshake Rules You Shouldn't Break," *Innovative Communication and Training Solutions*, ICTS, 3 July 2012, Web, 26 July 2013.

> According to authorities, we even quickly forget most of the message in formal oral communications (such as speeches), remembering only a fourth of the information after two days.

remembering Remembering what we hear is the third activity involved in listening. Unfortunately, we retain little of what we hear. We remember many of the comments we hear in casual conversation for only a short time—perhaps for only a few minutes or hours. Some we forget almost as soon as we hear them. According to authorities, we even quickly forget most of the message in formal oral communications (such as speeches), remembering only a fourth of the information after two days.

Improving Your Listening Ability

Improving your listening is largely a matter of mental discipline—of concentrating on the activity of sensing. If you are like most of us, you are often tempted not to listen, or you just find it easier

not to listen. Listening may seem like a passive activity, but it can be hard work.

After you have decided that you want to listen better, you must make an effort to be alert and to pay attention to all the words spoken.

Active listening is one technique individuals can use successfully. It involves focusing on what is being said and reserving judgment. Other components include sitting forward and acknowledging with "um-hm" and nodding.

Back-channeling (repeating what you think you heard) is an effective way to focus your attention, as is asking questions of a speaker. Communicators can use technologies such as chat and Twitter to comment on and enhance presentations

communication matters

The Ten Commandments of Listening

1. Stop talking. Unfortunately, most of us prefer talking to listening. Even when we are not talking, we are inclined to concentrate on what to say next rather than on listening to others. So you must stop talking before you can listen.

2. Put the talker at ease. If you make the talker feel at ease, he or she will do a better job of talking. Then you will have better input to work with.

3. Show the talker you want to listen. If you can convince the talker that you are listening to understand rather than oppose, you will help create a climate for information exchange. You should look and act interested. Doing things like reading, looking at your watch, and looking away distracts the talker.

4. Remove distractions. The things you do also can distract the talker. So don't doodle, tap with your pencil, shuffle papers, or the like.

5. Empathize with the talker. If you place yourself in the talker's position and look at things from the talker's point of view, you will help create a climate of understanding that can result in a true exchange of information.

6. Be patient. You will need to allow the talker plenty of time. Remember that not everyone can get to the point as quickly and clearly as you. And do not interrupt. Interruptions are barriers to the exchange of information.

7. Hold your temper. Anger impedes communication. Angry people build walls between each other; they harden their positions and block their minds to the words of others.

8. Go easy on argument and criticism. Argument and criticism tend to put the talker on the defensive. He or she then tends to "clam up" or get angry. Thus, even if you win the argument, you lose. Rarely does either party benefit from argument and criticism.

9. Ask questions. By frequently asking questions, you display an open mind and show that you are listening. And you assist the talker in developing his or her message and in improving the correctness of meaning.

10. Stop talking! The last commandment is to stop talking. It was also the first. All the other commandments depend on it.

Source: Anonymous.

Successful business people will have excellent speaking skills as well as excellent listening skills.

in real time, which helps keep people focused on what is being said.

In addition to working on the improvement of your sensing, you should work on the accuracy of your interpreting. To do this, you will need to think in terms of what words mean to the speakers who use them rather than what the dictionary says they mean or what they mean to you. You must try to think as the speaker thinks—judging the speaker's words by considering the speaker's knowledge, experiences, culture, and viewpoints. Like improving your sensing, improving your ability to hear what is intended requires conscious effort.

Accurately remembering what you hear requires making a conscious effort, such as taking notes. By taking care to hear what is said and by improving your ability to interpret messages, you increase your chance of accurately understanding what is said.

CONDUCTING AND PARTICIPATING IN MEETINGS

From time to time, you will participate in business meetings. These will range from extreme formality to extreme informality. On the formal end will be conferences and committee meetings, while discussions with fellow workers will be at the informal end. Whether formal or informal, the meetings will obviously involve communication, and the quality of the communication will determine their success.

Your role in a meeting will be that of either leader or participant. Of course, the leader's role is the primary one, but good participation is also vital. The following paragraphs review the techniques of performing well in either role.

Techniques of Conducting Meetings

How you conduct a meeting depends on the formality of the occasion. Meetings of such groups as formal committees, boards of directors, and professional organizations usually follow generally accepted rules of conduct called **parliamentary procedure**. These very specific rules are too numerous and detailed for review here, but when you are involved in a formal meeting, you can study one of the many books and Web sites covering parliamentary procedure before the meeting so that you know, for example, what it means to make a motion or call for a

from the tech desk

Collaborative Tools Support Virtual Meetings

It used to be that long-distance meetings or other types of collaboration required sophisticated teleconferencing or videoconferencing equipment. Today, however, anyone with Internet access can use virtual meeting software to collaborate with anyone anywhere in the world. Businesses routinely take advantage of this technology to save time and money while enhancing productivity.

In fact, you may have already used some of these online meeting tools, such as Skype, to communicate with your friends, family, and classmates. If not, becoming familiar with these tools is a good idea if you plan to enter the business world.

What can you do with a virtual meeting tool? If you're using Skype, for example, depending on the type of account you have, you can send large files to your group members (free); hold conference calls (free for Skype members); call landlines, mobile phones, or online numbers;

and use Skype Manager to create and manage accounts for your team.

So the next time you're working on a group project and need to meet, why not try a virtual meeting tool, save some travel time, and have

the meeting from the comfort of your own office, home, or dorm?

Source: Jennifer Caukin, "Workspace Blog: Tips and How-tos for Using Skype in the Workplace," *Skype*, Skype, 4 April 2012, Web, 26 July 2013.

Good talking is the foundation for effective oral communication in the workplace.

items come up during the meeting, you can take them up at the end or perhaps postpone them for a future meeting.

move the discussion along Another job you have as the leader is to control the agenda. After one item has been covered, bring up the next item. When the discussion moves off subject, move it back on subject. In general, do what is needed to proceed through the items efficiently, but do not cut off discussion before all the important points have been made. You will have to use your good judgment to permit complete discussion on the one hand and to avoid repetition, excessive details, and off-topic comments on the other.

control those who talk too much Keeping certain people from talking too much is likely to be one of your harder tasks. A few people usually tend to dominate the discussion, and one of your tasks as the leader is to control them. Of course, you want the meeting to be democratic, so you will need to let these people talk as long as they are contributing to the goals of the meeting. However, when they begin to stray, duplicate what's already been said, or bring in irrelevant matter,

vote. For less formal meetings, you can depart somewhat from parliamentary procedure, but keep in mind that every meeting has goals, and some structure is needed to help participants meet them. The following practices will help you conduct a successful meeting.

plan the meeting A key to conducting a successful meeting is to plan it thoroughly. For informal meetings, just knowing your plan may be sufficient, but before conducting formal or complex meetings, you may want to prepare an **agenda** (a list of topics to be covered). To prepare an agenda, select the items that need to be covered to achieve the goals of the meeting. Then arrange these items in the most logical order. Items that explain or lead to other items should come before the items that they explain or lead to. After preparing the agenda, make it available to those who will attend. Exhibit 10-1 shows an agenda created for a student organization meeting, but you can tailor an agenda to whatever will help you accomplish your goals. Word processing programs also have templates that may be helpful.

follow the plan You should follow the plan for the meeting item by item. In most meetings the discussion tends to stray, and new items tend to come up. As the leader, you should keep the discussion on track. If new

▼ **EXHIBIT 10-1** Example of a Meeting Agenda

Agenda

International Association of Business Communicators (IABC)
Executive Board Meeting
February 23, 2014
Clearwater Room, 7 p.m.

I. **Officer Reports**
    A.   President
    B.   Vice president
    C.   Secretary
    D.   Treasurer

II. **Committee Reports**
    A.   Public Relations Committee
    B.   Web Development Committee
    C.   Social Committee

III. **Old Business**
    A.   Bake sale fundraisers
    B.   Community service project

IV. **New Business**
    A.   Election of new officers
    B.   Attendance at exec board meetings
    C.   Hot chocolate promo

V. **Adjournment**

you should step in. You can do this tactfully by asking for other viewpoints or by summarizing the discussion and moving on to the next topic.

encourage participation from those who talk too little
Just as some people talk too much, some talk too little. In business groups, those who say little are often in positions lower than those of other group members. You should encourage these people to participate by asking them for their viewpoints and by showing respect for the comments they make.

control time
When your meeting time is limited, you need to determine in advance how much time will be needed to cover each item. Then, at the appropriate times, you should end discussion of the items. You may find it helpful to announce the time goals at the beginning of the meeting and to help the group members keep track of the time during the meeting. You might also consider including the time limits next to each item on the agenda so that readers know the time constraints before they attend the meeting and can plan their contributions accordingly.

written notes. The minutes of a meeting usually list the date, time, and location along with those persons who attended and those who were supposed to attend but were absent; some minutes may also note excused or unexcused absences. If there is an agenda, the minutes will usually summarize the discussion of each agenda topic. Exhibit 10-2 provides an example of minutes based on the meeting agenda presented in Exhibit 10-1. Generally, the person who takes minutes sends them to those who attended the meeting and requests corrections, or he or she presents the minutes at the next meeting and asks for changes or corrections at that time. Group members may also vote on whether to accept the minutes as they were recorded.

Techniques for Participating in a Meeting
From the preceding discussion of the techniques that a leader should use, you can infer the expectations for participating in a meeting. The following section reviews these expectations.

> [To ensure you have an accurate, objective account of the topics covered and decisions made at a meeting, assign the task of recording the meeting events (taking minutes) to someone.]

summarize at appropriate places
After a key item has been discussed, you should summarize what the group has covered and concluded. If a group decision is needed, the group's vote will be the conclusion. In any event, you should formally conclude each point and then move on to the next one. At the end of the meeting, you can summarize the progress made. You also should summarize whenever a review will help the group members understand what they agreed upon.

take minutes
People at meetings may hear or interpret what is said differently. In addition, you may need to refer to the discussions or to the decisions made at a meeting long after the meeting when people's memories are even less reliable. To ensure you have an accurate, objective account of the topics covered and decisions made at a meeting, assign the task of recording the meeting events (taking **minutes**) to someone. In particularly contentious or detailed discussions, it is important that everyone have a shared understanding of what has transpired.

The format of meeting minutes will depend on the nature of the meeting, group preferences, and company requirements. Some minutes are highly formal, with headings and complete sentences, while others might simply resemble casually

follow the agenda
When an agenda exists, you should follow it. Specifically, you should not bring up items not on the agenda or comment on such items if others bring them up. When there is no agenda, you should stay within the general limits of the goals for the meeting.

participate
The purpose of meetings is to get the input of everybody concerned. Your participation, however, should be meaningful. That is, you should participate when your input helps move the meeting toward its goals.

do not talk too much
As you participate in the meeting, you should speak up whenever you have something to say, but as in all matters of etiquette, always respect the rights of others to have the opportunity to speak, too. As you speak, ask yourself whether what you are saying really contributes to the discussion. Not only is the meeting costing you time, but it is costing other people's time and salaries, as well as the opportunity costs of other work they might be doing.

cooperate
Respect the leader and her or his efforts to make progress. Respect the other participants, and work with them in every practical way. You can demonstrate your willingness to

Minutes
International Association of Business Communicators (IABC)
Executive Board Meeting
February 23, 2014
Clearwater Room, 7 p.m.

Attended: Jim Solberg, Aaron Ross, Linda Yang, Tyler Baines, Sara Ryan
Absent: Jenna Kircher (excused), Rebecca Anderson (unexcused)

I. Officer Reports

A. *President:* Jim Solberg. Jim received a message from the director of university programs reminding him to view IABC's officer roster. He reviewed it on February 16 and signed the required forms.

B. *Vice president:* Linda Yang. Linda compiled job descriptions for all officer positions. Jenna put them on the IABC Web site. The link to the description of the secretary's position was not working. She is contacting Jenna to fix the link.

C. *Secretary:* Aaron Ross. Minutes of the last meeting were read and approved. He sent a thank-you note to Village Pizza for letting us have our last social there. Our average general meeting attendance is 15 even though we have 27 people who have paid dues. At the next exec board meeting, we should discuss ways to improve attendance.

D. *Treasurer:* Rebecca Anderson. No report.

II. Committee Reports

A. *Public Relations Committee:* Tyler Baines. The committee wants to have a public relations campaign in place for next fall. He will be asking for volunteers at the next general meeting.

B. *Web Development Committee:* Jenna Kircher. No report.

C. *Social Committee:* Sara Ryan. The next social will be at the campus bowling alley on March 12 at 7 p.m.

III. Old Business

A. *Bake sale fundraisers:* The $76 we earned this time is less than the $102 we earned at the last one. We will discuss creative fundraising ideas at the next general meeting.

B. *Community service project:* Linda has the forms for participating in IABC's Relay for Life team on April 26–27. She will present them at the next general meeting and ask for volunteers.

IV. New Business

A. *Election of new officers:* Linda. Officers will be elected at the April meeting. We need to encourage people to run.

B. *Attendance at exec board meetings:* Jim. Rebecca has missed every exec board meeting this semester without an excuse. The bylaws state that anyone with more than three unexcused absences in an academic year can be removed from the exec board and office. Jim sent Rebecca an email reminding her of the bylaws, but she did not respond. The Executive Board voted unanimously to remove Rebecca from the board and office. Jim will send her a letter thanking her for her service and telling her she is off the board and no longer treasurer.

C. *Hot chocolate promo:* Sara. Sara requested $20 to buy supplies to serve hot chocolate on the quad from 7:30–9:30 a.m. on Monday, March 9, to promote IABC. Linda moved to spend the $20. Jim seconded the motion. Motion carried unanimously.

V. Adjournment

The meeting adjourned at 8:30 p.m. The next general meeting will be Monday, March 2, at 7 p.m. in the Alumni Room. The next exec board meeting will be Monday, March 9, at 7 p.m. in the Clearwater Room.

Respectfully submitted,

Aaron Ross

cooperate by attending all meetings, completing your assigned tasks, or volunteering to lead or serve on a committee—anything that outwardly shows your desire to help your group achieve its objectives.

LO 10-5   Describe good phone and voice mail techniques.

USING THE PHONE

At first thought, a discussion of business phone techniques may appear to be unnecessary. After all, most of us have had experience in using the phone and may feel that we have little to learn about it. However, using the phone in professional contexts differs in some ways from how we use the phone in social contexts with family and friends.

Professional Voice Quality

Keep in mind that a phone conversation is a unique form of oral communication because the speakers cannot see each other unless they have phones that allow

> " A phone conversation is a unique form of oral communication because the speakers cannot see each other unless they have phones that allow face-to-face conversation. "

face-to-face conversation. As a result, impressions are formed only from the words and the quality of the voices. Thus, when speaking by phone, you must work to make your voice sound pleasant and friendly.

One often-suggested way of improving your phone voice is to talk as if you were face-to-face with the other person—even smiling and gesturing as you talk if this helps you be more natural. In addition, you want to be aware of your voice quality, pitch, and speed, just as you would in a face-to-face setting. You may even want to record a phone conversation and then judge for yourself how you come across and what you need to do to improve.

Courtesy

As in written communication, your goal in oral communication is to build goodwill. One way to do this over the phone is to always be courteous to your listener.

When you initiate the call, introduce yourself immediately and then ask for the person with whom you want to talk:

> This is Tessa Werner of Altman Media. May I speak with Mr. José Martinez?

If you are not certain with whom you should talk, explain the purpose of your call:

> This is Tessa Werner of Altman Media. We have a question about next week's photo shoot. May I speak with someone who can help me?

If a call is coming directly to you, identify yourself.

> Bartosh Realty. Toby Bartosh speaking. May I help you?

If you are screening calls for others, first identify the company or office and yourself and then offer assistance:

> Rowan Insurance Company. This is Audrey Peters. How may I help you?

> Ms. Santo's office. Marco Alba speaking. May I help you?

If the person whose calls you're screening is not available, be helpful by saying, "Ms. Santo is not in right now. May I ask her to return your call?" You could also ask "May I tell her who called?" or "Can someone else help you?"

Assistants to busy executives often screen incoming calls. In doing so, they should courteously ask the purpose of the calls. The response might prompt the assistant to refer the caller to a more appropriate person in the company. If the executive is busy at the moment, the assistant should explain this and either suggest a more appropriate time for a call or promise a callback by the executive. However, promise a call back only if one will be made.

Whether in or out of the office, those using mobile phones should practice the same voice quality and courtesies as on landlines.

If the person being called is on another line or involved in some other activity, it may be desirable to place the caller on hold or ask if the caller would like to leave a message. But good business etiquette dictates that the choice should be the caller's. However, if you have to put a caller on hold, be sure the hold time is reasonable. Avoid the practice of having an assistant place a call for an executive and then put the person called on hold until the executive is free to talk.

Effective Phone Procedures

At the beginning of a phone conversation you have initiated, state the purpose of the call and then use your listener's time efficiently by sticking to your point. To stay on point, you may want to outline an agenda for your call beforehand.

Courteous procedure is much the same in a phone conversation as in a face-to-face conversation. You listen when the other person is talking, refrain from interrupting, and avoid dominating the conversation.

Effective Voice Mail Techniques

Sometimes when the person you are calling is not available, you will be able to leave a voice message in an electronic voice mailbox. Not only does this save you the time involved

If Ms. Wilson would like to join them, I will be glad to make a tee time for her. She can contact me at 940-240-1003 before 5:00 PM this Friday. We look forward to seeing her at our Chief Executive Round Table meeting next Wednesday. Thank you.

Courteous Use of Cell Phones

According to the Pew Research Center, 88 percent of all Americans own a cell phone.[2] To say the least, this technology has greatly expanded our ability to communicate. To ensure that you project your best professional image, you'll want to keep in mind these tips for using your cell phone—whether you're calling, texting, or using any of your phone's other features.

1. Turn off the ringer in meetings and other places where it would be disruptive.

2. Do not use the cell phone at social gatherings.

3. Do not place the phone on the table while eating.

4. Avoid talking whenever it will annoy others. Usually this means within earshot of others.

5. Avoid discussing personal or confidential matters when others can hear you.

6. Do not talk in an excessively loud voice.

> ## At the beginning of a phone conversation you have initiated, state the purpose of the call and then use your listener's time efficiently by sticking to your point.

in calling back the person you are trying to reach, but it also allows you to leave a more detailed message than you might leave with an assistant.

You begin the message nearly the same way you would a phone call. Be as courteous as you would on the phone and speak as clearly and distinctly as you can. Tell the listener your name and affiliation. Begin with an overview of the message and continue with details. If you want the listener to take action, call for it at the end. If you want the listener to return your call, state that precisely, including when you can be reached. Slowly give the number where your call can be returned. Close with a brief goodwill message. For example, as a program coordinator for a professional training organization, you might leave this message in the voice mailbox of one of your participants:

> This is Ron Ivy from Metroplex Development Institute. I'm calling to remind Ms. Melanie Wilson about the Chief Executive Round Table (CERT) meeting next week (Wednesday, July 20) at the Crescent Hotel in Dallas. We will begin with breakfast at 7:30 AM and conclude with lunch at noon. Some of the CERT members will play golf in the afternoon at Dallas Country Club.

7. Preferably call from a quiet place, away from other people.

8. If you must talk while around people, be conscious of them. Don't hold up lines or get in the way of others.

9. Avoid using the phone while driving (the law in some states).

GIVING SPEECHES AND PRESENTATIONS

While much of your oral communication in the workplace will be the more informal types described above, some of your communication will require you to give more formal speeches and presentations. Many people find delivering speeches difficult and uncomfortable. Having to give the speech online or in front of a video camera can increase the risk of awkwardness. But you can improve your public speaking and reduce any anxiety associated with it by learning what good speaking techniques are and then putting those techniques into practice.

Determining the Topic and Purpose

Electronic means of communication have greatly expanded the types of speeches that businesspeople give. Sometimes face-to-face delivery is still the best choice, but other times a video or a live online presentation may be more appropriate.

Whatever the medium, your first step in formal speaking is to determine the topic and purpose of your presentation. In some cases, you will be assigned a topic, usually one within your area of specialization. In fact, when you are asked to make a speech on a specified topic, it is likely because of your knowledge of the topic. In some cases, your choice of topic will be determined by the purpose of your assignment, as

> **Whatever the medium, your first step in formal speaking is to determine the topic and purpose of your presentation.**

when you are asked to welcome a group, introduce a speaker, demonstrate a product, or pitch a proposal.

In other cases, you'll be asked to speak on a topic of your choice. In your search for a suitable topic, you should be guided by four basic factors. The first is your background and knowledge. Any topic you select should be one with which you are comfortable—one within your areas of proficiency. The second basic factor is the interests of your audience. Selecting a topic that your audience can appreciate and understand is vital to the success of your speech. The third basic factor is the occasion of the speech. Is the occasion a meeting commemorating a historic event? A monthly meeting of an executives' club? The keynote address at a professional conference? Whatever topic you select should fit the occasion. And fourth, consider the medium you'll be using. Whether your speech is an in-person address, a video, or a virtual presentation with remote participants may well affect your topic, as well as the length and contents of your speech.

from the tech desk

Have You Met TED?

To see great presenters in action, visit TED.com, where the world's top thinkers on a wide range of topics share their insights. Because each speaker is held to an 18-minute time limit, these speeches are often models of efficiency and clarity. And many of the talks are supported by stunning visuals, which the speakers skillfully integrate into their talks.

TED.com is sponsored by the organization TED—whose name is an acronym for technology, entertainment, and design. As the site explains, TED is a nonprofit organization "devoted to Ideas Worth Spreading." It began as the name of a conference organized by Richard Saul Wurman (the inventor of the term *information design*) and associates,

which first took place in Silicon Valley in 1984. Since then, the U.S. conference has become an annual event, and there is a yearly global conference as well.

The talks are organized by themes—such as "The Rise of Collaboration," "Not Business as Usual," "What's Next in Tech," and "Bold Predictions, Stern Warnings." You can also search the site by topic.

Here's a sampling of the business-related videos you'll find:

- In "Presentation Innovation," Dr. Hans Rosling makes public-health statistics come alive in animated visuals.
- In "Why Work Doesn't Happen at Work," Jason Fried, software entrepreneur, explores

the counterproductive qualities of office buildings.
- In "Evan Williams on Listening to Twitter Users," the co-founder of Twitter talks about the unexpected uses of Twitter that have fueled its astronomical growth.
- In "Why We Have Too Few Women Leaders," Sheryl Sandberg, COO of Facebook, analyzes the striking shortage of women in high places and offers advice for women seeking such roles.

If you haven't visited TED.com already, be sure to do so before your next speech to see how the best and brightest do it.

Preparing the Presentation

After you have decided what to talk about, why, and in what medium, you will gather the information you need. This step may involve recalling relevant experiences or generating ideas, conducting research in a library, searching through company data, gathering information online, or consulting people in your own company or other companies.

When you have your information, you are ready to begin organizing your talk. Although variations are sometimes appropriate, you should usually follow the basic pattern of introduction, body, and conclusion. This is the order described in the paragraphs below.

Usually, your first words should be a greeting appropriate for the audience. A simple "good morning" or "good evening" may suffice. Some speakers eliminate the greeting and begin with the speech, especially in more informal and technical presentations. If you have not been introduced to your audience, be sure to introduce yourself.

introduction The introduction of a speech has much the same goal as the introduction of a written or oral report: to prepare the listeners (or readers) to receive the message. But the opening of a speech usually needs some kind of attention-gaining material as well. The situation is somewhat like that of the sales message;

Other effective attention getters are quotations and questions. By quoting someone the audience would know and view as credible, you build interest in your topic. You also can ask questions. One kind of question is the rhetorical question—a question meant to provoke thought rather than elicit an answer, such as "Who wants to be free of burdensome financial responsibilities?" Another kind of question gives you background information on how much to talk about different aspects of your subject. With this kind of question, you must follow through by basing your presentation on the response. If you asked "How many of you have IRAs?" and nearly everyone raised a hand, you wouldn't want to talk about the importance of IRAs. You could skip that part of your presentation, spending more time on another topic, such as managing an IRA effectively.

Yet another possibility is the startling statement. Illustrating this possibility is the beginning of a speech to an audience of merchants on a plan to reduce shoplifting: "Last year, right here in our city, in your stores, shoplifters stole over $3.5 million of your merchandise."

In addition to arousing interest, your opening should lead into the topic of your speech. In other words, it should set up your message as the examples above do.

Following the attention-gaining opening, it is appropriate to tell your audience what you'll be talking about. In fact, in cases

> ## "By quoting someone the audience would know and view as credible, you build interest in your topic."

at least some of the people with whom you want to communicate won't initially be interested in your talk. You will need an appropriate strategy to engage the audience right away.

One possibility is a human-interest story. For example, a speaker presenting a message about the opportunities available to people with original ideas might open this way: "Nearly 150 years ago, an immigrant boy of 17 walked the streets of our town. He had no food, no money, no belongings except the shabby clothes he wore. He had only a strong will to work—and an idea."

Humor is another widely used technique. If you know that someone will be giving you a glowing introduction, you might say, "Wow, after that introduction, I can hardly wait to hear what I'm going to say."

One communications expert advises against using prepared jokes, though, especially if you are not a good joke teller. The better course, in his view, is to interject humor that is "spontaneous, related to the event, and self-directed."[3] Whichever way you choose, make sure your humor is relevant and inoffensive.

where your audience will already have an interest in what you have to say, you can begin here and skip the attention-gaining opening. Presentations of technical topics to technical audiences typically begin this way.

If you have a particular, and perhaps not widely shared, opinion about your subject, you may prefer to move into your subject indirectly—to build up your case before revealing your position. This inductive pattern may be especially desirable for sales proposals and other persuasive speeches.[4] But in most business-related speeches you should indicate your attitude toward the topic early in the speech.

body Organizing the body of your speech is much like organizing the body of a written report. You take the whole and divide it into comparable parts. Then you take those parts and divide them, and you continue to divide as far as it is practical to do so. In speeches, however, you are more likely to use factors (subtopics) rather than time, place, or quantity as the basis of division because in most speeches your presentation is likely to be built around issues and questions that pertain to the subject. Even so, time, place, and quantity subdivisions are possibilities.

call for action. The following close of a speech advocating a new marketing strategy illustrates this point:

> In short, switching from print to email marketing will save money, generate more sales leads, and gain us a bigger share of the market.

LO 10-7 Identify and select appropriate presentation methods.

Choosing the Presentation Method

In addition to determining the speech's content and structure, you need to decide on your method of presentation—that is, whether to present the speech extemporaneously, memorize it, or read it.

presenting extemporaneously Extemporaneous presentation is by far the most popular and effective method. With this method, you first thoroughly prepare your speech, as outlined above. Then you prepare notes and present

Successful oral presentations to large audiences are the result of thorough preparation.

> You need to emphasize the transitions between the divisions because, unlike the reader who can see them in a written report, the listener may miss them if they are not stressed adequately.

You need to emphasize the transitions between the divisions because, unlike the reader who can see them in a written report, the listener may miss them if they are not stressed adequately. Without clear transitions, you may be talking about one point, and your listener may think you are still on the previous point.

conclusion Like most reports, the speech usually ends by drawing a conclusion or conclusions. The conclusion is the culmination of your report—the point at which you achieve your communication goal. You should consider including these three elements in your close: (1) a restatement of the main point, (2) a summary of the key points developed in the presentation, and (3) a request for action on the part of your audience.

Bringing the speech to a climactic close—that is, making the conclusion the high point of the speech—is usually effective. Present the concluding message in words that gain attention and will be remembered. In addition to concluding with a summary, you can give an appropriate quote, use humor, or

the speech from them. You usually rehearse, making sure you have all the parts clearly in mind, but you make no attempt to memorize. Extemporaneous presentations generally sound natural to the listeners, yet they are (or should be) the product of careful planning and practice.

memorizing The most difficult method is to present your speech from memory. If you are like most people, you find it hard to memorize a long succession of words. Also, if you memorize, you can get flustered if you miss a word or two during your talk. You may even become panic-stricken.

For this reason, few speakers who use this method memorize the entire speech. Instead, they memorize key passages and use notes to help them through the speech. A delivery of this kind is a cross between an extemporaneous presentation and a memorized presentation.

reading The third presentation method is reading. Unfortunately, most of us tend to read aloud in a monotone. We also miss punctuation marks, fumble over words, lose our place, and so on. But many speakers have learned to read a speech in an interesting, smooth way, and with effort you can, too. One effective way is to practice with a recorder and listen to yourself. Then you can be your own judge of what you must do to improve your delivery. You would be wise not to read speeches until you have mastered this presentation method.

In most business settings, it is considered inappropriate to read. Your audience will want more personal interaction than that. However, when you are acting as the official spokesperson for a company or organization—for example, when responding to a crisis or giving an important announcement—reading from a carefully prepared speech is appropriate and even expected. Many top executives today use teleprompters when delivering read speeches, and many of these speeches are well done.

LO 10-8   Choose the means of audience feedback.

Choosing the Means of Audience Feedback

Traditionally, formal speeches have been one-way communication, with the presenter delivering a well-prepared talk and restricting audience participation to applause or questions at the end. Today's audiences are likely to expect more two-way communication, even during the talk.

With webinars, or live presentations conducted via the Web, participants expect to be kept involved through polls, questions, and even live chat with the speaker. But even in real settings,

from the tech desk

Look Like a Pro with PowerPoint Keyboard Shortcuts

If you've ever attended a presentation during which the speaker had trouble finding the right view, going back to an earlier slide, or using other features of the software, you know how distracting that can be.

Familiarizing yourself with keyboard shortcuts can help you move around quickly and skillfully in your presentation. Here, for example, are some PowerPoint shortcuts for managing your slides like a pro. Most of them also work in PowerPoint 2013.

To Do This	Press
Start your presentation in Slide Show view.	F5
Show or hide the arrow pointer.	A or =
Change the pointer to a pen (to draw on your slides while presenting).	CTRL + P
Change the pen back to an arrow.	CTRL + A
Show or hide the ink markup.	CTRL + M
Perform the next animation or advance to the next slide.	N, Enter, Page Down, Right Arrow, Down Arrow, or Spacebar
Perform the previous animation or advance to the next slide.	P, Page Up, Left Arrow, Up Arrow, or Backspace
Go to a certain slide.	Slide number + Enter
Return to the first slide.	Press and hold Right and Left Mouse buttons for 2 seconds
Display a blank black slide, or return to the last-viewed slide from a blank black slide.	B or Period
Display a blank white slide, or return to the last-viewed slide from a blank white slide.	W or Comma
Stop or restart an automatic presentation.	S
End a presentation.	Esc or Hyphen

Source: Microsoft, "Keyboard Shortcuts for Use While Delivering a Presentation in PowerPoint 2010," *Microsoft.com*, Microsoft Corporation, n.d., Web, 31 July 2013.

audiences often expect to be invited to contribute as well—whether by raising their hands, asking questions, voting with a clicker,[5] or tweeting their answers or comments.[6]

Even if you decide not to use back-channeling or audience feedback during your talk, you may want to incorporate audience participation at some points in your presentation to break up your speech and keep the attention level high. How you do so will depend on the media you will use to deliver your talk and your relationship with your audience.

Whether or not you invite audience feedback during your talk, consider inviting the audience to email you afterward with their questions and comments. Many webinar presenters follow up their online sessions with evaluation forms that they email to the participants. This kind of post-presentation interaction can help you improve your presentations, acquire positive statistics to show to your boss, and build productive relationships with those who heard you speak.

LO 10-9 Plan visuals to support oral communication.

SUPPORTING YOUR TALK WITH VISUALS

Audiences expect that most presentations will include a visual component to help them follow along and stay interested.

Plus, as professional trainers know, visuals also aid immensely in retention of the presented information. According to a U.S. government Web site that provides guidelines for safety and health training, an audience who has heard an oral presentation with no visuals recalls about 10 percent of the presentation. If the audience has been given only a visual with no oral component, that number increases to 35 percent. Combining the two media to present the information raises the number to 65 percent.[7]

Planning the most effective visuals to go with your oral report or speech should thus be integral to planning the talk overall. Chapter 3 also provides tips for choosing the best visuals for the types of data you may need to present in your oral report or speech.

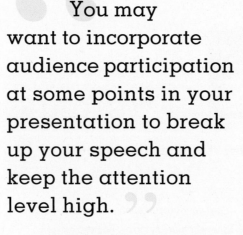

> You may want to incorporate audience participation at some points in your presentation to break up your speech and keep the attention level high.

What Kinds of Information to Present Visually

The purpose and content of your talk will determine the points at which a visual might be appropriate, but the following guidelines can help.

- Most presenters who will be using supporting visuals in electronic form begin their talk with a *title slide* that conveys the point of the talk, the speaker's name, and the name and logo of the sponsoring company (if any).

- Providing an *outline* of what you will cover can help your listeners comprehend you better as you proceed through your talk.

- *Charts, tables,* and *line art* or *diagrams* will help your listeners understand statistical information or a process.

- *Photographs* help your listeners form a concrete image of something you're discussing, whether a happy customer, your products, or a vision of the future.

- If not overdone, *animation* (zooming in and out, bars or columns that grow, moving text or objects, photographs that change) can reinforce a point you're making or help you show a process.

- Most popular digital presentation tools make it easy to incorporate *multimedia* into your show. Consider embedding

In oral presentations, appropriately handled visuals can help you communicate clearly.

audio or video files in your slides to bring in an expert opinion, share a story, or show an example. If you know you will have an Internet connection, consider embedding hyperlinks to get audience feedback (e.g., via Twitter).

When your talk will not be electronically supported—and such talks still take place frequently—you should consider incorporating visuals (except animation and multimedia, of course) into a print handout for your listeners. Many of these types of visuals can also be adapted to a flip chart or whiteboard. Be sure to consult the advice in Chapter 3 when designing your visuals, including how to cite the sources for any copyrighted visuals you'll be using.

Techniques for Using Visuals

Any visuals you use will need to be skillfully incorporated into your talk. Here are some basic dos and don'ts:

- Make certain that everyone in the audience can see the visuals. Too many or too-faint lines on a chart, for example, can be hard to see. An illustration that is too small can be meaningless to people far from the speaker. Even fonts must be selected and sized for visibility.

PowerPoint—the first widely used, and still the most popular, presentation software. But cloud applications (those hosted by a service provider, not installed on your computer) and apps for mobile devices have really broadened your options.

For example, you may have tried Prezi, a cloud application (available in a free version) that enables you to zoom in and out as you move around an online canvas. SlideRocket is another popular PowerPoint alternative. Even its free version comes with design, timing, and multimedia options—such as adding a live Twitter feed to a slide—that PowerPoint doesn't offer. And of course there's Google Drive's presentation tool, which has almost as many features as PowerPoint, along with the added benefit of enabling on-line collaboration without the need for additional software.

Whichever tool you use, stay clear of the following pitfalls:

- *Putting too much on a slide.* This is one of the most common errors presenters make. Remember that your talk, not your slides, should convey most of your detailed information. Making a slide or Prezi object too crowded is a surefire way to make your audience stop looking at it. For PowerPoint, we recommend approximately six brief

> ## " An illustration that is too small can be meaningless to people far from the speaker. "

- If necessary, explain the visual. Remember that the visual is there to help you communicate content, not just to add visual interest.

- When discussing a visual, refer to it with physical action and words. Use a laser pointer or slide animations to emphasize each point you're making.

- Talk to the audience—not to the visuals. Look at the visuals only when the audience should look at them. When you want the audience to look at you, you can regain their attention by covering the visual or making the screen white or black (in PowerPoint, toggle the W or B keys).

- Avoid blocking the listeners' views of the visuals, and make certain that the listeners' views are not blocked by lecterns, pillars, or furniture.

Use of Presentation Software

When most people think of designing visuals to go with a report or speech, they automatically think of

bullet points per slide and no more than two levels of bullet points. Exhibit 10-3 provides a checklist and an example for placing text on a slide.

- *Making the contents on the slide too small.* This is probably the second most common error. When you're designing your slides, you're looking at them on your computer, so small type and visuals are legible. But your audience for an in-person speech will be trying to read your slides from

▼**EXHIBIT 10-3**   An Attractive, Readable Text Slide in PowerPoint 2013

Placing Text on a Slide

- Choose an appropriate font size
  - Headings: 30 pt. minimum
  - Text: 24 pt. minimum
- Keep text minimal
  - Approximately six lines of bulleted text per slide
  - Four to six words per line
- Use one or two bullet levels
- Use dark text on a light background

many feet away from the screen, and even those viewing webinar slides on their computers won't like tiny content. We recommend 30-point type or larger for headings and 24-point type or larger for the text (Exhibit 10-3).

- *Using an inappropriate theme or unreadable color combinations.* Students tend to want to choose elaborate and flashy designs for their slides. But many templates included with the software are too busy for most presentation purposes, and certain text and background color combinations, such as red text on a blue background, make the slides difficult to look at, let alone read. Err on the side of conservatism. Black text on a white or pale background with the company logo discretely in the corner can make for easily readable slides as well as a clean, professional look.

- *Using too much animation.* Another tendency amateurs have is to go wild with the options for transitions between slides and with other dynamic features. Prezi is particularly susceptible to this problem; all of the zooming in and out

Make your decisions given the purpose and occasion of your talk. If you're presenting information that your listeners will want to be able to review later, give them complete copies of your slides or even a complete copy of your report or proposal. If you think your listeners will want to take notes as you go, give a handout at the start of the talk that has room for notes; otherwise, save the handout until the end so that your speech can benefit from the element of surprise. If the information you're covering can be distilled into a useful quick-reference sheet or a list of resources, handing that out at the end of your talk is sufficient. As you can see, there is no overarching rule; what's best will depend on the situation.

If possible, always bring one complete print copy of your slides or a detailed outline of your talk to the presentation venue. Computers crash; projector bulbs go out; Internet connectivity sputters. If technological disaster strikes, it's likely that the venue will still have a copier, enabling you to deliver your talk with print support.

> "The affordable costs make Web technology attractive to both large and small businesses for presentations to both large and small audiences."

can actually make viewers dizzy. Again, err on the side of conservatism. Remember that the goal is not to dazzle the audience with visual activity but to achieve your communication purpose.

- *Being inconsistent across slides.* Your whole presentation should have a consistent look. This means that similar slides should be formatted similarly (e.g., for text-based slides, use the same font throughout for the slide titles, as well as the same body font) and that no slide should look as though it belongs in a different presentation.

- *Reading verbatim what's on the slides (or, the opposite, not integrating what's on the slide into your talk).* Visuals should be included to help you present—not the other way around. Strike a good balance between helping the audience read what's on the screen and keeping the focus on you and what you have to say.

Use of Handouts

Should you supplement your presentation with handouts? If so, what kind? And at what point in the talk? These are good questions. You know from experience that some speakers provide no handouts, while others provide complete copies of their slides. If they have handouts, some speakers distribute them at the beginning of the talk, while others wait until the end.

LO 10-10   Plan and deliver effective Web-based presentations.

DELIVERING WEB-BASED PRESENTATIONS

Live Web presentations—commonly called *webinars* or *Web events*—have become a popular genre of business communication. They eliminate the speakers' and participants' travel expenses, and they can reach huge audiences. Plus, many powerful, easy-to-use applications for conducting such events are available. WebEx, once the undisputed leader in this area, now has competitors with such products as Citrix's GoToMeeting, Microsoft's Lync Online, Adobe's Acrobat ConnectPro, Dimdim, omNovia, and more. The affordable costs make this technology attractive to both large and small businesses for presentations to both large and small audiences.

Varieties of Web Presentations

While the terms *webcast*, *Web meeting*, and *webinar* are sometimes used interchangeably, and one Web-based application might support all three types of communication, the consensus of those in the industry seems to be that the terms have—or should have—different meanings.

A **webcast** typically consists of live video being "broadcast" to the audience. Like a television or radio show, it provides no means of audience participation. A **Web meeting** is a Web-based get-together, usually for a small group of people who have chosen to conduct their interactions online rather than in a conference room. Of the three terms, **webinar** is the most synonymous with *presentation*. Here, a main speaker or speakers present on their topic of expertise, but the audience almost always has an opportunity to participate.

Special Guidelines for Web Presentations

To deliver a live virtual presentation effectively, you'll need to prepare for certain preliminary, delivery, and closing activities.

Your first step is to choose a user-friendly technology that supports the type of webinar you want to give. Then you'll prepare and send out announcements of the presentation along with a note encouraging the audience to pretest their systems before the designated start time for the presentation. A day or two before and the day of the presentation, most presenters send email reminders of the event.

It is a good idea to line up a technical person to troubleshoot during the presentation, since some participants will have trouble connecting, others will fall behind, and software or Internet glitches can occur. Also, you'll need to arrange ahead of time for an assistant if you need one. An assistant can help you keep track of time, take over if necessary, and provide other help to keep the presentation going smoothly.

You may want to create something for early arrivers to view in the first 5 to 10 minutes before you start. This could be an

from the tech desk

Virtual Presentations: The Next Best Thing to Being There

Web-based presentation tools offer many options for participant interaction. Featured here is the interface of omNovia, a popular Web conferencing platform.

The participant area (circled in red) shows who is logged in. The area below it (circled in blue) shows who is currently presenting. The various options for presenters and participants (circled in green) let people easily navigate among functions. In omNovia, participants can chat privately or publicly, participate in Q&A sessions, and take surveys and polls. Presenters can show PowerPoint presentations, create outlines or illustrations on a virtual white board, and share the content of their desktops with participants. And if presenters have handouts or other documentation for the audience, they can create libraries in their conference rooms where participants can access pdf, HTML, and Word files. In

addition, presentations can be recorded and archived. Integration of social media tools such as Twitter and YouTube videos is also possible.

In other words, virtual presentation platforms can afford the same (and sometimes

more) opportunities for communication as face-to-face presentations, making them viable options when face-to-face interactions are not possible or desired.

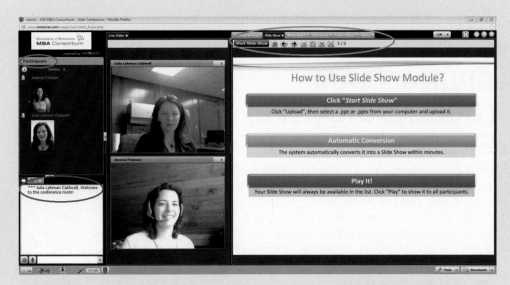

Source: omNovia, "Characteristics and Features of the omNovia Web Conference," *omNovia.com*, omNovia Technologies, n.d., Web, 20 June, 2013.

> You will need to take special care to plan the presentation—to determine the sequence of the presentation as well as the content of each team member's part.

announcement, news of an upcoming presentation, or information about your products and services. At the end of the talk, you will want to tell participants where to access additional information, including your slides, video recording of the presentation, and other business links.

The delivery of your presentation will be much like that for other presentations, except you will be doing it from your desktop using a microphone and perhaps a Web camera (if the talk will include video). You may want to use the highlighter, drawing tools, or animation features of PowerPoint or the webinar application to help you emphasize key points that you would otherwise physically point to in a face-to-face presentation. You will want to plan breaks during which you will poll or quiz the audience or handle questions that have come in through the chat tool.

In the closing, you will want to allow time to make any final points and answer any remaining questions. Watching your time is critical because some systems will drop you if you exceed your requested time.

Overall, presenting virtually requires the same keys to success as other presentations—careful planning, attentive delivery, and practice.

presentation as well as the content of each team member's part. You also will need to select supporting examples and design any visual components carefully to build continuity from one part of the presentation to the next.

Groups should plan for the physical aspects of the presentation, too. You should coordinate the type of delivery, use of notes, and attire to present an impression of competence and professionalism. You should also plan the transitions from one presenter to the next so that the team will appear coordinated.

Another presentation aspect—physical staging—is important as well. For face-to-face presentations, team members should decide where they will sit or stand, how visuals will be presented, how to change or adjust microphones, and how to enter and leave the speaking area. For videotaped and virtual presentations that will show the speakers and their settings, you will need to determine what kind of background to have and what distance the presenter will sit from the camera.

Attention to the close of the presentation is especially important because you'll need to ensure that all the pieces delivered by the team members are brought together into a coherent, complete conclusion. Teams need to decide who will present the close

---

LO 10-11 Work effectively with a group to prepare and deliver a team presentation.

GIVING TEAM (COLLABORATIVE) PRESENTATIONS

Team presentations are a common school assignment, and they're common in business as well. To give this type of presentation, you can apply all the preceding advice about giving individual presentations and speeches. You can also use much of Chapter 8's advice on preparing written reports collaboratively. But you will need to adapt the ideas to an oral presentation setting.

First, you will need to take special care to plan the presentation—to determine the sequence of the

It is a good idea to rehearse a team presentation with some colleagues before delivering it to an important audience.

and what will be said. If a summary will conclude the talk, the member who presents it should attribute key points to appropriate team members. If there is to be a question-and-answer session, the team should plan how to conduct it. For example, will one member take the questions and direct them to a specific team member? Or will the audience be permitted to direct questions to specific members? Some type of final note of appreciation or thanks needs to be planned, with all the team nodding in agreement or acknowledging the final comment in some way.

Teams should also allow for plenty of rehearsal time. They should practice the presentation in its entirety several times as a group before the actual presentation. During these rehearsals,

individual members should critique each other's contributions, offering specific ways to improve. After first rehearsal sessions, outsiders (nonteam members) might be asked to view the team's presentation and critique the group. Moreover, the team might consider videotaping a rehearsal of the presentation so that all members can evaluate it.

As you can see, an effective report or speech takes knowledge, preparation, and skill. But the rewards justify the effort. By following this chapter's advice and using good judgment, you can communicate orally in any medium or situation. This ability will make you a valued asset to your employer and further your professional success.

Communicating
in the Job Search

eleven

Did you know that 80 percent of open jobs are not advertised? Or that the average number of applicants for a job is 118, with only 20 percent getting an interview? Or that the average job interview is 40 minutes? According to Jacquelyn Smith of *Forbes,* these statistics mean that finding a job requires more than preparing a résumé and cover letter and preparing answers to interview questions. In addition to considering advertised jobs, today's job seekers must get their information out through several channels, including their personal and professional networks and social media outlets. In addition, applicants must arrive at interviews prepared not only to answer questions but also to present their own stories and ask questions.[1]

Whether it involves an internship, your first job, or a job further down your career path, job seeking is directly related to your success and your happiness. This chapter will help you conduct your search and prepare the documents essential for your success. ■

LEARNING OBJECTIVES

LO 11-1 Develop and use a network of contacts in your job search.

LO 11-2 Assemble and evaluate information that will help you select a job.

LO 11-3 Identify the sources that can lead you to an employer.

LO 11-4 Compile résumés for print and electronic environments that are strong, complete, and organized.

LO 11-5 Write targeted cover messages that skillfully sell your abilities.

LO 11-6 Explain how you can participate effectively in an interview.

LO 11-7 Write application follow-up messages that are appropriate, friendly, and positive.

LO 11-8 Maintain your job-search activities.

workplace scenario

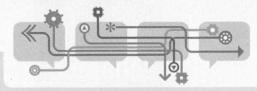

Finding Your First Post-College Job

Introduce yourself to this chapter by assuming a role similar to one you are now playing. You are Julia Alvarez, a student at Wilmont University. In a few months, you will complete your studies for work in human resources management.

You believe that it is time to begin seeking the job for which those studies have been preparing you. But how do you do this? Where do you look? What does the search involve? How should you present yourself for the best results? Once you get an interview, how should you prepare for it? After the interview, what should you say in a thank-you note to the employer?

And what about other correspondence you might have to write to accept a position, decline a position, or perhaps resign from your current job? The answers to these and related questions are reviewed in the following pages.

CONDUCTING THE JOB SEARCH

Of all the things you do in life, one of the most important is getting a job. Therefore, you must be careful and diligent in your job search. The following review of job-search strategies should help you succeed.

LO 11-1   Develop and use a network of contacts in your job search.

Building a Network of Contacts

You can begin the job search long before you are ready to find employment. In fact, you can do it now by building a **network of contacts**. More specifically, you can build relationships

your work ethic and your ability in the classroom is a great way to get your professors to know you and help you. Take advantage of opportunities to meet your professors outside the classroom, especially the professors in your major field.

Obviously, meeting business professionals can also lead to employment contacts. You may already know some through family and friends. But broadening your relationships among businesspeople would be helpful. You can do this in various ways, especially through college professional groups such as the Association for Information Technology Professionals, Delta Sigma Pi, and the Society for the Advancement of Management. By taking an active role in the organizations in your field of study, especially by working on program committees and by becoming an officer, you can get to know the executives who serve as guest speakers.

> The wider your circle of friends and acquaintances, the more likely you are to make employment contacts.

with people who can help you find work when you need it. Such people include classmates, professors, and businesspeople.

Right now, your classmates are not likely to be holding positions in which they make or influence hiring decisions, but some of them may know people who can help you. In the future, when you want to make a career change, they may hold such positions. The wider your circle of friends and acquaintances, the more likely you are to make employment contacts.

Knowing your professors and making sure that they know you can also lead to employment contacts. Because professors often consult for business, they may know key executives and be able to help you contact them. Professors sometimes hear of position openings and can refer you to the hiring executives. Demonstrating

You also might meet businesspeople online. If you share a particular interest on a blog or are known as one who contributes valuable comments to others' blogs, you may get some good job leads there. Likewise, you can network on social media sites such as LinkedIn by starting a group or actively contributing to one.

In addition to these more common ways of making contacts, you can use some less common ones. By working in community organizations (charities, community improvement groups, fundraising groups), you can meet community leaders. By attending meetings of professional associations (every field has them), you can meet the leaders in your field. In fact, participation in virtually any activity that provides contacts with businesspeople can be mutually beneficial, both now and in the future.

Obtaining an Internship

Internships are a wonderful way to network with people in your field, gain professional knowledge and experience, or simply learn whether your current field is where you want to build a career.

One source estimates that as many as 75 percent of employers look for internships on students' résumés.[2] According to a recent survey by the National Association of Colleges and Employers, "graduates who took part in a paid internship were more likely to get a job offer, have a job in hand by the time they graduated, and receive a higher starting salary offer than their peers who undertook an unpaid internship or no internship at all."[3] While many students may find paid internships more desirable, unpaid internships can also be valuable. Many students can earn college credits toward their degree (and perhaps get tuition for these credits waived or reduced) or receive a stipend, and the U.S. Department of Labor has laws governing the fair use of unpaid interns' work.[4] The key is to make sure that an unpaid internship provides the opportunity to gain real-world, marketable skills and to make sure the employer does not unfairly take advantage of an intern's unpaid time, expertise, and labor.

Though a quick Web search for internships will net several links to internship types, the first step in finding an internship simply may be to contact your school's career services office. You may also want to use many of the networking strategies discussed in the previous section.

LO 11-2 Assemble and evaluate information that will help you select a job.

Identifying Appropriate Jobs

To find the right job (or internship), you need to investigate both internal and external factors. The best fit occurs when you have carefully looked at yourself: your education, personal qualities, experience, and any special qualifications. However, to be realistic, these internal qualities need to be analyzed in light of the external factors. Some of these factors may include the current and projected job market, economic needs, location preferences, and family needs.

analyzing yourself When you are ready to search for a job, you should begin the effort by analyzing yourself. In a sense, you should look at yourself much as you would look at a product or service that is for sale. After all, when you seek employment, you are really selling your ability to work—to do things for an employer. A job is more than something that brings you money. It is something that gives equal benefits to both parties—you and your employer. Thus, you should think about the personal qualities you have that enable you to be an accountable and productive worker that an employer needs. This self-analysis should cover the following categories.

> "The best fit occurs when you have carefully looked at yourself: your education, personal qualities, experience, and any special qualifications."

Education. Perhaps you have already selected your career area such as accounting, organizational or technical communication, economics, finance, information systems, international business, management, or marketing. If you have, your task is simplified, because your specialized curriculum has prepared you for your goal. Even so, you may be able to note special points—for example, electives that have given you additional skills or that show something special about you (such as psychology courses that have improved your human relations skills, communication courses that have improved your writing and speaking skills, or foreign language courses that have prepared you for international assignments).

If you have pursued a more general curriculum, such as one in general business or liberal arts, you will need to examine what

The Web sites of universities' career centers, like the one shown here, can be a great place for students to start their job search.
Source: Used by permission from the University of Cincinnati.

communication matters

The Where, What, and Whys of Hiring

Surveying 225 employers from a population of 10,000 U.S. companies, Millennial Branding and Experience Inc. examined the role of internships in securing an entry-level position, qualities employees look for, and where employers find their new hires. Highlights of the study include the following:

- Internships. Even though employers are split 50/50 on whether they hire interns, 91 percent say students should have one or two internships before graduating; 87 percent say that the internships should last at least three months.
- Key Skills. Communication skills (98 percent), a positive attitude (97 percent), and teamwork skills (92 percent) were rated by employers as "important" or "very important" in people hired for entry-level positions.
- Employers' Hiring Resources. Employers use social networks to varying degrees to find employees. Few use social networking all or most of the time (16 percent). Instead, many use job boards (48 percent) and employee referrals (44 percent), a more traditional form of networking. However, many more do use social networks for background checks (35 percent). Within that 35 percent, LinkedIn (42 percent) and Facebook (40 percent) are the most popular, while Google + (15 percent) and Twitter (2 percent) are less popular.
- How to Stand Out. Managers rank the following as "important" or "very important" when they review a candidate's qualifications: relevant course work (69 percent); referrals from a boss or professor (65 percent); leadership positions in campus organizations (50 percent); and entrepreneurship (29 percent).
- How to Fail. Respondents rank unpreparedness for an interview (42 percent) and a bad attitude (26 percent) as major turn-offs in potential or new employees.

Source: Dan Schwabel, "Student Employment Gap Study," *Millennial Branding*, Millennial Branding and Experience, Inc., 14 May 2012, Web, 1 July 2013.

> " Your part-time server job or summer construction job may not seem like a big deal to you, but these jobs provide you with assets that any employer in any company can use. "

skills and knowledge will transfer to the workplace. Perhaps you will find an emphasis on computers, written communication, human relations, or foreign languages—all of which are highly valued by businesses. Or perhaps you will conclude that your training has given you a strong general base from which to learn specific business skills.

In analyzing your education, you should look at the quality of your record—grades, projects, honors, and any special recognitions you can highlight in an application.

Personal qualities. Your self-analysis also should cover your personal qualities. Employers often use **personality tests** such as the Myers-Briggs to screen new hires, and you can take them online as well as at most campus career centers. Qualities that relate to working with people are especially important. Qualities that indicate an aptitude for leadership or teamwork ability are also important.

Of course, you may want to check with friends to see whether they agree with your assessments. You also may need to check your record for concrete evidence supporting your assessments. For example, organization membership and participation in community activities are evidence of people and teamwork skills. Holding an office in an organization is evidence of leadership ability. Participation on a debate team, college bowl, or collegiate business policy team is evidence of communication skills.

Work experience. Work experience in your major field deserves the most emphasis, but work experience not related to the job you seek also can tell something important about you.

Your part-time server job or summer construction job may not seem like a big deal to you, but these jobs provide you with assets that any employer in any company can use, such as attention to detail, initiative, team skills, communication skills, and the ability to work well under pressure. You don't want to undersell this experience.

Special interests. As we've mentioned, your self-analysis should include special qualifications that might be valuable to an employer. Your analysis can also include personal interests that may be relevant to types of positions you are seeking. To illustrate, athletic experience might be helpful for work with a sporting goods distributor, a hobby of automobile mechanics might be helpful for work with an automotive service company, and an interest in music might be helpful for work with a radio station or an online music Web site.

You also might take an **interest inventory** such as the Strong Campbell Interest Inventory or the Minnesota Vocational Interest Inventory. These tests help match your interests to those of others successful in their careers. Most college counseling and career centers make these tests available to their students, and some are available online. Getting good help in interpreting the results is critical to providing you with valuable information.

Career fairs and job boards are good places to look for announcements of job openings.

analyzing outside factors After you have analyzed yourself, you need to combine this information with the work needs of business and other external influences to give realistic direction to your search for employment. Is there a demand for the work you want to do? Are jobs readily available? Can you make a living at doing what you want to do? Where is the kind of work you are seeking available? Are you willing to move? Is such a move compatible with others in your life—your partner, your children, your parents? Does the location meet your lifestyle needs? Although the job market may drive the answer to some of these questions (e.g., relocating in order to get the job you want), you should answer them on the basis of what you know now and then conduct your job search accordingly. Finding just the right job should be one of your most important goals. Reading about various careers in the *Occupational Outlook Handbook* at www.bls.gov/ooh/ is one way to learn about the nature of the jobs you seek as well as salary range and demand.

LO 11-3 Identify the sources that can lead you to an employer.

Finding Your Employer

You can use a number of sources in your search for an employer with whom you will begin or continue your career. Your choice of sources will probably be influenced by the stage of your career.

career centers If you are seeking an internship or just beginning your career, one good possibility is the career center at your school. Most schools have career centers, and these attract employers who are looking for suitable applicants. Many centers offer excellent job-search counseling and maintain databases on registrants' school records, résumés, and recommendations that prospective employers can review. Most have directories listing major companies with contact names and addresses. And most provide interviewing opportunities. Campus career centers often hold **career fairs**, which are an excellent place to find employers who are looking for new graduates or interns as well as to gather information about the kinds of jobs different companies offer. By attending them early, you often find out about internships and summer jobs or gather ideas for selecting courses that might give you an advantage when you do begin your career search.

network of personal contacts As we have noted, the personal contacts you make can be the leading means of finding employees. Business acquaintances may provide job leads outside those known to your friends.

classified advertisements Help-wanted advertisements in newspapers and professional journals, whether online or in print, provide good sources of employment opportunities for many kinds of work. Keep in mind that they provide only a partial list of jobs available. Many jobs are snapped up before they reach the classifieds, so be sure you are part of the professional grapevine in your field.

online sources In addition to finding opportunities in classifieds, you also will find them in online databases. Monster.com, for example, lists jobs available throughout the United States and beyond, with new opportunities posted regularly. Many companies post job openings on the Web, some with

from the tech desk

Make Your LinkedIn Profile Work for You

By one study's estimate, of the 45 percent of companies using social media to recruit employees, nearly 90 percent use LinkedIn to find and screen potential applicants—far more than they use Facebook (46 percent) or Twitter (22 percent). Whether you're new to LinkedIn or a LinkedIn veteran, Victor Reklaitis provides tips from industry experts for creating a personal brand that employers will respond well to:

1. Include a photo. A LinkedIn study shows that profiles with photos are seven times more likely to get responses.

2. Have at least 50 connections. You'll be 12 times more likely to get information about opportunities.

3. Provide a complete work history.

4. Customize your account settings so that the URL for your profile is www.linkedin .com/YourName. It's easier to remember than the longer default URL, and you can easily include it on a business card.

5. Tell an interesting and unique story in the summary that lets an employer learn more about you.

6. Avoid buzzwords to describe yourself; they don't allow you to stand out. According to LinkedIn, words in the list of top ten overused words of 2011 included *creative, effective, innovative, dynamic, motivated,* and *extensive experience.*

7. Make sure you give as well as receive. That is, in addition to using LinkedIn to get attention from employers, share your expertise (e.g., post an article you found on an industry trend).

8. Write a headline for your profile that accurately and precisely describes who you are.

Sources: Derek Thompson, "Here Come the Raises," *The Atlantic*, The Atlantic Monthly Group, 1 Mar. 2012, Web, 1 July 2012; Victor Reklaitis, "Manage Your Personal Brand Online: Tips for LinkedIn," *Investors. com,* Investor's Business Daily, Inc., 29 June 2012, Web, 20 June 2013.

---

areas dedicated to new college graduates. If you are working now, you may want to check your company's intranet for positions. And professional associations often maintain job databanks. Furthermore, you could use social media such as blogs, Facebook, LinkedIn, and Twitter to search for jobs or post queries about job openings that readers might know of.

employment agencies Companies that specialize in finding jobs for employees can be useful. Of course, such companies charge for their services. The employer sometimes pays the charges, usually if qualified applicants are scarce. **Executive search consultants** (headhunters) are commonly used to place experienced people in executive positions.

Employment agencies can also help job seekers gain temporary employment. Temping can lead to permanent employment. It allows the worker to get a feel for the company and the company to observe the worker before making a job commitment. You can also gain valuable on-the-job training in a temporary assignment.

prospecting Some job seekers approach prospective employers directly, by either personal visit, mail, or email. Personal visits are effective if the company has an employment office or if a personal contact can set up a visit. Mail contacts typically include a résumé and a cover letter. An email contact can include a variety of documents and be sent in various forms. The construction of these messages is covered later in the chapter.

PREPARING THE APPLICATION DOCUMENTS

How you pursue the employment opportunities that your research yields depends on your circumstances. When it is convenient and appropriate to do so, you make contact in person. It is convenient when the distance is not great, and it is appropriate when the employer has invited such a contact. When a personal visit is not convenient and appropriate, you may have to apply online or by mail, email, or fax.

Whether or not you apply in person, you are likely to need some written material about yourself. If you apply in person, you probably will take a résumé with you to leave as a record of your qualifications. If you do not apply in person, of course, the application is conducted completely in writing. Typically, it consists

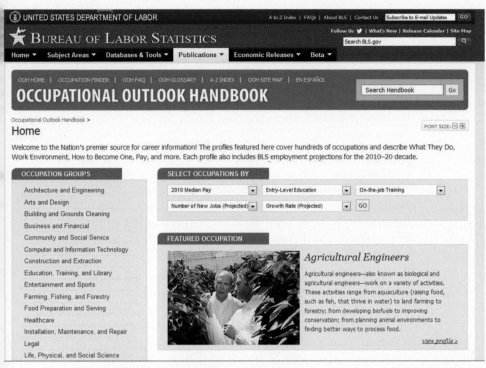

The Occupational Outlook Handbook *is one of the best resources for finding out about a wide variety of jobs, including their education requirements, expected earnings, job duties and conditions, and more.*

use a detailed cover message alone, while others include all three documents. You should use whatever will best support your case.

LO 11-4 Compile résumés for print and electronic environments that are strong, complete, and organized.

CONSTRUCTING THE RÉSUMÉ

Writing a résumé requires you to use many of the same strategies we've discussed in previous chapters for writing business documents. To plan your résumé, you'll analyze your goals, purpose, and audience. Then you will draft your résumé and revise and edit it until it represents your best professional work. As with any other business document, you also need to consider your channel of communication—whether your audience will read your résumé as a printed (hard copy) document or an electronic document. Constructing the content for each type is similar, but because you are likely to need both printed and electronic copies of your résumé, you will need to be familiar with the special considerations for each type.

> Designed for quick reading, the résumé lists facts that have been arranged for the best possible appearance.

of a résumé and a cover message. At some point in your employment efforts, you are likely to use each of these documents.

Approach preparing a résumé and cover letter as you would when preparing a sales campaign. Take a personal inventory, performing the self-analysis discussed earlier in the chapter, and list all the information about you that you believe an employer would want to know. Then learn as much as you can about the company—its plans, its policies, its operations. You can study the company's Web site, read its annual report and other publications, find any recent news articles about the company, and consult a variety of business databases. You should also learn as much as you can about the requirements of the work the company wants done.

With this preliminary information assembled, you are ready to plan the application. First, you need to decide what your application will consist of. Will it be just a cover message; a cover message and a résumé (also called a *vita, curriculum vita, qualifications brief,* or *data sheet*); or a cover message, résumé, and reference sheet? Though most select the combination of cover message and résumé, some people prefer to

In this section, we first present common content and organizational models for résumés in general. This is followed by a discussion of special considerations for formatting résumés to be read in print versus formatting résumés to be read in an electronic environment as email attachments and webpages. Lastly, we present guidelines for developing scannable résumés, which may be either print or electronic documents.

Résumé Content

Your résumé should include the information that your cover letter reviews plus supporting details. Designed for quick reading, the résumé lists facts that have been arranged for the best possible appearance. It should also be tailored to the position for which you are applying.

The arrangements of résumés differ widely, but the following process represents how most are written:

* Logically arrange information on education (institutions, dates, degrees, major field); information on employment (dates, places, firms, job titles, and accomplishments);

personal details (memberships, interests, and achievements—but not religion, race, and gender); and skills or specialized knowledge.

- Place your name and contact information at the top of the résumé and create subheadings for the main parts (e.g., objective, education, employment, extracurricular activities).

- Arrange the data for best visual appeal, making the résumé look balanced—without too much white space or too much text.

selecting the background facts Your first step in preparing the résumé is to review the facts you have assembled about yourself and then select the ones you think will help your reader evaluate you. You should include the information covered in the accompanying cover message because this is the most important information. In addition, you should include significant supporting details not covered in the accompanying cover message.

arranging the facts into groups After selecting the facts you want to include, you should sort them into logical

in the rest of the document so that your name stands out. If an employer remembers only one fact from your résumé, that fact should be your name, as in this example:

Julia M. Alvarez

The next level of headings might be *Objective, Education, Experience*, and *Skills*. These headings can be placed to the left or centered above the text that follows.

Consider using more descriptive headings that tell the nature of what follows. For example, instead of using the head *Education* and listing under that heading the software skills you acquired while a student, you might use *Computer Skills* and list your skills there. These heads better indicate the information covered, and they help the reader interpret the facts that follow.

As you can see from the exhibits in the chapter, the headings are distinguished from the other information in the résumé by the use of different sizes and styles of type. The main heading should appear to be the most important of all (larger and heavier). Your goal is to choose forms that properly show the relative importance of the information and are pleasing to the eye.

> [If an employer remembers only one fact from your résumé, that fact should be your name.]

groups. Many grouping arrangements are possible. The most conventional is the three-part grouping of *Education, Experience*, and *Skills* or *Interests*. Another possibility is grouping by job functions or skills, such as *Selling, Communicating*, and *Managing*. You may be able to think of other logical groups.

You also can derive additional groups from the three conventional groups mentioned above. For example, you can have a group of *Achievements*. Such a group would consist of special accomplishments taken from your experience and education. Another possibility is to have a group consisting of information highlighting your major *Qualifications*. Here you would include information drawn from the areas of experience, education, and skills or personal qualities. Illustrations of and instructions for constructing groups such as these appear later in the chapter.

constructing the headings With your information organized, a logical next step is to construct the headings for the résumé. Exhibit 11-1 provides a list of category headings to consider.

In a way, your name could be considered the main heading. It should be presented in type that is larger and bolder than that

including contact information Your address, phone number, and email address are the most likely means of contacting you. Most authorities recommend that you display them prominently somewhere in the résumé. You also may

Writing a résumé requires careful planning, data gathering, and organization.

Academic Achievements	Credentials	Professional Affiliations
Academic History	Degree(s)	Professional Affiliations & Awards
Academic Honors	Designations	Professional Employment
Academic Training	Dissertation	Professional Experience
Accomplishments	Education	Professional Leadership
Activities	Education Highlights	Professional Memberships
Additional Experience	Education & Training	Professional Organizations
Additional Professional Training	Educational Background	Professional Objective
Additional Training	Employment	Professional Qualifications
Affiliations	Employment History	Professional Seminars
Appointments	Employment Objective	Professional Summary
Areas of Expertise	Exhibitions & Awards	Publications
Associations	Experience(s)	Published Works
Athletic Involvement	Experience Highlights	Qualifications
Awards	Extracurricular Involvement	References
Awards & Distinctions	Field Placement	Related Course Work
Background and Interests	Foreign Language	Related Experience
Business Experience	Graduate School	Relevant Course Work
Career Goal	Graduate School Activities	Research Experience
Career Highlights	Graduate School Employment	Seminars
Career History	Hardware/Software	Skill(s) Summary
Career Objective	Highlights of Qualifications	Skills & Attributes
Career Profile	Honors	Skills & Qualifications
Career-Related Experience	Honors, Activities, & Organizations	Special Abilities
Career-Related Fieldwork	Honors and Awards	Special Awards
Career-Related Workshops	International Experience	Special Awards & Recognitions
Career-Related Training	International Travel	Special Courses
Career Skills & Experience	Internship Experience	Special Interests
Career Summary	Internship(s)	Special Licenses & Awards
Certificate(s)	Job History	Special Projects or Studies
Certifications	Languages	Special Skills
Classroom Experience	Leadership Roles	Special Training
Coaching Experience	License(s)	Strengths
Coaching Skills	Major Accomplishments	Student Teaching
College Activities	Management Experience	Student Teaching Experience
Communication Experience	Memberships	Study Abroad
Community Involvement	Memberships & Activities	Summary
Computer Background	Military Experience	Summary of Experience
Computer Experience	Military Service	Summary of Qualifications
Computer Knowledge	Military Training	Teaching Experience
Computer Languages	Objective	Teaching & Coaching Experience
Computer Proficiencies	Occupational History	Teaching & Related Experience
Computer Skills	Other Experience	Thesis
Computer Systems	Other Skills	Travel Abroad
Consulting Experience	Overseas Employment	Travel Experience
Cooperative Education	Overseas Experience	Volunteer Experience
Cooperative Education Experience	Planning & Problem Solving	Work Experience
Course Highlights	Portfolio	Work History
Course Work Included	Position Objective	Workshops & Seminars
Courses of Interest	Practicum Experience	

Source: "UW-Eau Claire, Sample Résumé Headings & Titles," *Career Services*, UW-Eau Claire, n.d., Web, 20 June 2013.

want to display your Web site address or addresses for your social networking sites. The most common location for displaying contact information is at the top under your name.

When it is likely that your contact information will change before the job search ends, you may want to include current and permanent contact information.

The logic of making the contact information prominent and inclusive is to make it easy for the employer to reach you. However, in the interest of privacy, you may want to include only an email address created specifically for job searches and a phone number for résumés shared through the Internet. For business use, a professional email address is always preferable to an informal one such as surferchick@yahoo.com.

including your objective

Although not a category of background information, a **statement of your objective** is appropriate in the résumé. Headings such as *Career Objective, Job Objective*, or just *Objective* usually appear at the beginning after your name.

Not all authorities agree on the value of including the objective, however. Some argue that the objective includes only obvious information that is clearly suggested by the remainder of the résumé. Moreover, they point out that an objective limits the applicant to a single position and eliminates consideration for other jobs that may be available.

Those who favor stating the objective reason that it helps the recruiter see quickly where the applicant might fit into the company. Since this argument appears to have greater support, you should include the objective. When your career goal is unclear, you may use broad, general terms, but when you are considering a variety of employment possibilities, you may want to have different versions of your résumé for each possibility.

Primarily, your objective should describe the work you seek. When you know the exact job title of a position you want at the targeted company, use it.

> Objective: To obtain a marketing research internship in the arts and entertainment industry.

Another technique includes using words that convey a long-term interest in the targeted company, as in the following example. However, using this form may limit you if the company does not have the career path you specify.

> Objective: To secure a full-time sales representative position for McGraw-Hill leading to sales management.

Also, wording the objective to point out your major strengths can be effective. It also can help set up the organization of the résumé.

> Objective: To apply three years of successful e-commerce accounting experience to a larger company with a need for careful attention to transaction management and analysis.

presenting the information

After crafting your job objective, you need to determine what information to present under the rest of your headings. Though the order of the headings will largely depend on your organizational strategy (see "Organizing for Strength," page 316), the information under each heading generally appears as follows.

Work experience. The description of your work experience should contain your job title/position, company name, location, and dates of employment. You should also include your job duties and the skills you acquired, especially those that relate to the position for which you are applying. Consider the following example:

> Marketing and Public Relations Intern
> Alliant Health Plans, Incorporated, Boston, MA
> Jan. 2014–May 2014
>
> - Created a webpage, brochure, and press release for a community wellness program
> - Interviewed and wrote about physicians, customers, and community leaders for newsletter articles
> - Worked with a team of interns in other departments to analyze and update the company's Web site

Note in the above example that in addition to the basic information regarding the job, the writer lists duties that anyone in a marketing or public relations field would likely use in a related position. The duties are represented with **action-oriented, past tense verbs** (e.g., *interviewed, wrote*) because the position has ended. If the intern were still in the position, he or she would have used **simple present tense verbs** for duties he or she currently performs (e.g., *interview, write*). The use of these action verbs strengthens a job description because verbs are the strongest of all words. If you choose them well, you will do much to sell your ability to do the jobs you are targeting. A list of the more widely used action verbs appears in Exhibit 11-2.

Note also that the writer uses both months and dates. This is especially important when you consider that simply saying "2012" doesn't let the reader know how long the internship was. In another example, if a reader sees dates of employment as 2012–2013, he or she does not know if that included a full year of employment or if the writer started on December 31, 2012, and quit on January 1, 2013. Including the months along with the dates is the clearest and most ethical way to represent your employment timeline.

Education. Because your education is likely to be your strongest selling point for your first job after college, you will cover it in some detail. (Unless it adds something unique, you usually do not include your high school education once you have enrolled in college. Similarly, you also minimize the emphasis on all your education as you gain experience.) At a minimum, your coverage of education should include institutions, dates, degrees, and areas of study. For some jobs,

EXHIBIT 11-2   A List of Action Verbs That Add Strength to Your Résumé

Communication Skills

Address
Arbitrate
Arrange
Author
Correspond
Develop
Direct
Draft
Edit
Enlist
Formulate
Influence
Interpret
Lecture
Mediate
Moderate
Motivate
Negotiate
Persuade
Promote
Publicize
Recruit
Spoke
Translate

Management Skills

Administer
Analyze
Assign
Attain
Chair
Contract
Consolidate
Coordinate
Delegate
Develop
Directed
Evaluate
Execute
Improve
Increase
Organize
Plan
Prioritize
Produce
Recommend
Review
Schedule

Strengthen
Supervise
Wrote

Teaching Skills

Adapt
Advised
Clarify
Coach
Communicate
Coordinate
Develop
Enable
Encourage
Evaluate
Explain
Facilitate
Guide
Inform
Initiate
Instruct
Persuade
Setting goals
Stimulate

Creative Skills

Act
Conceptualize
Create
Design
Develop
Direct
Establish
Fashion
Found
Illustrate
Institute
Integrate
Introduce
Invent
Originate
Perform
Plan
Publish
Revitalize
Shape

Helping Skills

Assist
Assess

Clarify
Coach
Counsel
Demonstrate
Diagnose
Educate
Expedite
Facilitate
Familiarize
Guide
Refer
Rehabilitate
Represent
Service
Support
Tend

Research Skills

Clarify
Collect
Critique
Diagnose
Evaluate
Examine
Extract
Identify
Inspect
Interpret
Interview
Invented
Investigate
Organize
Review
Summarize
Survey
Systematize

Financial Skills

Administer
Allocate
Analyze
Appraise
Audit
Balance
Budget
Calculate
Compute
Develop
Forecast

Manage
Market
Plan
Project
Research

Administrative Skills

Approve
Arrange
Catalogue
Classify
Collect
Compile
Dispatch
Execute
Generate
Implement
Inspect
Monitor
Operate
Organize
Prepare
Process
Purchase
Record
Retrieve
Screen
Specify
Systematize
Tabulate
Validate

Information Skills

Catalogue
Clarify
Classify
Compile
Compose
Convey
Copy
Correct
Define
Document
Gather
Inform
Kept records
Memorize
Proofread

Question
Review
Specify
Study
Survey
Tabulate
Test
Verify

Leadership Skills

Appoint
Approve
Arrange
Assess
Assign
Authorize
Carry out
Chair
Coach
Complete
Consult
Delegate
Demonstrate
Determine
Devise
Direct
Enlist
Facilitate
Head
Initiate
Launch
Motivate
Negotiate
Nominate
Preside
Set goals
Start

Problem-Solving Skills

Analyze
Apply
Calculate
Compile
Consult
Correct
Create
Critique
Design
Develop

Diagnose
Discover
Dissect
Examine
Explore
Problem solve
Propose
Research
Resolve
Revise
Search
Study
Track
Troubleshoot
Uncover

Technical Skills

Assemble
Built
Calculate
Compute
Design
Devise
Engineer
Fabricate
Maintain
Operate
Overhaul
Program
Remodel
Repair
Solve
Train
Upgrade

Teamwork Skills

Accomplish
Assist
Collaborate
Coordinate
Corroborate
Dispatch
Encourage
Explain
Follow
Help
Share
Team built
Volunteer

Source: UW-Eau Claire, "Sample Résumé Headings & Titles," *Career Services*, UW-Eau Claire, n.d., Web, 20 June 2013.

you may want to list and even describe specific courses, especially if you have little other information to present or if your course work has uniquely prepared you for those jobs. In particular, if you are applying for an internship, you may want to list your course work as an indication of your current level of academic preparation as it relates to the requirements of the position. If your grade-point average (GPA) is good, you may want to include it. Remember, for your résumé, you can compute your GPA in a way that works best for you as long as you label it accurately. For example, you may want to select just those courses in your major, labeling it "Major GPA." Or if your last few years were your best ones, you may want to present your GPA for just that period. In any case, include a GPA when it works favorably for you and be clear what grades or time period you are using to calculate it.

Personal information. What personal information to list is a matter for your best judgment. In fact, the trend appears to be toward eliminating such information. If you do include personal information, you should omit race, religion, gender, age, and marital status because current laws prohibit hiring based on them.

Personal information that is generally appropriate includes all items that tell about your personal assets. Information on your organization memberships, civic involvement, and social activities is evidence of experience and interest in working with people. Such information can be quite useful to some employers, especially when personal qualities are important to the work involved. Exhibit 11-3 presents a résumé that emphasizes education and experience.

References. Even though employers can (and do) check your social media sites, they are also likely to check specific references that you provide. Generally, you do not need to include references on or with your résumé unless the job posting asks you to. If the job posting does not require references, you can simply have a list of references ready for the point in the interview process when the employer requires it. Primary reasons for not including your references unless asked are that (1) references added to a résumé take up space that you could use to sell your skills, and (2) references included on a separate sheet, while not harmful, are not likely necessary if the employer has not asked for them. However, if you have a particularly impressive reference or a reference whom your audience might know, including your references can be a good idea. If you do include your references, it is usually best to put them on a separate **reference sheet**.

The type size and style of the main heading of this sheet should match that used in your résumé. It may say something like "References for [*your name*]." Below this heading is a listing of your references, beginning with the strongest one. In addition

> **In any case, include a GPA when it works favorably for you and be clear what grades or time period you are using to calculate it.**

to solving the reference dilemma, use of this separate reference sheet allows you to change both the references and their order for each job. A sample reference sheet is shown in the example in Exhibit 11-4 on page 318.

How many and what kinds of references to include will depend on your background. If you have an employment record, you should include one for every major job you have held—at least for recent years. You should include references related to the work you seek. If you base your application heavily on your education or your personal qualities, or both, you should include references who can vouch for these areas: professors, clergy, community leaders, and the like. Your goal is to list those people who can verify the points on which your appeal for the job is based. At a minimum, you should list three references. Five is a good maximum.

Your list of references should include accurate mailing addresses with appropriate job titles. Also useful are phone and fax numbers as well as email addresses. Job titles (officer, manager, president, supervisor) are helpful because they show what the references are able to tell about you. It is appropriate to include forms of address: Mr., Ms., Dr., and so on.

Some résumé writers may be tempted to put "references available upon request" at the bottom of their résumés. However, this expression is outdated and serves no purpose. When you think about it, of course you would always make your references available at the employer's request, which means you're stating the obvious. Though some may argue that including this statement shows a willingness to provide the information, you can show your willingness by including a separate references sheet. You may want to use the space you would devote to this statement by adding another line to your job duties or other experience to sell your skills and abilities.

When you do list someone as a reference, business etiquette requires that you ask for permission first. Although you will use only those who can speak highly of you, asking for your reference's permission beforehand helps that person prepare better. And, of course, it saves you from unexpected embarrassment if a reference does not remember you, is caught by surprise, or has nothing to say.

organizing for strength After you have identified the information you want to include on your résumé, you will want to organize or group items to present yourself in the best possible light. Three strategies for organizing this information are the reverse chronological approach, the functional or skills approach, and the accomplishments/achievements or highlights approach.

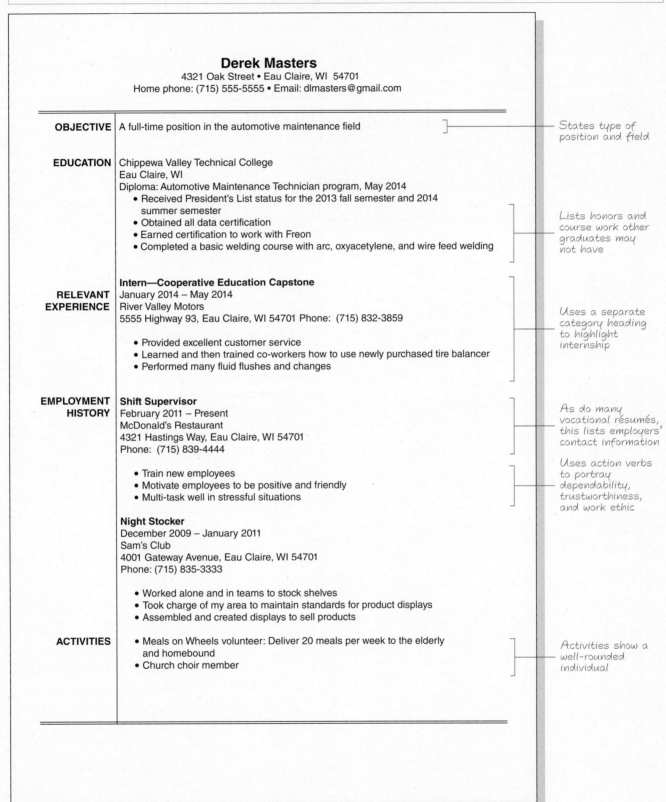

Derek Masters
4321 Oak Street • Eau Claire, WI 54701
Home phone: (715) 555-5555 • Email: dlmasters@gmail.com

OBJECTIVE | A full-time position in the automotive maintenance field

States type of position and field

EDUCATION | Chippewa Valley Technical College
Eau Claire, WI
Diploma: Automotive Maintenance Technician program, May 2014
- Received President's List status for the 2013 fall semester and 2014 summer semester
- Obtained all data certification
- Earned certification to work with Freon
- Completed a basic welding course with arc, oxyacetylene, and wire feed welding

Lists honors and course work other graduates may not have

RELEVANT EXPERIENCE | **Intern—Cooperative Education Capstone**
January 2014 – May 2014
River Valley Motors
5555 Highway 93, Eau Claire, WI 54701 Phone: (715) 832-3859

- Provided excellent customer service
- Learned and then trained co-workers how to use newly purchased tire balancer
- Performed many fluid flushes and changes

Uses a separate category heading to highlight internship

EMPLOYMENT HISTORY | **Shift Supervisor**
February 2011 – Present
McDonald's Restaurant
4321 Hastings Way, Eau Claire, WI 54701
Phone: (715) 839-4444

- Train new employees
- Motivate employees to be positive and friendly
- Multi-task well in stressful situations

As do many vocational résumés, this lists employers' contact information

Uses action verbs to portray dependability, trustworthiness, and work ethic

Night Stocker
December 2009 – January 2011
Sam's Club
4001 Gateway Avenue, Eau Claire, WI 54701
Phone: (715) 835-3333

- Worked alone and in teams to stock shelves
- Took charge of my area to maintain standards for product displays
- Assembled and created displays to sell products

ACTIVITIES |
- Meals on Wheels volunteer: Deliver 20 meals per week to the elderly and homebound
- Church choir member

Activities show a well-rounded individual

EXHIBIT 11-4 Thoroughness and Good Arrangement for a Reference Sheet

This reference sheet presents Julia Alvarez's references completely.

References for Julia M. Alvarez

3177 North Hawthorne Boulevard
St. Louis, MO 63139
314.967.3117
jmalvarez358@hotmail.com

Heading format matches résumé

Mr. John Gibbs
Human Resources Director
Upton Industries
7114 East 71st Street
St. Louis, MO 63139
Phone: 314.342.1171
Email: John.Gibbs@upton.com

Professor Helen K. Robbins
Department of Management
Wilmont University
St. Louis, MO 63139
Phone: 314.392.6673
Email: Helen.Robbins@wilmont.edu

Mr. Todd E. Frankle, Store Manager
The Gap, Inc.
Four Points Mall
St. Louis, MO 63139
Phone: 314.466.9101
Email: tfrankle@gap.com

Professor Carol A. Cueno
Department of Psychology
Wilmont University
St. Louis, MO 63139
Phone: 314.392.0723
Email: Carol.Cueno@wilmont.edu

Complete information and balanced arrangement

The **reverse chronological organizational layout** (Exhibit 11-5) presents your education and work experience from the most recent to oldest. It emphasizes the order and time frame in which you have participated in these activities. It is particularly good for those who have progressed in an orderly and timely fashion through school and work. The reverse chronological format is the most common way of organizing a résumé; therefore, it is likely the best choice for most job seekers (including college students), as it provides a format familiar to employers.

A **functional or skills layout** (Exhibit 11-6) organizes the résumé's contents around three to five areas particularly important to the job you want. Rather than showing that you developed one skill on one job and another skill on another job, this organizational plan groups related skills. It is particularly good for those who have had many jobs, have taken nontraditional career paths, or are changing fields. Creating this kind of résumé takes much work and careful analysis of both jobs and skills to show the reader that you are a good match for the position. If you use a functional résumé, be sure that readers can see from the other sections—such as employment and education—where you likely developed the skills that you are emphasizing. Enabling your readers to make these connections lends credibility to your claims to have such skills.

An **accomplishments/achievements layout** (Exhibit 11-7) foregrounds the most impressive factors about you.

It features a *Highlights* or *Summary* section that includes key points from the three conventional information groups: education, experience, and personal qualities. This information comes near the beginning of the résumé, usually following the objective. Typically, this layout emphasizes the applicant's most impressive background facts that pertain to the work sought, as in this example:

Summary

- **Experienced:** Three years of full-time work as programmer/analyst in designing and developing financial databases for the banking industry.

- **Highly trained:** BS degree with honors in management information systems.

- **Self-motivated:** Proven record of successful completion of three online courses.

Keep in mind that this section should not repeat sections of the résumé; it should highlight strengths. If your résumé is short and your summary is just a repeat of other content, you may choose not to include this section. In the rest of the résumé,

accomplishments should stand out. For example, rather than listing them within other categories, such as employment or education, you can put them in a separate section as illustrated by the example on page 322.

writing impersonally and consistently Because the résumé is a listing of information, you should write without personal pronouns (no *I's, we's, you's*). You should also write all equal-level headings and the parts under each heading in the same parallel grammatical form. For example, if one major heading in the résumé is a noun phrase, all the other major headings should be noun phrases. The following four job duties illustrate the point. All but the third are verb phrases. The error can be corrected by making the third a verb phrase, as in the examples to the right:

Not Parallel	Parallel
Greeted customers	Greeted customers
Processed transactions	Processed transactions
Data entry in Excel spreadsheets	Entered data in Excel spreadsheets
Balanced a cash drawer	Balanced a cash drawer

The following items illustrate grammatical inconsistency in the parts of a group:

Have fluency in Spanish

Active in sports

Ambitious

The understood word for the first item is *I* and for the second and third, the understood words are *I am*. Any changes that make all three items fit the same understood words would correct the error (e.g., fluent in Spanish, active in sports, ambitious).

Printed (Hardcopy) Résumés

As we've mentioned, one aspect of résumé writing you must consider is the environment in which your audience will read your document. Knowing whether your reader will read a printed or an electronic document helps you create résumés that are easily accessible to readers and ensure that you present a professional image.

A printed résumé is meant to be sent in hardcopy format through the mail and used in face-to-face interviews. Generally, a printed résumé is one or two pages. Whether a résumé is one or two pages is determined by the skills and other information that best demonstrate your qualifications for a position. A two-page résumé is not

Large size emphasizes name

Manny Konedeng
5602 Montezuma Road • Apartment 413 • San Diego • California • 92115
Phone: (619) 578-8508 • Email: mkonedeng@yahoo.com

Includes complete contact information

OBJECTIVE	A financial analyst internship with a brokerage firm

Uses descriptive, specific statement

EDUCATION	**Bachelor of Science Degree in Business Administration**, May 2014, San Diego State University, Finance Major

Dean's List
Current GPA: 3.32/4.00
Related Courses:
- Business Communication
- Investments
- Tax Planning
- Estate Planning
- Risk Management
- Business Law

Computer Skills:
- Excel, Word, PowerPoint, Access, QuickBooks
- Statistical Software: SAS, SPSS
- Web-based Applications—Surveymonkey, Blogger, GoToMeeting
- Research Tools—Center for Research in Securities Prices (CRSP) stock price database and the Standard and Poor's Research Insight database of corporations' financial statements

Expands and emphasizes strongest points through precise detail

WORK EXPERIENCE

Sales and Front Desk, Powerhouse Gym, Modesto, CA 95355   Summer 2013
- Sold memberships and facilitated tours for the fitness center
- Listened to, analyzed, and answered customers' inquires
- Accounted for membership payments and constructed sales reports
- Trained new employees to understand company procedures and safety policies

Relay Operator, MCI, Riverbank, CA 95367   Summers 2011 & 2012
- Assisted over 100 callers daily who were deaf, hard of hearing, or speech disabled to place calls
- Exceeded the required data input of 60 wpm with accuracy
- Multitasked with typing and listening to phone conversations
- Was offered a promotion as a Lead Operator

Co-founder and Owner, Fo Sho Entertainment, Modesto, CA 95355   Aug. 2009–May 2011
- Led promotions for musical events in the Central Valley
- Managed and hosted live concerts
- Created and wrote proposals to work with local businesses
- Collaborated with team members to design advertisements

Emphasizes position held rather than place or date

Uses descriptive action verbs

UNIVERSITY INVOLVEMENT

Communications Tutor, San Diego State University, San Diego, CA 92182   Spring 2014
- Critiqued and evaluated the written work for a business communication course
- Set up and maintained blog for business communication research

Recruitment Chair, Kappa Alpha Order Fraternity, San Diego, CA 92115
- Supervised the selection process for chapter membership   Fall 2013
- Individually raised nearly $1,000 for chapter finances
- Organized recruitment events with business sponsors, radio stations, and special guests

Includes items that will set him apart from other applicants

Carolynn W. Workman

12271 69th Terrace North
Seminole, FL 33772
727.399.2569 (Voice/Message)
cworkman@msn.com

Emphasizes tight organization through use of horizontal ruled lines

---

Objective   An entry-level tax accounting position with a CPA firm

---

Education   Bachelor of Science: University of South Florida, December 2014
Major: Business Administration
Emphasis: Accounting
GPA: 3.42 with Honors

Emphasizes degree and GPA

Uses internal bullets to increase readability

Accounting-Related Course Work:
Financial Accounting ❖ Cost Accounting and Control ❖ Accounting Information Systems ❖ Auditing ❖ Concepts of Federal Income Taxation ❖ Financial Policy ❖ Communications for Business and Professions

Emphasizes key skills relevant to objective

Activities:
Vice-President of Finance, Beta Alpha Psi
Editor, Student Newsletter for Beta Alpha Psi
Member, Golden Key National Honors Society

Skills
Computer

▶ Assisted in installation of small business computerized accounting system using QuickBooks Pro.
▶ Prepared tax returns for individuals in the VITA program using specialty tax software.
▶ Mastered Excel, designing data input forms, analyzing and interpreting results of most functions, generating graphs, and creating and using macros.

Accounting

▶ Reconciled accounts for center serving over 1,300 clients.
▶ Prepared income, gift, and estate tax returns.
▶ Processed expense reports for twenty professional staff.
▶ Generated financial statements and processed tax returns using Great Plains and Solomon IV.

Varies use of action verbs

Business
Communication

▶ Conducted client interviews and researched tax issues.
▶ Communicated both in written and verbal form with clients.
▶ Delivered several individual and team presentations on business cases, projects, and reports to business students.

Work History
Administrative
Assistant

Office of Student Disability Services, University of South Florida
Tampa, FL. Spring 2014.

Tax Assistant   Rosemary Lenaghan, Certified Public Accountant. Seminole, FL Jan. 2011–May 2014.

EXHIBIT 11-7 Traditional Print Résumé Using Highlights of Qualifications and Accomplishments Sections

Kimberly M. VanLerBerghe

2411 27th Street
Moline, IL 61265
309.764.0017 (Mobile)
kmv@yahoo.com

JOB TARGET Trainer/translator for a large, worldwide company with operations in Spanish-speaking markets

HIGHLIGHTS OF QUALIFICATIONS

Emphasizes those qualifications most relevant to position sought

- Experienced in creating and delivering multimedia PowerPoint presentations.
- Enthusiastic team member/leader whose participation brings out the best in others.
- Proficient in analytical ability.
- Skilled in gathering and interpreting data.
- Bilingual—English/Spanish.

EDUCATION

Presents the most important items here

DEGREE	B.A. English, June 2014, Western Illinois University	
EMPHASIS	Education and Spanish	MAJOR GPA: 3.87/4.00
HONORS	Dean's List, four semesters	Chevron Scholarship, Fall 2012
MEMBER	Mortar Board, Women's Golf Team	

EMPLOYMENT

Identifies most significant places of work and de-emphasizes less important work

DEERE & COMPANY, INC.
Student Intern, Summer 2013

CONGRESSMAN J. DENNIS HASTERT
Volunteer in Computer Services, Fall 2013

Several years' experience in the restaurant business including supervisory positions.

ACCOMPLISHMENTS

Presents only selected accomplishments from various work and volunteer experience that relate to position sought

- Trained executives to create effective cross-cultural presentations.
- Developed online training program for executive use of GoToMeeting.
- Designed and developed a database to keep track of financial donations.
- Coded new screens and reports; debugged and revised screen forms for easier data entry.
- Provided computer support to virtual volunteers on election committee.
- Provided translation services for Hispanic employees.

TheLadders, a company that matches career-seeking professionals with recruiters, recently published a study that used eye-tracking technology to measure how long recruiters looked at résumés before determining whether a candidate was a "fit" or "no fit."

How long do you think they took? Five minutes? Ten minutes? They took just six seconds. Of that six seconds, they spent 80 percent of that time looking at the candidate's name; current title/company; previous title/company; previous position start and end dates; current position start and end dates; and education.

When you think about the volume of résumés an employer might review every day, it's not surprising that an initial review takes only seconds. This means that job seekers must organize their content and format their résumés so that key skills and qualifications stand out.

Source: Will Evans, "Keeping an Eye on Recruiter Behavior," *TheLadders.com*, TheLadders, 2012, Web, 6 July 2013.

---

more impressive than a one-page résumé if the information on the second page is irrelevant. On the other hand, you should not feel pressured to have a one-page résumé and then cram two pages of content onto one page; doing so will crowd the page and make important information less obvious to your reader. Printed résumés should be formatted so that they are visually appealing and easily read. Be sure that your name, the word *résumé*, and the page number appear on page two as a page header.

The attractiveness of your résumé will say as much about you as the words. The appearance of the information that the reader sees plays a part in forming his or her judgment. While using a template is one solution, it will make you look like many other applicants. A layout designed with your reader and your unique data in mind will probably do a better job for you. Not only will your résumé have a distinctive appearance, but the design should sell you more effectively than one where you must fit your data to the design. A sloppy, poorly designed presentation, on the other hand, may even ruin your chances of getting the job. Thus, an attractive physical arrangement is a must.

There is no one best arrangement, but a good procedure is to approach the task as a graphic designer would. Your objective is to design an arrangement of type and space that appeals to the eye.

Margins look better if at least an inch of space is left at the top of the page and on the left and right sides of the page and if at least 1½ inches of space are left at the bottom of the page. Your listing of items by rows and columns appears best if the items are short and if they can be set up in two uncrowded columns, one on the left side of the page and one on the right side. Longer items of information are more appropriately set up in lines extending across the page. In any event, avoid long and narrow columns of data with large sections of wasted space on either side. Arrangements that give a heavy crowded effect also offend the eye. Extra spacing between subdivisions and indented patterns for subparts add visual appeal.

As you set up columns and bulleted lists, you will want to use tables rather than your space bar or tab key. This will allow you more control over your formatting and ensure that the alignment of the information in your columns is consistent and perfect.

While layout is important in showing your ability to organize and good spacing increases readability, other design considerations such as font and paper selection affect attractiveness almost as much. Commercial designers say that type size for headings should be at least 12 to 14 points and for body text, 10 to 12 points. The size of your name at the top of your resume, however, may be as large as 16 or 18 points (or more), depending on the font and the size of your other headings. They also recommend using no more than two font styles. Some word processing programs have a "shrink to fit" feature that allows the user to fit information on one page. It will automatically adjust font sizes to fit the page. Be sure the resulting type size is both appropriate and readable.

Another factor affecting the appearance of your application documents is the paper you select. The paper should be appropriate for the job you seek. In business, erring on the conservative side is usually better. The most traditional choice is white, 100 percent cotton, 20- to 28-lb. paper. Of course, reasonable variations can be appropriate. When you mail your printed résumé, do not fold it; mail it in a 9 × 12 envelope and be sure to type (not handwrite) the mailing labels.

contrasting bad and good examples The following two résumés are at opposing ends of the quality scale. The first one, scant in coverage and poorly arranged, does little to help the applicant. Clearly, the second one is more complete and better arranged.

Résumé with poor arrangement and incomplete information. Shortcomings in the first example (next page) are obvious. First, the form is not pleasing to the eye. The weight of the type is heavy on the left side of the page. Failure to indent wrapped lines makes reading difficult.

Traditional Print Résumé with Wording Errors, Poor Use of Space, and Insufficient Detail
This incomplete and poorly written résumé does not effectively present Julia Alvarez's qualifications.

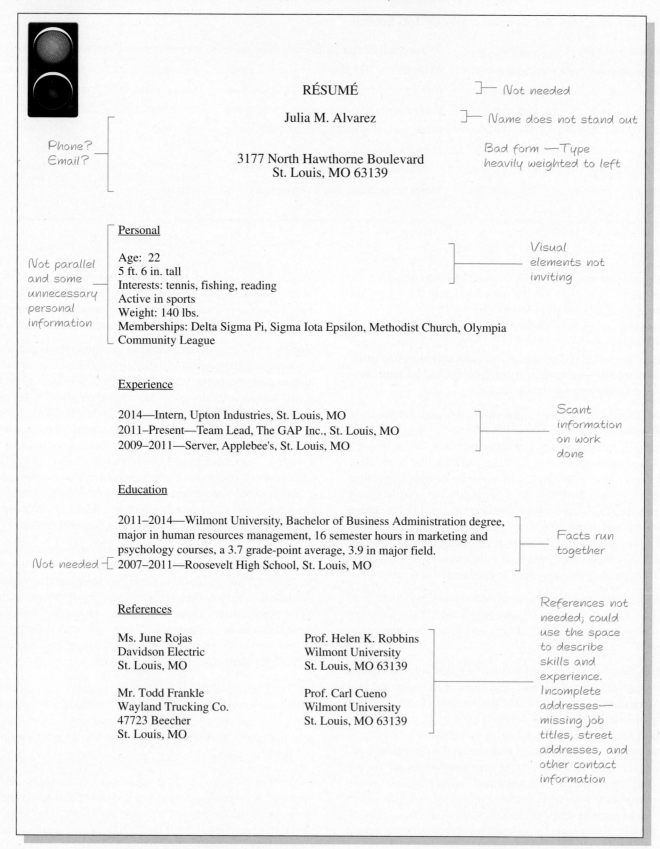

RÉSUMÉ

⎤— Not needed

Julia M. Alvarez

⎤— Name does not stand out

Phone?
Email?

3177 North Hawthorne Boulevard
St. Louis, MO 63139

Bad form —Type
heavily weighted to left

Personal

Not parallel
and some
unnecessary
personal
information

Age: 22
5 ft. 6 in. tall
Interests: tennis, fishing, reading
Active in sports
Weight: 140 lbs.
Memberships: Delta Sigma Pi, Sigma Iota Epsilon, Methodist Church, Olympia
Community League

Visual
elements not
inviting

Experience

2014—Intern, Upton Industries, St. Louis, MO
2011–Present—Team Lead, The GAP Inc., St. Louis, MO
2009–2011—Server, Applebee's, St. Louis, MO

Scant
information
on work
done

Education

2011–2014—Wilmont University, Bachelor of Business Administration degree,
major in human resources management, 16 semester hours in marketing and
psychology courses, a 3.7 grade-point average, 3.9 in major field.

Not needed —⎡ 2007–2011—Roosevelt High School, St. Louis, MO

Facts run
together

References

Ms. June Rojas
Davidson Electric
St. Louis, MO

Prof. Helen K. Robbins
Wilmont University
St. Louis, MO 63139

Mr. Todd Frankle
Wayland Trucking Co.
47723 Beecher
St. Louis, MO

Prof. Carl Cueno
Wilmont University
St. Louis, MO 63139

References not
needed; could
use the space
to describe
skills and
experience.
Incomplete
addresses—
missing job
titles, street
addresses, and
other contact
information

Traditional Print Résumé with Thoroughness and Good Arrangement
This complete and reverse chronologically organized résumé presents Julia Alvarez's case effectively.

Julia M. Alvarez

3177 North Hawthorne Boulevard
St. Louis, MO 63139
314.967.3117 (Voice/Message)
jmalvarez358@hotmail.com

Presents contact data clearly

Objective

To obtain a full-time position in human resources management specializing in safety and OSHA compliance.

Education

Bachelor of Business Administration
Wilmont University, St. Louis,
MO University—May 2014
GPA: 3.7/4.0

Major: Management:
             Human Resources Emphasis
Minor: Psychology
Certificate: Advanced Business
                Communication

Layout emphasizes key educational facts

Related Coursework:

Highlights most relevant courses and subjects

- Compensation Theory and Administration
- Organizational Change and Development
- Industrial Relations
- Advanced Human Resource Management

- Managerial Accounting
- Training and Human Resource Development

SHRM Certification

- Passed the Society for Human Resource Management (SHRM) Certification Examination

Internship

Human Resource Intern, Upton Industries, St. Louis, MO, June 2014–August 2014
- Consulted with management to ensure compliance with OSHA and state safety regulations
- Analyzed data from Upton's human resource database and created reports to guide management decisions
- Researched industry trends and reported them to management
- Participated in personnel-related tasks such as recruiting, reviewing candidates' résumé, conducting interviews, completing new hires' paperwork, training both new and current employees, determining compensation and benefits, and facilitating exit interviews
- Prepared timesheets for payroll

Special Project: Created and maintained an "HR News and Updates" page for the company intranet

Employment

Team Lead, The Gap, Inc., St. Louis, MO, October 2011–Present
- Was promoted to Team Lead after two months of employment
- Was named top store sales associate four of eight quarters
- Create merchandise displays
- Train new sales associates
- Participate in interviews and hiring
- Set the weekly schedule

Action verbs portray an image of a hard worker with good interpersonal skills

Host and Food Server, Applebee's, St. Louis, MO, September 2009 – September 2011
- Provided exceptional customer service
- Worked well as part of a team to seat and serve customers quickly and efficiently

Activities

Includes only most relevant information

Delta Sigma Pi (professional); Sigma Iota Epsilon (honorary), served as treasurer and president; Board of Stewards for church; Society for Human Resources Management (SHRM), served as chapter vice president

This résumé also contains numerous errors in wording. The headings are not parallel in grammatical form. All are in topic form except the first one. The items listed under *Personal* are not parallel either and contain irrelevant and inappropriate personal information. Throughout, the résumé coverage is scant, leaving out many of the details needed to present the best impression of the applicant. Under *Experience*, little is said about specific tasks and skills in each job; and under *Education*, high school work is listed unnecessarily. The references are incomplete, omitting street addresses and job titles.

Traditional print résumé with thoroughness and good arrangement. The revised résumé for Julia Alvarez (page 325) appears better at first glance, and it gets even better as you read it. It is attractively arranged. The information is neither crowded nor strung out. The balance is good. Its content is also superior to that of the other example. Additional words show the quality of Ms. Alvarez's work experience and education, and they emphasize points that make her suited for the work she seeks. This résumé excludes inappropriate personal information and has only the facts that tell something about Ms. Alvarez's personal qualities. A bulleted list of duties under each job describes the skills and qualities Ms. Alvarez brings to a human resources position. A separate references sheet (see Exhibit 11-4

readers who will view it electronically, how you use software to format your résumé may affect how the document looks when the reader views it on a computer screen.

creating an unformatted résumé As you will learn in upcoming paragraphs, before you think about sending your résumé electronically, you'll want to have an **unformatted (plain-text) version** of your résumé to work with. A plain-text résumé contains no formatting (e.g., bold type, italics, tables, lists, bullets, centered text), which, as we discuss below, is useful when sending a résumé over email or submitting an application online. To create a plain-text résumé, just open your formatted résumé in Microsoft Word, select File > Save As > Save as Type dropdown list > "Plain text." You'll see that the file now has a .txt file extension. If you close the file and reopen it by clicking the file itself, the document will likely open in Notepad or WordPad. If you want to open the .txt file in Word, open Word first and then go to File > Open.

sending your résumé via email If an employer requests that you send your résumé by email, he or she may specify how to submit it. Always follow those directions. In the absence

> ## In the absence of directions, you can place the résumé text in the body of the email, attach your formatted résumé to the email, or do both.

on page 318) with complete contact information permits the reader to contact the references easily. Job titles tell how each is qualified to evaluate the subject. Note that moving the references to a separate page frees space in the résumé for presenting Ms. Alvarez's qualifications.

Electronic Résumés

While some employers require printed résumés, others will require that you submit your résumé electronically via email or perhaps through an online application form or database. You might even want to create your own Web site to showcase your résumé or post a profile using sites such as LinkedIn (see Exhibit 11-8).

Preparing a résumé for electronic use requires more attention to software than it might if you were preparing a résumé to be read in printed form. For example, when you format your printed résumé, you can manipulate the software any way you wish to achieve the desired look or effect—and this is all right because all your reader will see is the final printed copy; he or she does not need to know what you did to get your document to look a particular way. However, if you send this formatted résumé to

of directions, you can place the résumé text in the body of the email, attach your formatted résumé to the email, or do both.

One reason to copy the text of your résumé into an email is that you will know your reader received the information. Another is that the reader may find it helpful to have the text in the body of the email—especially if he or she is using scanning software to screen candidates. If you're copying the text of your résumé into the body of the email, always copy from an unformatted (plain-text) résumé. You cannot be sure that your reader is able to read HTML or formatted messages; he or she may be only able to read plain-text messages. Copying from the unformatted document ensures that your text appears in plain format. It also saves you time from having to "unformat" anything you would have copied from your formatted résumé.

While copying the text of a résumé into the body of an email is functional, it does not leave you with a résumé as nice as the one you've formatted. For this reason, many job candidates will attach a copy of the formatted résumé to the email. It is especially important in formatted résumés sent as attachments that you do not format lists and blocks of text using the space bar or tab key. Many times when you do this, the attachment will look

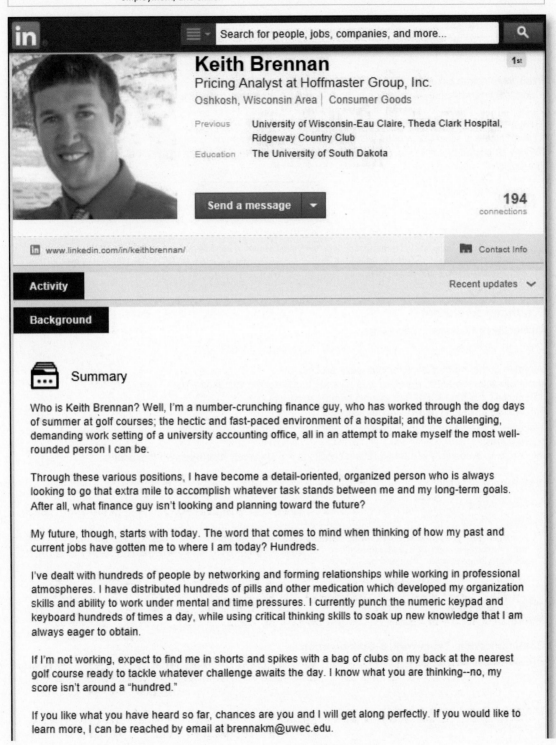

EXHIBIT 11-8 (continued)

 Experience

Pricing Analyst

Hoffmaster Group, Inc.

July 2012 – Present (1 year 5 months) | Oshkosh, Wisconsin Area

Work directly with buyers and sales reps to make pricing decisions.
Complete financial analysis of products and send out quotes to prospective buyers.
Accept or decline quotes coming in from sales reps.
Deal with the contract rebating process.
Communicate product information with other areas of the company.
Work with customer service to view, analyze, confirm, and send out orders to the plant for final production.

▼ 1 recommendation

I am writing this letter of recommendation with the highest level of confidence on behalf of Keith Brennan. Keith is a very driven, selfstarter, motivated, and dedicated employee. He has always demonstrated these attributes by always giving a... View ↓

Financial Assistant

University of Wisconsin-Eau Claire

February 2011 – May 2012 (1 year 4 months) | University Accounting

Assisted in the monthly bank reconciliation process.
Inputted data for budget, expense, and revenue transfers into Microsoft Excel.
Reconciled the 2011 fiscal year university grant accounts.
Reconciled university general ledger accounts.
Completed daily spreadsheets involving credit card transactions, ACH deposits, and Sallie Mae transactions.
Worked with thousands of student refund checks to ensure timely disbursement to students.

▼ 1 recommendation

Keith is a very good worker and is willing to take on any task asked of him. He is detail orientated and does not have any problem asking questions if he is not sure how to do something. Keith is also very pleasant to have in the office and has... View ↓

 Certifications

Communication, Teamwork, & Collaboration

MRA-Institute of Management

July 2013 – Present

Microsoft Excel 2010- Level 2

New Horizons of Wisconsin

March 2013 – Present

Microsoft Excel 2010- Level 3

New Horizons of Wisconsin

April 2013 – Present

EXHIBIT 11-8 (continued)

Courses

The University of South Dakota

- Quantitative Analysis (BADM 720)
- Managerial Economics (ECON 782)
- Managerial Marketing (BADM 770)

Skills & Expertise

Most endorsed for...

6	Time Management	+
4	Customer Service	+
4	Financial Analysis	+
4	Microsoft Excel	+

Education

The University of South Dakota
Master of Business Administration (MBA), Finance, General
2013 – 2015 (expected)

▸ 3 courses

University of Wisconsin-Eau Claire
Bachelor of Business Administration (BBA), Finance; IS
2008 – 2012

Activities and Societies: Financial Management Association, Intramural Basketball, Intramural Volleyball, Intramural Soccer

Additional Info

Personal Details

Birthday

fine, but occasionally, the reader will open the file only to see text scattered across the page, as the tabs and the spacing were not preserved. To ensure that your text and lists stay aligned, use the ruler or a table (just be sure to turn off "View Gridlines" before emailing the document). Also be sure to use a standard font so that the font you use is the font the reader sees. For example, if you use Calibri but your reader is using an older version of Word that does not have Calibri, Word will substitute a new font. If you use Arial, a font all versions of Word have, you know your reader will see your document as you created it.

As a courtesy to your reader, be sure to use a common file format for attached documents. Because many people use Word, a .doc file or .docx file will work for most readers. In addition, include your name in the file name. Instead of a generic "résumé.doc," use "karen_jones_résumé.doc" so the reader can easily identify your résumé.

Lastly, before sending your résumé to an employer, always test it by sending it and any attachments to yourself and a few others to ensure that the formatting will be preserved. Another option for ensuring that the résumé you create in Microsoft Word or other program looks the same to both you and your reader is to

presented earlier in this chapter. Whether you are creating your own webpage or using a template, you have a variety of options for creating a visually appealing, easily navigable Web résumé. The tips for Web writing presented in Chapter 2 can guide your design and presentation decisions; guidelines presented earlier in this chapter will help you develop the content.

Scannable Résumés

Some companies may require either in print or via email a résumé that can be scanned into a database and retrieved when a position is being filled. Since the objective is getting your résumé reviewed in order to be interviewed, you should use the following strategies to improve your chances of having it retrieved.

include keywords One strategy, using **keywords**, is often recommended for use with scannable résumés. These keywords are usually nouns or concrete words that describe skills and accomplishments precisely. Instead of listing a course in comparative programming, you would list the precise languages compared, such as PHP, C++, and Java. Instead of saying you would like a job in information systems, you would

> [Some companies may require either in print or via email a résumé that can be scanned into a database and retrieved when a position is being filled.]

save the résumé as a PDF and attach it to your email. Word's "Save As" feature will let you save files as a PDF. If you are not using Word, free PDF converters are available online.

submitting an online job application If you're applying online, you may be asked to attach a résumé in the same way you would attach a file to an email. If this is the case, the same guidelines discussed above apply. However, applying online may require that you enter information into text boxes or fields. Because these fields are not likely to allow text formatting, you will want to copy the information from your unformatted (plain-text) résumé into the fields. Many times if you copy from a formatted résumé, you have to delete tabs or extra spaces, so copying from an unformatted résumé will save time. If you want to use any kind of formatting, you can always capitalize headings or use an asterisk (*) or hyphen (-) in place of a bullet.

using a webpage or social networking site Social networking sites such as LinkedIn provide a convenient way to present your professional qualifications. Exhibit 11-8 presents an example of a well-designed LinkedIn page. Also, as we have mentioned, several free sites such as Weebly and Yola provide professional templates for creating personal webpages. All you need to do is input the content, following the guidelines

name specific job titles such as systems analyst, network specialist, or application specialist. Using industry-specific terminology is also persuasive.

Some ways to identify the keywords in your field are to read ads, listen to recruiters, and listen to your professors. Start building a list of words you find used repeatedly. From this list, choose those words most appropriate for the kind of work you want to do. Increase your use of abbreviations, acronyms, and jargon appropriate to the work you want to do.

Some experts recommend using a separate keyword section at the beginning of the résumé, loading it with all the relevant terms. If you use this technique, include the heading "Keywords" before your objective and follow it with 6 to 12 keywords. However, many résumé writers are well aware of the importance of using keywords and consciously work to integrate them into their résumés. Remember, though, to be ethical in your use of keywords from the job posting by using only those keywords that actually represent your qualifications.

choose words carefully Unlike the traditional résumé, the scannable résumé is strengthened not by the use of action verbs but rather by the use of nouns. Informal studies have shown that those retrieving résumés from such databases tend to use precise nouns when searching the database.

You may, however, prefer to combine the use of precise nouns with strong action verbs. The nouns will help ensure that the résumé gets pulled from the database, and if a recruiter pulls your résumé for further review, the verbs will help the recruiter see the link to the kind of work you want to do.

keep the design simple Since you want your résumé to be read accurately, you will use a font most scanners can read easily, such as Helvetica, Arial, and Times Roman. Most scanners can easily handle fonts between 10 and 14 points. Although many handle bold, when in doubt use all caps for emphasis rather than bold. Also, because italics often confuse scanners, avoid them. Underlining is best left out as well. It creates trouble with descending letters such as *g* or *y* when the line cuts through the letter. Also, avoid graphics and shading; they just confuse the software. Use white paper to maximize the contrast, and always print in the portrait mode. The Manny Konedeng résumé in Exhibit 11-9 (page 332) is a scannable résumé employing these guidelines.

Today companies accept résumés by mail, fax, Web, and email. Be sure to choose the channel that serves you best. Some employers give the option to the sender. Obviously, when speed gives you a competitive advantage, you'll choose the fax, Web, or email options. If you elect to print and send a scannable résumé, do not fold it. Just mail it in a 9 × 12 envelope. For

exists but would like to investigate the possibility of employment with a company. Generally, a cover letter is organized according to the following plan:

- An introduction that gets the reader's attention and provides just a brief summary of why you are interested or qualified or previews the information in the body of the letter. If you are writing a solicited letter, you will also mention where you learned of the position.

- A body that matches your qualifications to the reader's needs. You should also use good sales strategy, especially the you-viewpoint and positive language.

- A conclusion that requests action such as an interview and provides contact information that makes a response easy.

Exhibits 11-10 through 11-13 provide examples of effective cover letters.

gaining attention in the opening As in sales writing, the opening of the cover message has two requirements: It must gain attention and it must set up the information that follows.

Gaining attention is especially important in prospecting messages. Such letters are likely to reach busy executives. Unless the résumé gains favorable attention right away, the potential employer probably will not read it. Even invited messages must

> As in sales writing, the opening of the cover message must gain attention and set up the information that follows.

a little extra cost, you will help ensure that your résumé gets scanned accurately rather than wondering if your keywords were on a fold that a scanner might have had difficulty reading.

LO 11-5 Write targeted cover messages that skillfully sell your abilities.

WRITING THE COVER MESSAGE

You should begin work on the cover message by fitting the facts from your background to the work you seek and arranging those facts in a logical order. Then you present them in much the same way that a sales writer would present the features of a product or service, carefully managing the appeal. Wherever possible, you adapt the points you make to the reader's needs.

Cover Letters

Cover letters come in two types: **solicited (invited)** and **unsolicited (prospecting)**. As their names suggest, a solicited letter is written in response to an actual job opening, and an unsolicited letter is written when you don't know whether a job

gain attention because they will compete with other invited messages. Invited messages that stand out favorably from the beginning have a competitive advantage.

As the cover message is a creative effort, you should use your imagination in writing the opening, but the work you seek and your audience should guide your imagination. Take, for example, work that requires an outgoing personality and a vivid imagination such as sales or public relations. In such cases, you would do well to show these qualities in your opening words. At the opposite extreme is work of a conservative nature, such as accounting or banking. Openings in such cases should normally be more restrained.

In choosing the best opening for your case, you should consider whether you are writing a prospecting or an invited message. If the message has been invited, your opening words should refer to the job posting and begin qualifying you for the advertised work, as in these examples:

> Will an honors graduate in accounting, with experience in tax accounting, qualify for the work you listed in today's *Times*?

> Because of my specialized training in accounting at State University and my practical experience in cost-based accounting, I believe I have the qualifications you described in your *Journal* advertisement.

Manny Konedeng
5602 Montezuma Road
Apartment 413
San Diego, California 92115
Phone: (619) 578-8058
Email: mkonedeng@yahoo.com

Avoids italics and underlines yet is arranged for both scanner and human readability

KEYWORDS

Finance major, bachelor's degree, leadership skills, ethics, communication, teamwork

OBJECTIVE

A financial analyst internship with a broker-dealer where both analytical and interpersonal communication skills and knowledge are valued

Uses all caps and spacing for enhanced human readability

EDUCATION

Bachelor of Science Degree in Business Administration, May 2014
San Diego State University, Finance Major

Dean's List
Current GPA: 3.32/4.00

Related Courses

Business Communication
Investments
Tax planning
Estate Planning
Risk Management
Business Law

All items are on one line and tabs avoided for improved comprehension

Computer Skills

Statistical Software: SAS, SPSS Excel, Word, PowerPoint, Access, QuickBooks
Web-based Applications: Surveymonkey, Blogger, GoToMeeting
Research Tools: Center for Research in Securities Prices (CRSP) stock price database and the Standard and Poor's Research Insight database of corporations' financial statements

Accomplishments

Published in Fast Company Magazine and the San Diego Union Tribune
Won Greek scholarship
Finished in top five mathematics competition

WORK EXPERIENCE

Powerhouse Gym, Sales and Front Desk, Summer 2013 Modesto, CA 95355

Sold memberships and facilitated tours for the fitness center
Listened to, analyzed, and answered customers' inquiries
Accounted for membership payments and constructed sales reports
Trained new employees to understand company procedures and safety policies

Integrates precise nouns and industry-specific jargon as keywords

MCI, Relay Operator, Summer 2011 & 2012, Riverbank, CA 95367

Assisted over 100 callers daily who were deaf, hard of hearing, or speech disabled to place calls
Exceeded the required data input of 60 wpm with accuracy
Multitasked with typing and listening to phone conversations
Was offered a promotion as a Lead Operator

EXHIBIT 11-9   (continued)

Fo Sho Entertainment, Co-founder and Owner, Aug. 2009-May 2011
Modesto, CA 95355

Led promotions for musical events in the Central Valley
Managed and hosted live concerts
Created and wrote proposals to work with local businesses
Collaborated with team members to design advertisements

UNIVERSITY EXPERIENCE

Information Decision Systems, Communications Tutor, Spring 2014,
San Diego, CA 92182
Critiqued and evaluated the written work for a business
communication course

Set up and maintained blog for business communication research
Kappa Alpha Order Fraternity, Recruitment Chairman, Fall 2013,
San Diego, CA 92115
Supervised the selection process for chapter membership
Individually raised nearly $1,000 for chapter finances
Organized recruitment events with business sponsors, radio stations,
and special guests

ACTIVITIES AND SERVICE

Campus Leadership

Recruitment Chair, Kappa Alpha Order Fraternity
Supervised all new member recruitment
Coordinated fundraisers for chapter finances
Organized recruitment events with business sponsors, radio stations,
and special guests
Advocated Greek Freshmen Summer Orientation and Greek life

Correspondent for External Chapter Affairs, Kappa Alpha Order
Fraternity
Communicated with chapter alumni and National Office to fulfill
chapter obligations

Upsilon Class Treasurer, Kappa Alpha Order Fraternity
Managed chapter budgets and expenditures
Held several Interfraternity Council Roles

Member, Fraternity Men against Negative Environments and Rape
Situations

Cochairman, Greek Week Fundraiser

Candidate, IFC Treasurer

Professional and Community Service

Member, Finance & Investment Society
Presenter, Peer Health Education
Marshal, SDSU New Student & Family Convocation
Volunteer, Muscular Dystrophy national philanthropy
Volunteer, Service for Sight philanthropy
Volunteer, Victims of Domestic Violence philanthropy
Volunteer, Camp Able philanthropy
Associated Students' Good Neighbor Program volunteer
Volunteer, Designated Driver Association
Volunteer, Beach Recovery Project

Avoids graphics and extra lines

Adds other relevant information since there is no physical page limit

Uses black on white contrast for improved scanning accuracy

From: Molly Everson <mheverson@creighton.edu>
To: Marlene O'Daniel <modaniel@cic.org>

SUBJECT: Application—Communications Specialist Position

Dear Ms. O'Daniel:

One of your employees, Victor Krause, suggested that I apply for the communications specialist position you have open. My résumé is attached for your review.

Gains attention with an associate's name

Presently, I am a communications intern for Atlas Insurance. My work consists primarily of writing a wide variety of documents for Atlas policyholders. This work has made me an advocate for well-crafted business communication, and it has sharpened my writing skills. More importantly, it has taught me how to get and keep customers for my company through writing well.

Employs a conservative style and tone

Shows the writer knows the skills needed for the job

Additional experience working with businesspeople has given me insight into the communication needs of business. This experience includes planning and presenting a communication improvement course for Atlas employees.

Uses subtle you-viewpoint— implied by the writer's understanding of the work

My college training provided a solid foundation for work in business communication. Advertising and public relations were my areas of concentration for my BS degree from Creighton University. As you will see on the enclosed résumé, I studied all available writing courses in my degree plan. I also studied writing through English and journalism courses.

References the résumé

Brings the review to a conclusion— fits the qualifications presented to the job

My education and experience have prepared me for work as your communication specialist, as the attached writing samples show. I know business writing, and I know how it can be used to your company's advantage. May we discuss my qualifications in an interview? You can reach me at 402-786-2575 to arrange a convenient time and place to meet.

Moves appropriately for action

Sincerely,

Molly H. Everson

4407 Sunland Avenue
Phoenix, AZ 85040-9321

July 8, 2014

Ms. Anita O. Alderson, Manager
Tompkins-Oderson Agency, Inc.
3901 Tampico Avenue
Los Angeles, CA 90032-1614

Dear Ms. Alderson:

Uses reader's words for good attention gainer

Marketing student … interest in advertising … team skills….

Demonstrates ability to write advertising copy through writing style used

These keywords in your July 6 advertisement on State University's Career Finder student Web site describe the intern you want, and I believe I am that person.

Shows clearly what the writer can do on the job

I have gained experience in every area of retail advertising while working for the *Lancer*, our college newspaper, selling advertising, planning layouts, and writing copy. During the last two summers, I obtained firsthand experience working in the advertising department of Wunder & Son. My main responsibility was to write copy, some of which is enclosed for your inspection; you will find numerous other examples on my blog at http://janekbits.blogspot.com.

Shows strong determination through good interpretation

In my major, I am studying marketing with a specialization in advertising and integrated marketing communications. My course work has provided several opportunities to participate in real-world projects, including one where my teammates and I used a variety of media to raise money for schools in Louisiana, Texas, and Mississippi's hurricane damaged areas. Understanding the importance of being able to get along well with people, I actively participated in Sigma Chi (social fraternity), the Race for the Cure (breast cancer fundraising event), and Pi Tau Pi (honorary business fraternity). The experience gained in these associations makes me confident that I can fit in well at Tompkins-Oderson.

Provides good evidence of social skills

Leads smoothly to action

I ask that you review my qualifications and contact me for an interview. You can email me at janek@hotmail.com or call and text message me at 602-713-2199 to arrange a convenient time to talk about my joining your team.

Sincerely,

Michael S. Janek

Michael S. Janek

enclosures

12712 Sanchez Drive
San Bernadino, CA 92405

April 9, 2014

Mr. Conrad W. Butler
Office Manager
Darden, Inc.
14326 Butterfield Road
San Francisco, CA 94129

Dear Mr. Butler:

Gains attention with question

Can Darden, Inc., use a hardworking Grossmont College business administration major seeking an internship in management or office administration? My experience, education, and personal qualities have prepared me to contribute to your operations.

Sets up rest of letter

As the attached résumé indicates, I have worked as a receptionist for the past three summers. My duties have included managing a multi-line telephone system, using Excel and Access to monitor office traffic, and greeting visitors. I am excited about the possibility of developing more skills through an internship with Darden.

Brings out highlights with review of experience

Complementing my work experience are my studies at Grossmont College. In addition to studying the prescribed courses in my major field of business office technology, I have completed electives in Dreamweaver, QuickBooks, and professional speaking to help me in my career objective. In spite of full-time employment through most of my time in college, I was awarded the Associate of Arts degree last May with a 3.3 grade point average (4.0 basis). But most important, I am learning from my studies how office work can be done efficiently.

Interprets education facts for the reader

In addition, I have the personal qualities that would enable me to fit smoothly into your organization. I like people, and through work and academic experiences, I have learned how to work with them as both a team player and a leader.

Sets up action and uses adaptation in concluding statement

May I meet with you to talk about interning for Darden? Please call me at 714-399-2569 or email me at jgoetz@gmail.com to arrange an interview.

Requests action clearly and appropriately

Sincerely,

Jimmy I. Goetz

Jimmy I. Goetz

Enc.

MARY O. MAHONEY

May 17, 2014

Mr. Nevil S. Shannon
Director of Personnel
Snowdon Industries, Inc.
1103 Boswell Circle
Baltimore, MD 21202

Dear Mr. Shannon:

Effective attention-getting question — Will you please review my qualifications for work in your management trainee program? My education, work attitude, and personal skills qualify me for this program. — *Good setup for paragraphs that follow*

Good interpretation of education — My education for administration consists primarily of four years of business administration study at State University. The Bachelor of Business Administration degree I will receive in June has given me a broad foundation of business knowledge. As a general business major, I studied all the functional fields (management, marketing, information systems, finance, accounting) as well as the other core business subjects (communications, statistics, law, economics, production, and human resources). I have the knowledge base that will enable me to be productive now, and I can build upon this base through practical experience.

In addition, I completed an internship last summer in the claims department of Advantage Insurance. My duties included accurately processing a high volume of claims and documenting interactions with both our customers and other claimants. At the end of the internship, my supervisors commented on my ability to diffuse tense situations and work with people of all types and backgrounds. In addition, they noted the creative thinking and team skills I demonstrated as we all worked to make the claims process more efficient. — *Emphasizes skills transferable to the position applied for.*

Good use of fact to back up personal qualities — Throughout college, I developed my personal skills. As an active member of the student chapter of the Society for the Advancement of Management, I served as treasurer and program chairperson. I participated in intramural golf and volleyball, and I was an active worker in the Young Republicans, serving as publicity chairperson for three years. All this experience has helped me to acquire the balance you seek in your administrative trainees.

Good request for action — These highlights and the additional evidence presented in the enclosed résumé present my case for a career in management. May I have an interview to continue my presentation? You can reach me at 301.594.6942 or marymahoney@yahoo.com. Thank you for your consideration. — *Clear request for action flows logically from preceding presentation*

Sincerely,

Mary O Mahoney

Mary O. Mahoney

Enclosure

1718 CRANFORD AVENUE • ROCKWELL, MD • 20854
VOICE/MESSAGE/FAX: 301.594.6942 • EMAIL: MARYMAHONEY@YAHOO.COM

Developing a Professional Portfolio

Imagine yourself in an interview. The interviewer says, "This position requires you to use PowerPoint extensively. How are your presentation and PowerPoint skills?" What do you say? Of course you say your skills are excellent. And so does everyone else who interviews.

One way you can set yourself apart from other applicants is to take a professional portfolio to an interview to demonstrate your qualifications.

A portfolio may contain a title page, your résumé, references list, cover letter, a transcript, a program description, copies of licenses and certifications, work samples, letters of recommendation, personal mission statements—whatever creates your best professional image. All you need to do is put your documents in sheet protectors in a professional looking three-ring binder for easy editing and updating and create

tab dividers for the sections of the portfolio, and you're on your way. One note of advice, though: Protect your information by including a confidentiality statement; removing any student ID numbers, SSN numbers, or other private information from your documents; and using copies rather than originals of any licenses or certificates.

Source: "Portfolios," *University of Wisconsin–Eau Claire Career Services*, University of Wisconsin–Eau Claire, 9 Mar. 2010, Web, 8 July 2012.

You can gain attention in the opening of an unsolicited letter in many ways. One way is to use a topic that shows understanding of the reader's operation or of the work to be done. Employers are likely to be impressed by applicants who have made the effort to learn something about the company, as in this example:

> Now that Taggart, Inc., has expanded operations to Central America, can you use a broadly trained international business major who knows the language and culture of the region?

Another way is to make a statement or ask a question that focuses attention on a need of the reader that the writer seeks to fill. The following opening illustrates this approach:

> How would you like to hire a University of Cincinnati business major to fill in for your vacationing summer employees?

Sometimes you will learn of a job possibility through a company employee. Mentioning the employee's name can gain attention, as in this opening sentence:

> At the suggestion of Mr. Michael McLaughlin of your staff, I am sending the following summary of my qualifications for work as your loan supervisor.

Many other possibilities exist. In the final analysis, you will have to use what you think will be best for the particular job you're seeking. But try to avoid the overworked beginnings that were popular a generation or two ago such as "This is to apply for . . ." or writer-centered beginnings such as "I am writing to apply for . . ." or the tentative "I would like to apply for. . . ."

selecting content Following the opening, you should present the information about your

qualifications for the work. Begin this task by reviewing the job requirements. Then select the facts about you that qualify you for the job.

If your application has been invited, you may learn about the job requirements from the source of the invitation. If you are answering an advertisement, study it for the employer's requirements. If you are following up on an interview, review

Companies often describe themselves and their career opportunities on their Web sites.
Source: http://corporate.target.com/ Reprinted with permission of Target.

the interview for information about job requirements. If you are prospecting, your research and your logical analysis should guide you.

In any event, you are likely to present facts from three background areas: education, experience, and skills and/or personal details.

How much you include from each of these areas and how much you emphasize each area should depend on the job and on your background. Most of the jobs you will seek as a new college graduate will have strong educational requirements, and your education is likely to be your strongest selling point at this stage of your career. Thus, you should stress your education. When you apply for work after you have accumulated experience, you will probably need to stress experience. As the years go by, experience becomes more important and education less important. Your personal characteristics are important for many jobs, especially jobs that involve working with people.

If a résumé accompanies the cover message, do not rely on it too much. Remember that the message does the selling, and the résumé lists the significant details. Thus, the message should contain the major points around which you build your case, and the résumé should include these points plus supporting details. As the two are parts of a team effort, somewhere in the message you should refer the reader to the résumé.

organizing for persuasion You will want to present the information about yourself in the order that is best for you. In general, the plan you select is likely to follow one of three general orders. The most common order is a logical

> " Your personal characteristics are important for many jobs, especially jobs that involve working with people. "

from the tech desk

Web Sites Offer Valuable Interview Advice

The Web is a rich resource for help with interviewing. Sites such as Monster.com and many of the other online job database sites offer tips on all aspects of interviewing. You can get ideas for questions to ask interviewers, techniques for staying calm, and methods of handling a phone interview. They even include practice interactive virtual interviews, with immediate feedback on your answers as well as suggestions

and strategies for handling difficult questions. The Monster site includes a planner listing a host of good commonsense tips from polishing your shoes to keeping an interview folder to keep track of all written and oral communication. Using these sites when preparing for interviews will help you not only feel more confident and interview more effectively but also be ready to evaluate the company as well.

Source: Used by permission of Monster.com.

> ## "Merely presenting facts does not ensure conviction. You will also need to present the facts in words that make the most of your assets."

grouping of the information, such as education, experience, and skills and/or personal details. A second possibility is a time order. For example, you could present the information to show a year-by-year preparation for the work. A third possibility is an order based on the job requirements. For example, selling, communicating, and managing might be the requirements listed in an advertised job.

Merely presenting facts does not ensure conviction. You will also need to present the facts in words that make the most of your assets. Think of this as the difference between showing and telling. You could tell the reader, for example, that you "held a position" as sales manager, but it is much more convincing to say that you "supervised a sales force of 14," which actually shows your ability. Likewise, you do more for yourself by writing that you "earned a degree

in business administration" than by writing that you "spent four years in college." And it is more effective to say that you "learned tax accounting" than to say that you "took a course in tax accounting."

You also can help your case by presenting your facts in reader-viewpoint language wherever this is practical. More specifically, you should work to interpret the facts based on their meaning for your reader. For example, you could present a cold recital like this one:

> I am 21 years old and have an interest in mechanical operations and processes. Last summer I worked in the production department of a container plant.

Or you could interpret the facts, fitting them to the one job:

> Last summer's experience working 10- and 12-hour days in the production department of Miller Container Company is evidence of my interest in mechanics and shows that I can and will work hard.

This prospecting message is dull, selfish, and poorly written.

Dear Mr. Stark:

This is to apply for a position in marketing with your company.

At present, I am completing my studies in marketing at Wilmont University and will graduate with a Bachelor of Business Administration degree with an emphasis in human resource management this May. I have taken all the courses in marketing available to me as well as other helpful courses such as statistics, organizational psychology, and ecommerce.

I have had good working experience as a host and food server, a sales associate, and an HR intern. Please see details on the enclosed résumé. I believe that I am well qualified for a position in human resource management and am considering working for a company of your size and description.

Because I must make a decision on my career soon, I request that you write me soon. For your information, I will be available for an interview on March 17 and 18.

Sincerely,

Julia M. Alvarez

Since you will be writing about yourself, you may find it difficult to avoid overusing *I*-references, but an overuse of *I*'s sounds egotistical and places too much attention on the often repeated word. This creates the impression that you are more focused on yourself than you are on the reader's needs. Some *I*'s, however, should be used. After all, the message sells your skills and abilities.

Overall, you are presenting your professional image, not only as a prospective employee but also as a person. Carefully shaping the character you are projecting is arguably just as important to the success of your cover message as using convincing logic.

driving for action in the close

The presentation of your qualifications should lead logically to the action that the close proposes. You should drive for whatever action is appropriate in your case. It could be a request for an interview or an invitation to engage in further communication (perhaps to answer the reader's questions). You are concerned mainly with opening the door to further negotiations.

Your action words should be clear and direct. As in the sales message, the request for action may be made more effective if it is followed by words recalling a benefit that the reader will get from taking the action. The following closes illustrate this technique:

The highlights of my education and experience show that I have been preparing for a career in human resources. May I now discuss beginning this career with you? You can reach me at 727-921-4113 or by email at owensmith@att.com

I am very much interested in meeting for an interview to discuss with you how my skills can contribute to your company's mission.

contrasting cover messages

The message on page 340 and the message below present the qualifications of Julia M. Alvarez, the job seeker described in the Workplace Scenario at the beginning of the chapter. The first message follows few of the suggestions given in the preceding pages, whereas the second message is persuasive and well organized and clearly asks for action in the closing.

a bland and artless presentation of information

The bad message on page 340 begins with an old-style opening. The first words are of little interest. The presentation of qualifications that follows is a matter-of-fact, uninterpreted review of information. Little you-viewpoint is evident. In fact,

This letter follows the chapter's advice for writing prospecting messages.

Dear Mr. Stark:

Is there a place in your Human Resource Department for someone who is well trained in the field and can talk easily and competently with employees? As a June graduate with a degree in human resource management and internship experience, I am well qualified to meet your needs.

My studies at Wilmont University were specially planned to prepare me for a career in human resource management. I have taken courses in compensation theory and administration, organizational change/development, and training and human resource development. As part of my degree requirements, I also passed the Society for Human Resource Management (SHRM) Certification Examination. In addition, I studied a wide assortment of supporting subjects: economics, business communication, information systems, psychology, interpersonal communication, and operations management. My studies have given me a solid foundation in HR work.

As my résumé shows, I have completed an HR internship, where I performed tasks ranging from researching and reporting data on company and industry trends to participating in the hiring process. I would welcome the opportunity to continue working in HR with your company.

My work experiences have also prepared me for a career in human resource management. While in college I worked as a server at Applebee's, where I developed customer service and team skills, and I continued to develop my skills as a team lead for The Gap, where I was the top seller for four of eight quarters. From these experiences, I have learned to understand human resource management and listen carefully to people.

These brief facts and the information in my résumé describe my diligent efforts to prepare for a position in human resource management. May I talk with you about beginning that position? You can reach me at 917.938.4449 to arrange an interview to discuss how I could fit in your Human Resource Department.

Sincerely,

Julia M. Alvarez

most of the message emphasizes the writer (note the *I*'s), who comes across as bored and selfish. The information presented is scant. The closing action is little more than an I-viewpoint statement about the writer's availability.

skillful selling of one's ability to work The better message on page 341 begins with an interesting question that sets the stage for the rest of the contents. The review of experience is interpreted, showing how the experience would help the applicant perform the job. The review of education is similarly covered. Notice how the interpretations show that the writer knows what the job requires. Notice also that reader-viewpoint is stressed throughout. Even so, a moderate use of *I* gives the letter a personal quality, and the details show the writer to be a thoughtful, engaged person. The closing request for action is a clear, direct, and courteous question. The final words recall a main appeal of the letter.

Email Cover Messages

Like other email messages, an email cover message needs a clear subject line; like print cover messages, it needs a formal salutation and closing. And its purpose is still to highlight your qualifications for the particular job you are applying for. It can be identical to one you might create for print, or you may opt to introduce yourself and your purpose in a short email message and attach your full cover letter. The primary job of the email cover message is to identify the job, highlight the applicant's strengths, and invite the reader to review the résumé.

Notice how the solicited cover message below quickly gains the reader's attention in the opening, highlights the skills in the body, and calls for action in the close.

To: Kate Troy <kate_troy@thankyoutoo.com>
From: Jessica Franklin <jessica_franklin@yahoo.com>
Date: October 1, 2014
Subject: Web Design Intern Position

Dear Ms. Troy:

Yesterday my advisor here at Brown University, Dr. Payton Kubicek, suggested that I contact you about the summer intern position in Web design you recently announced on your Web site.

At Brown I have taken courses that have given me a good understanding of both the design aspects and the marketing elements that a good Web site needs. Additionally, several of my course projects involved working with successful Web-based businesses, analyzing the strengths and weaknesses of their business models.

I would enjoy applying some of these skills to help build a successful site targeted toward Thankyoutoo.com's high-end customers. You will see from my webpage profile at www.jessicafranklin.com/ that my design preferences

and styles complement those of your company's Web site, allowing me to contribute almost immediately. I can be available for an interview at any time.

Sincerely,

Jessica Franklin

LO 11-6   Explain how you can participate effectively in an interview.

HANDLING THE INTERVIEW

Your initial contact with a prospective employer can be by mail, email, phone, or a personal (face-to-face) visit. If all goes well, your application will eventually involve a personal visit—an interview. Sometimes, before inviting candidates to a formal interview session, recruiters use phone interviews for preliminary screening.

In a sense, the interview is the key to the success of the application—the "final examination." You should carefully prepare for the interview, as the job may be lost or won in it. The following review of employment interview highlights should help you do your best in your interviews. You will find additional information about interviewing on the textbook Web site.

Some employers use phone interviews for initial screening of job candidates. Being both well rested and well prepared for those initial contacts will help ensure that one gets an opportunity for a face-to-face interview later.

communication matters

Investigating the Company

Before arriving for an interview, you should learn what you can about the company: its products or services, its personnel, its business practices, its current activities, its management. Such knowledge will help you talk knowingly with the interviewer. And perhaps more important, the interviewer is likely to be impressed that you took the time to investigate the company. That effort can give you an advantage.

Making a Good Appearance

How you look to the interviewer is a part of your message. Thus, you should work to present just the right image. Interviewers differ to some extent on what that image is, but you would be wise to present a conservative appearance. This means avoiding faddish, offbeat styles and preferring the conservative, conventional business colors such as black, brown, navy, and gray. Remember that the interviewer wants to know whether you fit into the role you are seeking. You should look like you are right for the job.

Some may argue that such an insistence on conformity in dress and grooming infringes on one's personal freedom. Perhaps it does, but if the people who can determine your future have fixed views on matters of dress and grooming, it is good business sense to respect those views.

Anticipating Questions and Preparing Answers

You should be able to anticipate some of the questions the interviewer will ask. Questions about your education (courses, grades, honors) are usually asked. So are questions about work experience, interests, career goals, location preferences, and activities in organizations. You should prepare answers to these questions in advance. Your answers will then be thorough and correct, and your words will display poise and confidence. Your preparation will also reflect your interest.

In addition to general questions, interviewers often ask more complicated ones. Some of these are designed to test you—to learn your views, your interests, and your ability to deal with difficult problems. Others seek more specific information about your ability to handle the job in question. Although such questions are difficult to anticipate, you should be aware that they are likely to be asked. Following are questions of this kind that one experienced interviewer asks:

What can you do for us?

Would you be willing to relocate? To travel?

Do you prefer to work with people or alone?

How well has your performance in the classroom prepared you for this job?

What do you expect to be doing in 10 years? In 20 years?

What income goals do you have for those years (10 and 20 years ahead)?

Why should I rank you above the others I am interviewing?

Why did you choose _____ for your career?

How do you feel about working overtime? Nights? Weekends?

Did you do the best work you are capable of in college?

Is your college record a good measure of how you will perform on the job?

What are the qualities of the ideal boss?

What have you done that shows leadership potential? Teamwork potential?

What are your beginning salary expectations?

Sometimes interviewers will throw in tough questions to test your poise. The Communication Matters information on page 342 provides some examples of these questions and provides tips for answering them.

Some questions, though, may not be legal regardless of the interviewer's intent, whether the interviewer is making small talk, is unaware the questions are illegal, plans to discriminate against you, or just wants to test whether you respond. How you respond is up to you; before you respond, you may want to ask how the question is relevant to the position, or you may politely decline to answer.

What religion do you practice?

How old are you?

Are you married?

Do you plan to have children?

Recently, the **behavioral interview style** has become popular with campus recruiters. Rather than just determining your qualifications for the job, interviewers are attempting to verify if you can do the work. They ask questions about what you would do in certain situations because how you behave now is likely to transfer to similar situations in another job. Here are a few examples of behavioral questions:

What major problem have you faced in group projects and how did you deal with it?

Can you tell us about a time when you had to choose between following the rules or stretching them?

Describe a conflict you had with someone and how you resolved it.

> "Interviewers ask questions about what you would do in certain situations because how you behave now is likely to transfer to similar situations in another job."

Keep your answers concise. Briefly state the situation, describe the steps you took, and summarize the results. You can also share what you learned from the experience. For more practice preparing for questions, check the resource links on the textbook Web site.

Putting Yourself at Ease

Perhaps it is easier to say than to do, but you should be calm throughout the interview. Remember that you are being inspected and that the interviewer should see a calm and collected person. Appearing calm involves talking in a clear and strong voice. It also involves controlling your facial expressions and body movements. Developing such controls requires self-discipline and reassuring self-talk. You may find it helpful to convince yourself that the stress experienced during an interview is normal. Or you may find it helpful to look at the situation realistically—as merely a conversation between two human beings. Practicing your answers to common

The job interview is the final stage of the job application. Appearing professional, calm, and enthusiastic will help you succeed.

communication matters

What's the Number One Interviewing Mistake?

According to Jessica Liebman, managing editor of the business-news site *Business Insider*, the number one interview mistake isn't wearing the wrong clothes, showing up with a cup of coffee, being late, having a limp handshake, or opening the interview by asking if you'll get a free iPad.

It's an action (or nonaction) that takes place after the interview—namely, failing to send a thank-you message. As Liebman says, even a brief email of thanks helps seal a positive impression, whereas no message indicates that "you don't want the job" or "you're disorganized and forgot about following up."

So the next time you have an interview, be sure you get everyone's name. As soon as possible after the interview (preferably the same day), send your interviewers a thank-you email. It could mean the difference between getting your job and continuing your search.

Source: Jessica Liebman, "The Number One Mistake People I Interview Are Making These Days," *Business Insider*, Business Insider, 24 Feb. 2013, Web, 20 June 2013.

> Your goal should be to make certain that both the interviewer and you get all the information you consider important.

interview questions out loud may be helpful. You may even want to record one of these practice sessions and analyze your performance. Your school's career services office may be able to help with this. Other approaches may work better for you. Use whatever approaches work. Your goal is to control your emotions so that you present the best possible appearance to the interviewer.

Helping to Control the Dialogue

Just answering the questions asked is often not enough. The questions you ask and the comments you make should bring up what you want the interviewer to know about you. Your self-analysis revealed the strong points in your background. Now you should make certain that those points come out in the interview.

How to bring up points about you that the interviewer does not ask is a matter for your imagination. For example, a student seeking a job in advertising believed that her teamwork skills should be brought to the interviewer's attention. So at an appropriate time in the interview, she asked, "How important is the ability to collaborate in this company?" The anticipated answer—"very important"—allowed her to discuss her skills. To take another example, a student who wanted to bring out his knowledge of the prospective employer's operations did so with this question: "Will your company's expansion in the Madison area create new job opportunities there?" How many questions of this sort you should ask will depend on your need to supplement your interviewer's questioning.

Although you want to ask questions that highlight your skills, you should also ask questions to determine if the company is a good fit for you such as "How would you describe the work environment here?" Your goal should be to make certain that both the interviewer and you get all the information you consider important.

LO 11-7 Write application follow-up messages that are appropriate, friendly, and positive.

FOLLOWING UP AND ENDING THE APPLICATION

The interview not the last step in the application process. A variety of other steps can follow. Sending a brief thank-you message email is an essential follow-up step. Not only does it show courtesy, but it can also give you an advantage because some of your competitors will not do it. If you do not hear from the prospective employer within a reasonable time, it is appropriate to inquire by telephone, email, or letter about the status of your application. You should certainly do this if you are under a time limit on another employer's offer. The application process may end with no offer (frequently with no notification at all—a discourteous way of handling applicants), with a rejection notice, or with an offer. How to handle these situations is reviewed in the following paragraphs.

Other Job-Search Messages

writing a thank-you message After an interview it is courteous to write a thank-you message, whether or not you are interested in the job. If you are interested, the message can help your case. It singles you out from the competition and shows your interest in the job, and as the Communication Matters feature on page 345 indicates, not sending a thank-you note could determine whether you are offered a job.

Such messages are usually short. They begin with an expression of gratefulness. They say something about the interview, the job, or the company. They take care of any additional business (such as submitting information requested). Then they end on a goodwill note—perhaps a hopeful look to the next step in the negotiations. While you can send your message by mail, an

Sending a thank-you note shows professionalism and good manners.

email will quickly convey your thanks. The following message illustrates:

> Dear Mr. Woods:
>
> Thank you for talking with me yesterday about the finance internship. I enjoyed learning more about Sony Corporation of America and the financial analyst position.
>
> As you requested, I have enclosed samples of the financial analysis I developed as a class project. If you need anything more, please let me know.
>
> I look forward to the possibility of discussing employment with you soon.
>
> Sincerely,

constructing a follow-up to an application When a prospective employer is late in responding or you receive another offer with a time deadline, you may need to write a **follow-up message**. Employers are often just slow, but sometimes they lose the application. Whatever the explanation, a follow-up message may help to produce action.

Such a message is a form of routine inquiry. It can use the need to make a job decision or some other good explanation as the reason for writing. The following message is an example:

> Dear Ms. Yang:
>
> Because the time is approaching when I must make a job decision, could you please tell me the status of my application with you?
>
> You may recall that you interviewed me in your office November 7. You wrote me November 12 indicating that I was among those you had selected for further consideration.
>
> SAIC remains one of the organizations I would like to consider in making my career decision. I would very much appreciate hearing from you by December 3.
>
> Sincerely,

planning the job acceptance Job acceptances are favorable response messages with an extra amount of goodwill. Because the message should begin directly, a "yes" answer in the beginning is appropriate. The remainder of the message should contain a confirmation of the starting date and place and comments about the work, the company, the interview—whatever you would say if you were face to face with the reader. The message need not be long. This one does the job well:

> Dear Ms. Garcia:
>
> Yes, I accept your offer of employment as a junior analyst. After my first interview with you, I was convinced that Allison-Caldwell was the organization for me. I am delighted that you think I am right for Allison-Caldwell.
>
> Following your instructions, I will be in your Toronto headquarters on May 28 at 8:30 AM ready to work for you. Thank you for this opportunity.
>
> Sincerely,

> ## As a matter of policy, some companies require a written resignation even after an oral resignation has been made.

writing a message refusing a job

Messages refusing a job offer follow the indirect refusal pattern. One good technique is to begin with a friendly comment—perhaps something about past relations with the company. Next, explain and present the refusal in clear yet positive words. Then end with a more friendly comment. This example illustrates the plan:

> Dear Mr. Chen:
>
> Meeting you and the other people at Northern was a genuine pleasure. Thank you for sharing so much information with me and for the generous job offer.
>
> I was impressed with all I learned about Northern, and as we discussed, a special interest of mine is to work abroad. After considerable thought, I have decided to accept an offer with a firm that has extensive opportunities along these lines.
>
> I appreciate the time and the courteous treatment you gave me.
>
> Sincerely,

writing a resignation

At some point in your career you are likely to resign from one job to take another. When this happens, you will probably inform your employer of your resignation orally. But when you find it more practical or comfortable, you may choose to resign in writing. In some cases, you may do it both ways. As a matter of policy, some companies require a written resignation even after an oral resignation has been made.

Your resignation should be as positive as the circumstances permit. Even if your work experiences have not been pleasant, you want to depart without a final display of anger. As an anonymous philosopher once explained, "When you write a resignation in anger, you write the best letter you will ever regret."

The indirect order is usually the best strategy for negative messages such as a resignation. But some are written in the direct order. They present the resignation right away, following it with expressions of gratitude or favorable comments about past working experiences. Either approach is acceptable. Even so, you would do well to use the indirect order because it is more likely to build the goodwill you want to leave behind you.

The example below illustrates the indirect order. It begins with a positive point—one that sets up the negative message. The negative message follows, clearly yet positively stated. The ending returns to positive words chosen to build goodwill and fit the case.

> Dear Ms. Shuster:
>
> Working as your assistant for the past five years has been a genuinely rewarding experience. Under your direction I have grown as an administrator, and I have learned a great deal from you about retailing.
>
> As you may recall from our past discussions, I have been pursuing the same career goals that you held early in your career. To achieve these goals I am now resigning to accept a store management position with Lawson's in Belle River. I would like my employment to end on the 31st, but I could stay a week or two longer if needed to help train my replacement.
>
> I leave with only good memories of you and the other people with whom I worked. Thanks to all of you for a valuable contribution to my career.
>
> Sincerely,

LO 11-8 Maintain your job-search activities.

Continuing Job-Search Activities

Continuously keeping your finger on the pulse of the job market is a good idea. Not only will it provide you with information about changes occurring in your field, but it will also keep you alert to better job opportunities as soon as they are announced.

maintaining your résumé

While many people intend to keep their résumés up to date, they just do not make it a priority. Some others make it easy by updating as changes occur. And a few update their résumés at regularly designated times, such as a birthday, New Year's Day, or even the anniversary of their employment. No matter what works best for you, updating your résumé as you gain new accomplishments and skills is important. Otherwise, you will be surprised to find how easily you can lose track of important details.

reading job ads/professional journals

Nearly as important as keeping your résumé updated is keeping up on your professional reading. Most trade or professional journals have job notices or bulletin boards you should check regularly. These ads give you insight into what skills are in demand, perhaps helping you choose assignments where you get the opportunity to develop new skills. Staying up to date in your field can be stimulating; it can provide both challenges and opportunities.

conduct a winning job campaign!

- Would you like to learn more about creating a LinkedIn profile?
- Have you ever taken a personality test to determine what careers might be right for you?
- Are you headed to a career fair and need tips on how to stand out?

Scan the QR code with your smartphone or use your Web browser to find out at www.mhhe.com/RentzM3e. Choose Chapter 11 > Bizcom Tools & Tips. While you're there, you can view exercises, PPT slides, and more to conduct a winning job campaign.

www.mhhe.com/RentzM3e

Endnotes

CHAPTER 1

1. The Conference Board, Corporate Voices for Working Families, the Partnership for 21st Century Skills, and the Society for Human Resource Management, *Are They Ready to Work? Employers' Perspectives on the Basic Knowledge and Applied Skills of New Entrants into the 21st Century Workforce*, 21, *Partnership for 21st Century Skills*, Partnership for 21st Century Skills, 2 Oct. 2006, Web, 22 Apr. 2013.

2. *NACE*, National Association of Colleges and Employers, 2011, Web, 22 Apr. 2013.

3. Shirley Taylor, "Why Are Communication Skills Important?," *ST Training Solutions*, ST Training Solutions Pte Ltd, n.d., Web, 22 Apr. 2013.

4. Jonathan Farrington, "The MOST Important Leadership Trait?—It's a 'No-Brainer,'" *Blogit*, Jonathan Farrington, 26 Sept. 2008, Web, 22 Apr. 2013.

5. Chuck Martin, "NFI Research Result: Wish List," *Forbes.com*, Forbes.com, 4 Feb. 2010, Web, 22 Apr. 2013.

6. Rich Maggiani, "The Costs of Poor Communication," *Solari*, Solari Communication, 2012, Web, 22 Apr. 2013.

7. SIS International Research, "SMB Communications Pain Study White Paper: Uncovering the Hidden Cost of Communications Barriers and Latency," *SIS International Research*, SIS International Research, *Market Intelligence Journal*, 10 Mar. 2009, Web, 22 Apr. 2013.

8. David Bollier, *The Future of Work: What It Means for Individuals, Businesses, Markets and Governments*, 15, *The Aspen Institute*, Aspen Institute, 2011, Web, 22 Apr. 2013.

9. Institute for the Future for Apollo Research Institute, *Future Work Skills 2020*, 8, *Apollo Research Institute*, Apollo Research Institute, 2011, Web, 22 Apr. 2013.

10. Bollier 19.

11. Institute for the Future for Apollo Research Institute, *Future of Work Report: Executive Summary*, 4, *Apollo Research Institute*, Apollo Research Institute, Mar. 2012, Web, 22 Apr. 2013.

12. Bollier 22.

13. Jim Keane, President, Steelcase Group, *Future of Work Webinar*, *Apollo Research Institute*, Apollo Research Institute, n.d., Web, 7 May 2012.

14. Institute for the Future for Apollo Research Institute, *Future Work Skills 2020*, 9.

15. According to Ross C. DeVol, chief research officer for the Milken Institute, one in five Americans will have hit 60 in 2030, and many of these will be staying in the workforce (*Future of Work Webinar*, *Apollo Research Institute*, Apollo Research Institute, n.d., Web, 7 May 2012).

16. Katherine Haynes Sanstad, Regional Executive Director, Diversity, Kaiser Permanente, *Future of Work Webinar*, *Apollo Research Institute*, Apollo Research Institute, n.d., Web, 7 May 2012.

17. Institute for the Future for Apollo Research Institute, *Future Work Skills 2020*, 9.

18. Sanstad.

19. Institute for the Future for Apollo Research Institute, *Future Work Skills 2020*, 4.

20. Institute for the Future for Apollo Research Institute, *Future Work Skills 2020*, 10.

21. Institute for the Future for Apollo Research Institute, *Future Work Skills 2020*, 8.

22. Bollier 8.

23. Bollier 3.

24. Institute for the Future for Apollo Research Institute, *Future of Work Report* 6.

25. See Edgar H. Schein, *Organizational Culture and Leadership*, 4th ed. (San Francisco: Jossey-Bass, 2010), print, which reviews the literature on this important concept.

26. For discussions of problem solving, see the following print resources: John R. Hayes, *The Complete Problem Solver*, 2nd ed. (Hillsdale, NJ: Lawrence Erlbaum, 1989); Morgan D. Jones, *The Thinker's Toolkit* (New York: Three Rivers Press, 1998); Janet E. Davidson and Robert J. Sternberg, eds., *The Psychology of Problem Solving* (Cambridge, UK: Cambridge University Press, 2003); Dan Roam, *The Back of the Napkin* (London: Portfolio, 2008); John Adair, *Decision Making and Problem Solving Strategies*, 2nd ed. (London: Kogan Page, 2010).

27. See research by Dorothy A. Winsor, especially *Writing Power: Communication in an Engineering Center* (Albany: SUNY Press, 2003), print.

CHAPTER 2

1. Paula Wasley, "Tests Aren't Best Way to Evaluate Graduates' Skills, Business Leaders Say in Survey," *The Chronicle of Higher Education*, The Chronicle of Higher Education, 23 Jan. 2008, Web, 26 May 2013.

2. JoAnne Yates, *Control through Communication: The Rise of System in American Management* (Baltimore: The Johns Hopkins UP, 1989) 95, print.

3. PRWeb, "Tablets Make Their Way into the Workplace: Email and Note Taking Are the Most Popular Business Uses, Says NPD In-Stat," *PRWeb*, Vocus, Inc., 14 Feb. 2012, Web, 22 May 2013.

4. Kristen Purcell, "Search and Email Still Top the List of Most Popular Online Activities," *Pew Internet & American Life Project*, Pew Research Center, 9 Aug. 2011, Web, 24 May 2013.

5. Govloop, "Does Email Help or Hinder Your Professional Productivity?" *AOL Government*, AolGov, 24 Apr. 2012, Web, 24 May 2013.

6. Heidi Schultz, *The Elements of Electronic Communication* (Boston: Allyn and Bacon, 2000) 43–47, print.

7. Erica Swallow, "How Recruiters Use Social Networks to Screen Candidates," *Mashable.com*, Mashable, Inc., 23 Oct. 2011, Web, 26 May 2013.

8. Janice Redish, *Letting Go of the Words: Writing Web Content That Works* (San Francisco: Elsevier, 2012) 20, print.

9. Jakob Nielsen, "Writing Style for Print vs. Web," *Alert Box*, Jakob Nielsen, 7 June 2008, Web, 24 May 2013.

10. Redish 102.

CHAPTER 4

1. "What Is Plain Language?," *PlainLanguage.gov*, Plain Language Action and Information Network (PLAIN), n.d., Web, 20 May 2013.

2. Quoted by Helen Sword, "Yes, Even Professors Can Write Stylishly," *The Wall Street Journal*, Dow Jones & Company, Inc., 6 Apr. 2012, Web, 21 May 2013.

3. "Disability in America Infographic," *Disabled World*, disabled-world.com, 1 Dec. 2011, Web, 23 May 2013. The graphic is based on U.S. census information.

CHAPTER 6

1. Valerie Creelman, "The Case for 'Living' Models," *Business Communication Quarterly* 75:2 (2012): 176–191, print.

2. Jennifer R. Veltsos, "An Analysis of Data Breach Notifications as Negative News," *Business Communication Quarterly*, 75:2 (2012): 192–203, print; Creelman.

3. Veltsos 203.

CHAPTER 7

1. *10 Best Practices for Email Marketing*, 2, *eMarketer*, eMarketer, May 2011, Web, 20 May 2013.

2. See Helen Rothschild Ewald and Roberta Vann, "'You're a Guaranteed Winner': Composing 'You' in a Consumer Culture," *Journal of Business Communication* 40 (2003): 98–117, print.

3. Charles A. Hill, "The Psychology of Rhetorical Images," *Defining Visual Rhetorics*, ed. Charles A. Hill and Marguerite Helmers (Mahwah, NJ: Lawrence Erlbaum, 2004) 30–38, print.

4. For further information, consult the Federal Trade Commission's publication *The CAN-SPAM Act: A Compliance Guide for Business* at http://business.ftc.gov/documents/bus61-can-spam-act-compliance-guide-business.

5. "Room for Improvement in Email Opt-Outs," *eMarketer*, eMarketer, 5 Apr. 2010, Web, 21 May 2013.

6. "Room for Improvement."

7. "What It Takes to Win," *CapturePlannning.com*, CapturePlanning.com, 2012, Web, 21 May 2013.

8. Carl Dickson, "What a Private Sector Company Can Learn From Government Proposals," *Captureplanning.com*, CapturePlanning.com, 2012, Web, 21 May 2013.

CHAPTER 8

1. "Search Engine Trends," *Experian Hitwise*, Experian Information Solutions, Inc., 2012, Web, 2 June 2013.

2. Leah Graham and Panagiotis Takis Metaxas, "'Course It's True; I Saw It on the Internet!': Critical Thinking in the Internet Era," *Communications of the AMC* 46.5 (2003): 73, print.

3. "Social Marketing Continues Meteoric Rise Among Local Businesses," *MerchantCircle*, Reply! Inc., 15 Feb. 2011, Web, 2 June 2013.

4. Tom Webster, "Twitter Use in America: 2010," *Edison Research*, Edison Research, 29 Apr. 2010, Web, 2 June 2013.

5. Clive Thompson, "The Early Years," *New York Magazine*, New York Media LLC, 12 Feb. 2006, Web, 3 June 2013.

6. "Blog Directory," *Technorati*, Technorati, Inc., n.d., Web, 3 June 2013.

7. "Technorati Authority FAQ," *Technorati*, Technorati, Inc., n.d., Web, 3 June 2013.

8. "Usability Basics," *Usability.gov*, U.S. Department of Health & Human Services, n.d., Web, 3 June 2013.

CHAPTER 9

1. For a fuller description and history, see Joyce Wycoff, "5-15 Reports: Communication for Dispersed Organizations," *Innovation Network*, InnovationNetwork, 2001, Web, 12 June 2013.

CHAPTER 10

1. Ronald B. Adler, Jeanne Marquardt Elmhorst, and Kristen Lucas, *Communicating at Work: Principles and Practices for Business and the Professions* (New York: McGraw-Hill, 2013) 247, print.

2. "Nearly Half of American Adults Are Smartphone Users," *Pew Internet*, Pew Internet and American Life Project, 1 Mar. 2012, Web, 25 July 2013.

3. Bill Lampton, "5 Reasons You Shouldn't Start Your Speech with a Joke," *Business Know-How*, Attard Communications, Inc., n.d., Web, 28 July 2013.

4. "The Secret Structure of Great Talks," a popular TED video by Nancy Duarte, recommends using a structure that moves back and forth between "what is" and "what could be," ending with the "call to action" (*TED*, Ted Conferences, LLC, Feb. 2012, Web, 29 July 2013). This would be an appropriate structure for talks intended to inspire, sell, or change people's minds. Its indirectness would not be appropriate for instructional or informational talks.

5. Jan Hoffman, "Speak Up? Raise Your Hand? That May No Longer Be Necessary," *The New York Times*, The New York Times Company, 30 Mar. 2012, Web, 29 July 2012.

6. See Kathy Reiffenstein's blog *Professionally Speaking* for the research behind Twitter's growing popularity as a presentation-enhancing tool ("Twitter in Presentations—Love It or Leave It?," 28 May 2009, Web, 29 July 2012).

7. United States, "Construction Safety and Health Outreach Program," Occupational Safety & Health Administration, Department of Labor, n.d., Web, 30 July 2013.

CHAPTER 11

1. Jacquelyn Smith, "7 Things You Probably Didn't Know about Your Job Search," *Forbes*, Forbes.com, LLC., 17 April 2013, Web, 20 June 2013.

2. National Association of Colleges and Employers, "Paid Internships Key to Job-Search Success for New College Grads," *NACE*, NACE, 6 Oct. 2011, Web, 1 July 2013.

3. National Association of Colleges and Employers.

4. Jean Chatzky, "Why Students Shouldn't Take Unpaid Internships," *Newsweek/The Daily Beast*, The Newsweek/Daily Beast Company, LLC, 21 Nov. 2011, Web, 1 July 2013.

BONUS CHAPTER C

1. Laurel Delaney, *The World Is Your Market: Small Businesses Gear up for Globalization*, *Scribd*, Scribd, 2004, Web, 2 May 2012.

2. *Economic News Release: Table A-7*, Bureau of Labor Statistics, US Department of Labor, 4 May 2012, Web, 6 May 2012.

3. Geert Hofstede, "National Cultures and Corporate Cultures," *Communication Between Cultures*, ed. Larry A. Samovar and Richard E. Porter (Belmont, CA: Wadsworth, 1984) 51, print.

4. Fons Trompenaars and Peter Woolliams, *Business Across Cultures* (London: Capstone, 2003) 53, print.

5. Thomas L. Friedman, *The World Is Flat: A Brief History of the Twenty-First Century* (New York: Farrar, Straus, and Giroux, 2005), print.

6. John Mattock, ed., *Cross-Cultural Communication: The Essential Guide to International Business,* rev. 2nd ed. (London: Kogan Page, 2003) 15–23, print.

7. Roger E. Axtell, *Gestures: The Do's and Taboos of Body Language around the World* (New York: John Wiley & Sons, 1998) 43, print.

8. Wang De-hua and Li Hui, "Nonverbal Language in Crosscultural Communication," *Sino-US English Teaching* 4.10 (2007): 67, *www.linguist.org.cn,* Web, 5 May 2013.

9. Allan Pease and Barbara Pease, *The Definitive Book of Body Language* (New York: Bantam, 2006) 111, print.

10. Iris Varner and Linda Beamer, *Intercultural Communication in the Global Workplace,* 5th ed. (New York: McGraw-Hill/Irwin, 2011) 101–102, print.

11. "Power Distance Index," *ClearlyCultural,* ClearlyCulural.com, n.d., Web, 2 May 2013.

12. Sejung Mariana Choi, Shu-Chuan Chu, and Yoojung Kim, "Culture-Laden Social Engagement: A Comparative Study of Social Relationships in Social Networking Sites among American, Chinese and Korean Users," *Computer-Mediated Communication across Cultures,* ed. Kirk St. Amant and Sigrid Kelsey, IGI Global, 2012, Web, 2 May 2013.

13. Kirk St. Amant, "Culture, Context, and Cyberspace: Rethinking Identity and Authority in the Age of the Global Internet," Association for Business Communication Southeast Regional Conference, St. Petersburg, FL, Mar. 2013, conference presentation.

14. "Business Culture in Spain," *WorldBusinessCulture,* Global Business Culture, 2012, Web, 5 May 2013.

15. Jensen J. Zhao, "The Chinese Approach to International Business Negotiation," *Journal of Business Communication* 37 (2000): 225, print.

16. Zhao 225.

17. Naoki Kameda, *Business Communication toward Transnationalism: The Significance of Cross-Cultural Business English and Its Role* (Tokyo: Kindaibungeisha Co., 1996) 34, print.

18. Mattock 14–15.

19. Danielle Medina Walker, Thomas Walker, and Joerg Schmitz, *Doing Business Internationally: The Guide to Cross-Cultural Success,* 2nd ed. (New York: McGraw-Hill/Irwin, 2003) 211, print.

20. Jean-Claude Usunier, "Ethical Aspects of International Business Negotiations," *International Business Negotiations,* ed. Pervez N. Ghauri and Jean-Claude Usunier, 2nd ed. (Amsterdam: Pergamon, 2003) 437–438, print.

BONUS CHAPTER E

1. Examples included here have been adapted to business communication from the *MLA Handbook for Writers of Research Papers,* 7th ed. (New York: MLA, 2009), print; Diana Hacker and Barbara Fister, *Research and Documentation Online,* 5th ed., Bedford/St. Martin's, n.d., Web, 19 July 2013; and Linn-Benton Community College, "MLA Citation Guide," *Scribd,* Scribd Inc., 2012, Web, 24 July 2013. When these sources did not agree, we used the model that seemed to fit best with the logic used for other entries.

index

active-review cards
Communicating in the Workplace

The following questions will test your take-away knowledge from this chapter. How many can you answer?

LO 1-1 In what ways is communication important to you and to business?

LO 1-2 What are the main challenges facing business communicators today?

LO 1-3 What are the three main categories of business communication?

LO 1-4 What are the two primary communication networks in a business, and how are they different?

LO 1-5 What factors can affect the type and amount of communicating that a business does?

LO 1-6 What characteristics make business communication a form of complex problem solving?

LO 1-7 What are the contexts that may influence any given act of business communication?

LO 1-8 What are the steps that two business communicators will usually go through in the process of communicating with each other?

Practical Application

Reread the Workplace Scenario that opens Chapter 1 (page 4). Let's assume that the young employee asked to join this important committee is Jenny Charles, a recent college graduate. Let's also assume that the committee has met to divide up the research, and Jenny has been given the task of gathering ideas from the customer service area in which she works and from young employees like herself throughout the company. After a few weeks, the chair of the committee, a high-ranking executive, emails each committee member to ask for a brief progress report. Here is Jenny's:

I've spoken to several people in customer service about the internal communication methods they think we should use. Frankly, they don't seem very interested in the topic. Most comment that they get too many emails, and they seem to want a better way of staying informed about the company, but that's about as much as they contribute. I think we're going to have to find out what works at other companies and go from there.

Jenny

Applying what you learned in this chapter about the workplace and about communication problem solving, explain why this is a poor handling of the situation.

Did your answers include the following important points?

LO 1-1
- Good communicators have an edge in the job market and are more likely to be promoted.
- Every business depends on the coordinating of people's activities through communication. Good communication skills support other important skills, such as problem solving and collaborating.

LO 1-2
- Staying abreast of changing information technologies
- Being able to communicate skillfully with those from other cultures and with co-workers who have different backgrounds (e.g., different ethnicities, different ages)
- Being able to analyze data, situations, and people effectively (with computational thinking, visual literacy, and interpretive skills)
- Maintaining high ethical standards in one's own communication and helping one's company be socially responsible in its communications

LO 1-3
- Internal-operational, external-operational, and personal

LO 1-4
- Virtually all organizations have both a formal and an informal network. The formal network uses official, approved communication channels and genres. The informal network (also known as the "grapevine") consists of personal channels that individuals in the company have created. The former is more stable and more business related. The latter has an ever-changing structure and contains a good deal of extraneous, even erroneous information—but it still helps achieve the work of the organization.

LO 1-5
- The nature of the business
- The nature of the business's environment (industry)
- The geographic dispersion of the operations of the business
- The people who make up the business
- The business's organizational culture

LO 1-6
- Because businesses are goal oriented, good business communications are also goal oriented. Like other problem-solving activities, business communication helps close the gap between a current situation and a more desirable one.
- Each problem the communicator faces is, in some ways, a unique problem requiring a unique solution. As with all ill-defined problems, solving business communication problems requires analysis, judgment, and creativity.
- There is no one best solution to business-communication problems, and even a carefully planned solution can fail. But it will have a much better chance of succeeding than a poorly planned one.

LO 1-7
- The larger external context (business-economic, sociocultural, historical)
- The communicators' relationship
- The communicators' individual contexts (organizational, professional, personal)

LO 1-8
- The initiating communicator senses a communication need, defines the situation, considers possible strategies, selects a course of action, composes the message, and sends the message.
- The recipient receives the message, interprets it, decides on a response, and replies, following the same problem-solving steps that the original communicator followed.

ANSWERS TO PRACTICAL APPLICATION   First, the "report" has almost no real content. The committee chair needs information, not excuses. If Jenny is having trouble getting the information she has been asked to provide, she needs to figure out a better way to communicate with other employees so that they will give her more input. It is also unprofessional of her to complain about her co-workers, and it is presumptuous to advise the chair on how to proceed, especially since she is a newcomer. The message thus fails to achieve its two main goals: to share data that will help the committee and to project a positive image of the writer.

active-review cards
Understanding the Writing Process and the Main Forms of Business Messages

The following questions will test your take-away knowledge from this chapter. How many can you answer?

LO 2-1 What are the main stages of the writing process and some effective strategies for managing that process?

LO 2-2 In what kinds of situations are business letters primarily used?

LO 2-3 What are the main parts of a memorandum, and when are memos primarily used?

LO 2-4 What are the main parts of an email message, and what are the differences in the levels of formality?

LO 2-5 When might it be appropriate to use text and instant messaging in business?

LO 2-6 How can social media be used in the workplace?

LO 2-7 What is the inverted pyramid, and why is it useful for writing Web documents?

Practical Application

Eleanor Hadley, a manager at a marketing research firm, has a new intern, Seth Haderberg. The first assignment she has given him is to recruit between 7 and 12 people for a focus group (a form of market research in which representatives of the target market offer reactions to a product that's about to be marketed). Using the phone numbers of potential participants, he needs to call until he gets enough people, find out which days and times are best for them to meet, and schedule the focus group. Then he needs to ask Eleanor's assistant, Alex, to send a letter to each of the participants telling them when and where to meet and how they'll be compensated. Eleanor has asked Seth to let her know when he has made these arrangements and what they are.

Seth is now sitting in an airport waiting to go on a quick trip to Disney World. He's done all Eleanor asked—but he realizes that he forgot to tell her so. With no computer at hand, he decides to send her a text message. It reads as follows:

> lined up focus group. c alex. going 2 fla. c u mon. will say hi 2 mickey 4 u lol

What advice would you give Seth about his writing process? About his use of text messaging?

Did your answers include the following important points?

LO 2-1
- The main stages of the writing process are planning, drafting, and revising.
- Spending about a third of the writing time on each stage is a good idea. Remembering that these stages are recursive, not strictly chronological, will also help you write more effective messages.
- The planning stage consists of collecting the information, analyzing and organizing it, and making basic decisions about form, channel, and format.
- While drafting, avoid perfectionism, keep moving forward at a reasonably steady pace, and use any other strategy that will help you write.
- When revising, act as your own critic, honestly assessing how the reader will be likely to respond to each part of what you've written. Using three "levels of edit"—revising, editing, and proofreading—will help you write a more successful, polished message.
- When appropriate, seek feedback from others, and receive their criticism with an open mind.

LO 2-2
- Business letters are usually written to external audiences and in more formal situations—though they still project a human voice.

LO 2-3
- A memorandum typically begins with the date, a "To" line, a "From" line, and a "Subject" line. Then comes the body of the message. There is usually no complimentary close or signature. Instead, the writer initials his/her name in the "From" line.
- Memos are usually written to internal audiences, and usually in situations when email is not practical. They tend to be informal but can be used in more formal situations.

LO 2-4
- The header part of an email typically contains these parts: To, Cc, Bcc, Subject, and Attachments. The body of the email usually starts with the reader's name (sometimes after a brief salutation), followed by the message itself and finally the writer's name (sometimes with such other elements as a complimentary close, the writer's title and company, and so forth).
- Depending on the situation, emails can use casual, informal, or formal language, but the most commonly appropriate style is the informal style.
- Emails should be concise, clear, courteous, and grammatically correct.

LO 2-5
- Text messaging allows you to use a mobile phone to send, receive, and view typed messages.
- It is most appropriate when used with a mobile workforce and when the message to be sent is very short.
- Text messages should be professional and efficient.
- Instant messaging is online chatting. It is essentially a typed telephone conversation, taking place in real time.
- On the job, instant messages should be kept relatively short to enable quick back-and-forth communication. They typically have a conversational, professional tone, but their level of formality should be determined by the relationship of the communicators.

LO 2-6
- Business professionals use social networking tools such as blogs, wikis, Facebook, and Twitter to communicate with internal and external audiences.
- Because information on these sites can be shared with anyone, you must make sure the content on your site is professional and appropriate.

LO 2-7
- Documents written in the inverted-pyramid style present the main point first. This strategy is especially useful in writing Web documents because many Web readers scan documents quickly and may not scroll to find the main point.

active-review cards
Communicating Effectively with Visuals

The following questions will test your take-away knowledge from this chapter. How many can you answer?

LO 3-1 What factors should guide the parts of your presentation that are supported by visuals?

LO 3-2 What factors are important in the construction of effective visuals?

LO 3-3 What are some text-based visuals a writer might consider?

LO 3-4 In what contexts would you use a bar chart? A pie chart? A line chart?

LO 3-5 What are some common errors and ethical problems constructing and using visuals? How can you avoid them?

LO 3-6 What are some guidelines for placing and interpreting visuals effectively?

Practical Application

Can you identify at least five ways to improve on the physical presentation of the visual below?

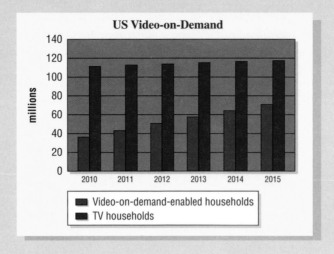

US Video-on-Demand

Legend:
- Video-on-demand-enabled households
- TV households

Did your answers include the following important points?

LO 3-1
- Plan the visuals for your document soon after you organize your findings.
- Your choice of visuals should be guided by your business and communication goals.
- Visuals should clarify, emphasize, add coherence to, or summarize data.
- Visuals should provide interest.
- Visuals should make complex information clearer.

LO 3-2
Several factors are important in constructing visuals to present information effectively:
- Size
- Orientation
- Type
- Rules and borders
- Colors and cross-hatching
- Clip art
- Background
- Numbering
- Titles and captions

LO 3-3
Text-based visuals include
- Tables
- Pull quotes
- Bulleted lists
- Flow charts and process charts

LO 3-4
- A bar chart is used to compare differences in quantities.
- A pie chart is used to compare subdivisions (pieces) of the whole (the pie).
- A line chart is use to indicate changes in information over time.

LO 3-5
Common errors that result in distorting information include
- Errors in graphing
- Errors of format
- Scale distortion
- Scale that does not begin at zero
- Accurate representation of the content

LO 3-6
Visuals must be placed appropriately and interpreted accurately.
- When possible, place a visual near the point in your presentation or document where you discuss it.
- Refer to the visual before the reader sees it in your presentation or document.
- After you present the visual, call your reader's attention to more specific points in the visual.

The ethical problem that results is the distortion of the meaning of the data. To avoid errors, writers should carefully choose content, format, and design for visuals.

The following questions will test your take-away knowledge from this chapter. How many can you answer?

LO 4-1 What are two ways to simplify your writing through word choice?

LO 4-2 Why should you use popular clichés and slang with caution?

LO 4-3 When is the use of technical terms and acronyms appropriate?

LO 4-4 What are three ways to make your wording more precise?

LO 4-5 What word pairs and word combinations can give writers trouble?

LO 4-6 What are three ways to make your verbs active?

LO 4-7 What are four kinds of discriminatory writing to avoid?

LO 4-8 What are two ways to keep your sentences short?

LO 4-9 What are two ways to use sentence structure to emphasize the most important information?

LO 4-10 Avoiding which problems will help give your sentences unity and clear logic?

LO 4-11 What are five good pieces of advice for writing clear, smooth paragraphs?

LO 4-12 Define "conversational style" and explain how to use it in formal situations.

LO 4-13 What is the "you-viewpoint" and why should you use it?

LO 4-14 What is the justification for using positive wording in your messages?

LO 4-15 Besides using polite expressions, what are three ways to achieve courtesy in your writing?

LO 4-16 What are the three major techniques for emphasizing the positive and deemphasizing the negative?

Practical Application

How many wording problems can you spot in the following message thanking a new customer for his business?

Dear Mr. Taylor,

You are welcomed as a new customer and thanked for your business.

It will be seen that we have a lot of wonderful, cheap products which have gained us notoriety in our industry. Plus, our CRM is second to none. And our customer-service girls will answer your questions ASAP, 24-7.

We are anxious to serve you further and look forward to a harmonious and mutually profitable collaborative experience.

Did your answers include the following important points?

LO 4-1
- Use familiar words; use short words.

LO 4-2
- They can quickly grow stale; they can "sound like a substitute for thinking"; they may not be understood, especially by nonnative speakers.

LO 4-3
- Such expressions are appropriate (1) when the reader has the specialized knowledge required to understand them or (2) when you define them for nonexperts.

LO 4-4
- Use concrete words; use specific words; use words with the appropriate connotation.

LO 4-5
- Commonly confused words (e.g., less/fewer, advice/advise); idioms (e.g., authority on, comply with).

LO 4-6
- Limit use of "to be" verbs; use active voice; avoid camouflaged verbs.

LO 4-7
- Sexist words
- Words that stereotype by race, nationality, or sexual orientation.
- Words that stereotype by age.
- Words that typecast those with disabilities.

LO 4-8
- Limit the content; use economical wording.

LO 4-9
- Put the more important points in the shorter sentences.
- State the important points as main clauses and subordinate the less important information.

LO 4-10
- Including unrelated ideas; including excessive detail; mixing two types of sentences; writing incomplete constructions; using modifiers illogically; putting like elements in nonparallel form.

LO 4-11
- Giving paragraphs unity.
- Keeping paragraphs short.
- Making good use of topic sentences.
- Leaving out unnecessary detail.
- Using coherence devices (such as repetition of key words and transitional words).

LO 4-12
- A conversational style avoids "rubber stamps" and other kinds of stiff, impersonal wording. In formal writing, you should avoid colloquialisms and contractions but still use language that sounds like one human being conversing with another.

LO 4-13
- The "you-viewpoint" is an attitude of mind that places the reader in the center of things. Using the you-viewpoint shows the reader that you have his/her needs and interests in mind.

LO 4-14
- Readers respond more favorably to positive words, and they help build goodwill.

LO 4-15
- Avoid blaming the reader; do no more than is expected; be sincere.

LO 4-16
- Put positive points in positions of emphasis in the message; put positive points in short sentences or main clauses; devote more space to positive points.

The following questions will test your take-away knowledge from this chapter. How many can you answer?

LO 5-1 How does assessing the reader's likely reaction to your message help you organize your message?

LO 5-2 What is the general plan for direct-order messages?

LO 5-3 How should you write clear, well-structured routine inquiries for information?

LO 5-4 How should you write direct, orderly, and friendly favorable responses?

LO 5-5 How should you write order acknowledgments and other thank-you messages that build goodwill?

LO 5-6 How should you write direct claims that objectively and courteously present the claim and explain the facts?

LO 5-7 How should you compose adjustment grants that regain any lost confidence?

LO 5-8 What are some strategies for writing clear and effective operational communications?

Practical Application

Can you suggest some ways to improve on the request below?

Every year our company rewards its top salespeople with a weekend retreat for them and their families. We enjoy getting together to have fun, relax, and enjoy one another's company. There are around 50 of us. We are looking at your resort and waterpark. Do you have openings for December 3–5? Please let me know as soon as possible. Please include a list of amenities.

Did your answers include the following important points?

LO 5-1
- If the reaction will be negative, indirect order is your likely choice.
- If it will be positive or neutral, you will want directness.

LO 5-2
- Begin with the objective.
- Provide any additional guidelines or explanation.
- End with adapted goodwill.

LO 5-3
- Begin it with a request—either (1) a request for specific information wanted or (2) a general request for information.
- Somewhere in the message explain enough to enable the reader to answer.
- If the inquiry involves more than one question, order questions in a logically ordered bulleted or numbered list.
- End with an appropriate friendly comment.

LO 5-4
- If the response contains only one answer, begin with it. If it contains more than one answer, begin with a major one or a general statement indicating you are answering.
- Identify the message being answered early, perhaps in a subject line.
- Arrange your answers (if more than one) logically. And make them stand out.
- If both good- and bad-news answers are involved, give each answer the emphasis it deserves, perhaps by subordinating the negative.
- For extra goodwill effect, consider doing more than was asked.
- End with appropriate cordiality.

LO 5-5
- Handle some by form messages or notes, but in special cases use individual messages.

- Begin such messages directly, telling the status of the goods ordered or
- In the remainder of the message, build goodwill, perhaps including some selling or reselling.
- Include an expression of appreciation somewhere in the message.
- End with an appropriate, friendly comment.

LO 5-6
- Even though they carry bad news, these claims are written in the direct order.
- Somewhere early in the message, identify the transaction. Then state the claim.
- Follow with a clear review of the facts without showing anger.
- You may want to suggest a remedy.
- End with cordial words.

LO 5-7
- Write them in direct order, realizing they differ from other direct order messages in that they involve a negative situation. You need to overcome the negative image in the reader's mind.
- You do this by first telling the good news—what you are doing to correct the wrong.
- In the opening and throughout, emphasize the positive. Avoid the negative—words such as *trouble, damage,* and *broken.*
- Try to regain the reader's lost confidence with an explanation or with assurance of corrective measures taken.
- End with a goodwill comment, avoiding words that recall what went wrong.

LO 5-8
- Operational messages are routine messages sent internally and most organized in direct order.
- Write the casual ones like good conversation, but make them clear and courteous.
- Organize them logically; strive for clarity.

ANSWERS TO PRACTICAL APPLICATION 1. Put the request for information in the beginning. 2. Order the questions logically and put them in an ordered list. 3. Give a date by which the information is needed. 4. Include more details regarding the information requested.

The following questions will test your take-away knowledge from this chapter. How many can you answer?

LO 6-1 In what situations might it be better to use the indirect order for negative-news messages?

LO 6-2 What is the general plan for writing indirect-order messages?

LO 6-3 How would you adapt the general plan to refused requests?

LO 6-4 When is a claim written in an indirect pattern?

LO 6-5 How would you adapt the general plan to adjustment refusals?

LO 6-6 How would you adapt the general plan to negative announcements?

Practical Application

Consider the following message sent to a company's employees announcing a new policy in the workplace. Using what you know from Chapter 6 about how to write a negative announcement, what should the writer do to improve the message?

Beginning immediately, space heaters are banned from use in your individual office or cubicle. If you are using one, you must discontinue using it immediately or face disciplinary action.

Yes, I realize that some offices are colder than others, and many of you have space heaters, but unfortunately, whenever you use your space heater in your office, you affect our heating plant's ability to regulate the temperature evenly in the building. In other words, you make the matter worse, and the heating plant can't fix the problem of heating all areas evenly. If you have a problem with this new policy, we suggest you bring a sweater or dress in warm clothing.

We apologize for the inconvenience, but for the good of everyone, you must remove space heaters from your office immediately or face the consequences. If you have any questions, please contact me.

Did your answers include the following important points?

LO 6-1

The indirect order is usually best when . . .

- You are delivering negative news and want to convince the reader that your position is correct, logical, and reasonable.
- You are delivering negative news and want to cushion the shock.

LO 6-2

- Use a strategic buffer.
- Set up the negative news.
- Present the bad news as positively as possible.
- Offer an alternative solution.
- End on a positive note.

LO 6-3

- Your buffer would indicate that you are responding to a request but would not yet give your response.
- Your setup for the news would be convincing reasons for having to say "no."
- The refusal itself, presented as positively as possible, would be short, leading quickly to what you can do or recommend.
- You would end on a forward-looking note—not with a reminder of the refusal (such as an apology).

LO 6-4

- The indirect pattern is used to write a claim when the writer anticipates resistance or a hostile reaction from the audience.
- Some of these hostile situations might involve large dollar or time requirements, unique occasions, tight practices, and more.

LO 6-5

- Your buffer would acknowledge the reader's request and indicate common ground between you and the reader (for example, that you both value quality or fairness).
- You should explain the refusal using logical reasons and objective language. Your setup for the news could explain what your company's policy is in such cases and show—without blaming—that, under this policy, the reader's case does not warrant an adjustment.
- You would say—perhaps as the statement of the refusal itself—what you can do to solve the reader's problem.
- Your ending would be positive and forward looking, with no reminder of the refusal (such as an apology).

LO 6-6

- If you determine that an indirect plan is best, you would start with a buffer introducing a topic related to the announcement.
- You would next explain the situation that requires the negative news.
- Your statement of the news would need to be thorough and clear, since people would not be expecting it and would be likely to have various questions about it.
- You would then help people see what to do next or, if there were no next steps, focus on the positive side of the announcement (for example, that it helps solve a problem that the reader cares about).
- Your close would help the reader feel as positively as possible about the situation and the writer/writer's company.

The following questions will test your take-away knowledge from this chapter. How many can you answer?

LO 7-1 What are important strategies for writing any persuasive message?

LO 7-2 What is the general plan for persuasive requests?

LO 7-3 What are some common unethical techniques used in sales messages?

LO 7-4 What are the planning steps for direct-mail or email sales messages?

LO 7-5 What are the techniques for writing an effective sales message?

LO 7-6 In order to be persuasive, proposals need to satisfy what three criteria on which the readers will be likely to judge the proposal?

Practical Application

When you're not doing your schoolwork, you volunteer for People Working Cooperatively (PWC), an Ohio-based nonprofit organization that provides free home repairs for low-income, elderly, or disabled homeowners. Repair Affair, the PWC's main spring event, is about to take place, and you've been asked to write an email message to persuade students at colleges and universities in your area to participate. Study the organization and this event at <http://www.pwchomerepairs.org/ohio.aspx> and make a list of all the potential benefits you can think of that might appeal to college students.

Did your answers include the following important points?

LO 7-1
- Know your readers.
- Choose and develop targeted reader benefits.
- Make good use of the three kinds of persuasive appeals: character-based, logical, and emotional.
- Make it easy for your readers to comply.

LO 7-2
- Open with words that (1) gain attention and (2) set up your chosen strategy.
- Develop your appeals using persuasive language and you-viewpoint.
- Make the request clearly and without negatives (1) at the end of the message or (2) followed by a last persuasive appeal.

LO 7-3
- Overloading people's email in-boxes with spam.
- Using deceptive wording and visuals.
- Omitting important information or otherwise impairing the readers' ability to make a reasoned judgment.

LO 7-4
- Learn all about your readers.
- Determine the central appeal that you will use in your message.
- Determine the makeup of the mailing.

LO 7-5
- Gain favorable attention on the envelope or in the subject line and hold it in the first paragraph.
- Build a persuasive case, using appeals based on what you are selling and on the traits of your readers.
- Use the you-viewpoint and positive wording throughout.
- Give your message visual appeal.
- Include all necessary information and any auxiliary pieces that support the message.
- Request action, making that action as easy to perform as possible and linking it to a reason to act now or to the main appeal.
- Possibly add a postscript.
- In email writing, offer to remove the reader's name from your mailing list.

LO 7-6
- Desirability of the solution (Do we need this? Will it solve our problem?)
- Qualifications of the proposer (Can the writer or his/her company really deliver, and on time and on budget?)
- Return on investment (is the expense, whether time or money, justified?)

ANSWERS TO PRACTICAL APPLICATION   Sample Answers: You will be helping the unfortunate. The people you'll help truly deserve it (which PWC can verify). The little bit you give can change a life. You'll get a chance to use any "handyman" skills you may have. You don't have to have any special skills. You can learn useful skills from licensed professionals. You'll have fun. You'll feel like part of a team. You'll meet other young professionals. It's only for one day. You'll be part of an award-winning organization. Your school may count the volunteer hours toward any "community service" hours you might need. You'll have something appealing to put on your résumé. It's a nice break from studying. It'll put things in perspective. You don't need to look any further for a good volunteer opportunity.

The following questions will test your take-away knowledge from this chapter. How many can you answer?

LO 8-1 What are problem and purpose statements, and why are they important?

LO 8-2 What kinds of factors may be involved in a report problem?

LO 8-3 What's the difference between primary and secondary research?

LO 8-4 What are the three main Boolean logic operators, and how can they help you conduct Internet research?

LO 8-5 What is an RSS feed, and how can it help you with research?

LO 8-6 What four features of a Web site can help you determine its reliability?

LO 8-7 What social networks can be used for gathering information?

LO 8-8 What are the three main types of library resources for conducting business research?

LO 8-9 What is the difference between probability and nonprobability sampling, and why is it important to have as representative a sample as possible?

LO 8-10 What are six guidelines for constructing a reliable and valid questionnaire?

LO 8-11 Describe the observation method of research. How does it differ from conducting an experiment?

LO 8-12 What are the differences between the before-after experiment and the controlled before-after experiment?

LO 8-13 What kinds of data do focus groups and interviews enable you to gather?

LO 8-14 What are two important guidelines for conducting ethical research?

LO 8-15 What are eight ways to avoid error when interpreting your data?

LO 8-16 What are some common patterns of organization for report outlines?

LO 8-17 What are five guidelines for turning an outline into an effective table of contents?

LO 8-18 What are five qualities of a well-written report?

LO 8-19 In what ways is writing a report collaboratively different from writing a report by oneself?

Practical Application

Assume that you help maintain your company's new blog. The traffic to the blog thus far has been disappointing, and your boss has asked you to identify ways to generate more visits and comments. What kinds of research could you do to solve this problem?

Did your answers include the following important points?

LO 8-1
- The problem statement describes the situation that created the need for the report.
- The purpose statement describes the specific goal of the research conducted to address the problem.
- Formulating these statements helps focus your research, and when stated in the final report, they help your reader understand what you did and why.

LO 8-2
- Subtopics for information reports.
- Hypotheses (possible explanations) for problems requiring a solution.
- Bases of comparison for problems requiring evaluation.

LO 8-3
Primary research produces new information through firsthand observation. Secondary research uses material that someone else has gathered and published.

LO 8-4
The three main operators are AND, OR, and NOT. They help you narrow your search, expand your search, or exclude particular topics.

LO 8-5
Subscribing to an RSS (Really Simple Syndication) feed from relevant Web sites will enable you to receive the most current information on the topics you're researching.

LO 8-6
Its apparent purpose, the qualifications of the authors, indicators of the contents' validity, and its structure.

LO 8-7
Facebook, Twitter, LinkedIn, wikis, blogs, listservs, and Web sites of professional organizations, and social-bookmarking Web sites.

LO 8-8
Library catalogs, databases, and reference materials.

LO 8-9
With probability sampling, all members of the population being studied have the same chance of being chosen to participate in the study. With nonprobability sampling, the participants are chosen on some basis (e.g., convenience, relevance to the research purpose, ability to be accessed). The more representative your sample is, the more valid your claims about the population or subsets of the population will be.

LO 8-10
- Avoid leading questions.
- Avoid absolute terms.
- Focus on one concept per question.
- Make the questions easy to understand.
- Avoid questions that touch on personal prejudices or pride.
- Ask only for information that can be remembered.

LO 8-11
The observation method may be defined as seeing with a purpose. It consists of watching events and systematically recording what is seen. The events observed are not manipulated as they are in an experiment.

LO 8-12
In a before-after experiment, one measures the variable one is interested in, introduces the experimental factor, and then measures the variable again. There is only one group being studied, and if the variable changes during the experiment, it is possible that other factors have caused the change.
In the controlled before-after design, there are two similar groups being compared: a control group and an experimental group. The variable of interest is measured in both groups, the experimental factor is introduced, and the variable is measured again in both groups. With this design, one can make a stronger claim that the factor actually caused the difference between the experimental group's before and after numbers.

LO 8-13
Focus groups and interviews are best for gathering qualitative data, such as beliefs, attitudes, and people's personal experiences.

LO 8-14
Treat research participants ethically; report information accurately and honestly.

LO 8-15
(1) Report the facts as they are; (2) draw conclusions only when the facts warrant it; (3) don't interpret a lack of evidence as proof to the contrary; (4) don't compare noncomparable data; (5) don't mistake correlations for cause-effect relationships; (6) beware of unreliable and unrepresentative data; (7) don't oversimplify; (8) tailor your claims to your data.

LO 8-16
Organization by time periods, places/locations, quantities, factors, or a combination.

LO 8-17
(1) Use a readable and appropriate format; (2) write helpful topic or talking heads; (3) make the headings that are on the same level grammatically parallel; (4) use concise wording; (5) avoid excess repetition of words.

LO 8-18
(1) A reader-centered beginning and ending; (2) objectivity; (3) a consistent time viewpoint; (4) smooth transitions from one part to the next; (5) language that is interesting without being distracting.

LO 8-19
When writing reports collaboratively, the first step is to determine who will be in the group. The group should then establish its ground rules, figure out how they will share their work and ideas, and make a project plan. When researching and writing the report, all team members should help define the problem and purpose, but different people will usually have responsibility for different parts of the report. All should help edit the report, though one person will usually have the primary responsibility for this task.

ANSWERS TO PRACTICAL APPLICATION Your first step would likely be an Internet search to find out what advice others might even have posted. Perhaps Google Scholar might even have some articles on the topic. You could also look at the blogs of your competitors and see what is working for them. You could query your Facebook and LinkedIn contacts or post a question on a professional listserv. Your local library may well have books and articles on the subject. You could also conceivably conduct a focus group or interviews with some of your company's customers or potential customers. Once you gather possible strategies, you could use user-testing or before-after experiments to see which strategies generated the most positive results.

The following questions will test your take-away knowledge from this chapter. How many can you answer?

LO 9-1 How do the length and formality of the report affect its makeup?

LO 9-2 What are four ways that the writing in short reports differs from the writing in long reports?

LO 9-3 What are the three main forms for short to mid-length reports?

LO 9-4 In terms of *purpose*, what are four common types of short reports?

Practical Application

Claudia Messer is the office manager for a small real estate firm. A couple of weeks ago, her boss dropped by her office and asked her to look into whether creating a company Facebook page might be a good idea.

Claudia has done the research and is now writing an email report to her boss. Here is the beginning of her report:

Subject: Facebook Report

It's amazing how much stuff is out there about company Facebook pages. It was almost overwhelming! I wound up focusing on *Facebook for Dummies*, consultants' Web sites, and articles from eMarketer.com. They had some very interesting things to say.

Given what you've read in Chapter 9, what advice would you give Claudia about the way she has started her report?

Did your answers include the following important points?

LO 9-1
- At their most lengthy and formal, reports include a title fly, a title page, a letter of transmittal, a table of contents, and an executive summary. As their length and formality decrease, reports lose more and more of these prefatory parts. The shortest and least formal reports are simple email or memo reports.

LO 9-2
- There is less need for introductory information in short reports.
- Short reports are almost always direct, whereas long reports may be organized indirectly (conclusions last).
- Short reports use a more personal writing style.
- Short reports do not require an elaborate coherence plan.

LO 9-3
- Short-report format
- Letter format
- Email or memo format

LO 9-4
- Routine operational reports
- Progress reports
- Problem-solving reports
- Meeting minutes

ANSWERS TO PRACTICAL APPLICATION ● Make the subject line refer more specifically to the topic and purpose of the report—perhaps something like "Potential Advantages of a Facebook Page for XYZ Realty." ● Open the first paragraph with a clearer indication of the context for the report, so help orient him/her at the start of the report ("As you requested, I . . ."). ● State the main point of the report in your introduction. Your boss wants an answer to the question, "Would creating and maintaining a Facebook page be worth the effort for our company?" There is no good reason in this case for the indirect order. ● You might want to indicate briefly what topics you'll discuss in the report (that is, give an overview of the report's structure). ● It is a good idea to identify your sources. However, that shouldn't be the focus of the first paragraph.

The following questions will test your take-away knowledge from this chapter. How many can you answer?

LO 10-1 What are some key elements of talking that you can use to improve your own talking?

LO 10-2 What can you do to improve the physical aspects of your oral communication such as your posture, walking, facial expressions, and gestures?

LO 10-3 What are some of the challenges of listening? How can you overcome them?

LO 10-4 What are some guidelines for conducting and participating in meetings?

LO 10-5 What are some examples of good phone and voice mail techniques?

LO 10-6 How do you determine an appropriate topic, purpose, and structure for a speech or presentation?

LO 10-7 How can you identify and select appropriate presentation methods?

LO 10-8 How can an audience provide feedback? What are some factors to consider when soliciting audience feedback?

LO 10-9 What can you do to use visuals to help communicate ideas?

LO 10-10 What factors should you consider when planning and delivering a Web-based presentation?

LO 10-11 What factors should you consider when preparing and delivering a team presentation?

Practical Application

Can you identify the strengths of the closing of the commencement speeches below?

"Who will create the Conscious Businesses of the 21st century—businesses that have deeper purpose and are managed consciously to create value on behalf of all of the stakeholders? Why not some of the Bentley graduates here today? Why not you?

I have personally found nothing more fun, more meaningful, or more rewarding than creating and growing the Whole Foods Market. It was what my heart called me to do, and I have followed that calling for 30 years now. To the Bentley graduates today I put forth this challenge: What is your own heart calling you to do? Whatever it is, have the courage to follow it. The grand adventure of your own life now lies open before you. Seize the day!

Thank you. I have greatly enjoyed being with you today."

—John Mackey, Commencement Speech at Bentley College, 2008

"And I hope you will come back here to Harvard 30 years from now and reflect on what you have done with your talent and your energy. I hope you will judge yourselves not on your professional accomplishments alone, but also on how well you have addressed the world's deepest inequities on how well you treated people a world away who have nothing in common with you but their humanity.

Good luck."

—Bill Gates, Commencement Speech at Harvard, 2007

"Stewart and his team put out several issues of *The Whole Earth Catalog*, and then when it had run its course, they put out a final issue. It was the mid-1970s, and I was your age. On the back cover of their final issue was a photograph of an early morning country road, the kind you might find yourself hitchhiking on if you were so adventurous. Beneath it were the words: "Stay Hungry. Stay Foolish." It was their farewell message as they signed off. Stay Hungry. Stay Foolish. And I have always wished that for myself. And now, as you graduate to begin anew, I wish that for you.

Stay Hungry. Stay Foolish.

Thank you all very much."

—Steve Jobs, Commencement Speech at Stanford, 2005

Did your answers include the following important points?

LO 10-1
- Talking is the oral expression of our knowledge, viewpoints, and emotions.
- It depends on four critical factors: voice quality, speaking style, word choice, and adaptation.

LO 10-2
- What listeners see and hear affects the communication.
- They see the physical environment (stage, lighting, background), personal appearance, posture, walking, facial expressions, and gestures.
- They hear your voice.
- For best effect, vary the pitch and speed.
- Give appropriate vocal emphasis.
- Cultivate a pleasant quality.

LO 10-3
- Listening is just as important as talking and oral communication, but it creates challenges.
- Listening involves how we sense, filter, and retain incoming messages.
- Most of us do not listen well, because we tend to avoid the hard work that listening requires.
- You can improve your listening with effort.
- Put your mind to it and discipline yourself to be attentive.
- Make a conscious effort to improve your mental filtering of incoming messages; discipline and practice to retain what you hear.
- Follow the practical suggestions offered in "The Ten Commandments of Listening."

LO 10-4
- In business, you are likely to participate in meetings, some formal and some informal.
- If you are in charge of a meeting, you should know parliamentary procedure for formal meetings, plan the meeting (develop an agenda and circulate it in advance), follow the plan, keep the discussion moving, control those who talk too much, encourage participation from those who talk too little, control time (making sure the agenda is covered), and summarize at appropriate times.
- If you are a participant at a meeting, you should stay with the agenda, participate fully, but do not talk too much, cooperate, and be courteous.

LO 10-5
- To improve your phone and voice mail techniques, you can cultivate a pleasant voice, talk as if you were in a face-to-face conversation, and follow courteous procedures.
- For good voice mail messages, follow these suggestions: (1) identify yourself by name and affiliation, (2) deliver a complete and accurate message, (3) speak naturally and clearly and (4) give important information slowly, and (5) close with a brief goodwill message.
- Demonstrate courtesy when using cell phones by turning off the ringer when and where it could disrupt others, avoiding use at social gatherings, keeping the phone off the table during meals, talking only in places where others won't be in earshot, avoiding talking about confidential or private business, keeping voice volume down, initiating calls in quiet places away from others, being conscious of others when you talk, and avoid talking while driving, especially if it is against the law.

LO 10-6
- Consider the following suggestions in selecting and organizing a presentation.
- Begin by selecting an appropriate topic—one in your area of specialization and of interest to your audience.
- Organize the message (probably by introduction, body, conclusion).
- Consider an appropriate greeting ("ladies and gentlemen," "friends").
- Design the introduction to meet these goals:
- Arouse interest with a story or humor.
- Introduce the subject (theme).
- Prepare the reader to receive the message.
- Use indirect order presentation to persuade and direct order for other cases.
- Organize like a report: divide and subdivide, usually by factors.
- Select the most appropriate ending, usually restating the subject and summarizing.
- Consider using a climactic close.

LO 10-7
- Choose the best manner of presentation.
- Extemporaneous is usually best.
- Memorizing is risky.
- Reading is difficult, unless you are skilled.

LO 10-8
- When you plan your presentation, determine whether you want audience feedback and if so, what type (applause, questions, comments, evaluation forms).
- In a webinar, people can participate in polls or chat with the speaker.
- You can solicit feedback during or after your feedback.

LO 10-9
- Use visuals whenever they help communicate.
- Select the types that do the best job.
- Blend visuals into your speech, making certain that the audience sees and understands them.
- Organize the visuals as part of your message.
- Emphasize the visuals by pointing to them.
- Do not block your audience's view of the visuals.
- Use presentation applications appropriately and in appropriate contexts.

LO 10-10
- Technology and affordable costs have made Web-based presentations popular with large and small businesses.
- Choose a user-friendly technology.
- Make sure you have technology support available.
- Let participants know where you have archived any handouts or other information they may want to retrieve.
- Deliver your presentation much as you would a face-to-face presentation with PowerPoint, a highlighter, or other tools.

LO 10-11
- Team presentations require extra planning to
- reduce overlap and provide continuity
- provide smooth transitions between presentations, and
- coordinate questions and answers.

ANSWERS TO PRACTICAL APPLICATION   Some strengths are these: 1. Successfully signals the end. 2. Summarizes in broad terms the ideas likely presented earlier. 3. Audience centered. 4. Inspiring

active-review cards
Communicating in the Job Search

The following questions will test your take-away knowledge from this chapter. How many can you answer?

LO 11-1 How can you use a network of contacts in the job search?

LO 11-2 What kind of information can you assemble and evaluate that will help you select the right job?

LO 11-3 What sources will lead you to an employer?

LO 11-4 How do you compile print and electronic résumés that are strong, complete, and organized?

LO 11-5 How can you write targeted cover messages that skillfully sell your abilities?

LO 11-6 How can you participate effectively in an interview?

LO 11-7 What other kinds of application follow-up messages are appropriate in business?

LO 11-8 How can you best maintain your job-search activities?

Practical Application

Can you identify some ways to improve the opening below of an email message regarding a summer internship sent to an employer?

I would like to apply for a summer internship at XYZ Company in Information Systems.

Can you identify some ways to improve the closing of an email regarding a summer internship sent to an employer?

Thank you. I look forward to hearing from you soon.

Did your answers include the following important points?

LO 11-1
- A good first step in your job search is to build a network of contacts.
- Get to know businesspeople who might help you later such as classmates, professors, and business leaders.
- Use them to help you find a job, particularly with tools such as LinkedIn.

LO 11-2
- When you're ready to find a job, analyze yourself and outside factors.
- Look at your education, personal qualities, and work experience.
- From this review, determine what work you are qualified to do.
- Then select the job that is right for you.

LO 11-3
- When you are ready to apply for a job, use the contact sources available to you.
- Check university career centers, personal contacts, advertisements, online sources, employment agencies, personal search agents, and web page profiles.
- If these do not produce results, prospect by mail, email, phone, or the Web.

LO 11-4
- In your application efforts, you are likely to use résumés and cover messages. Prepare them as you would written sales material.
- First, study your product—you.
- Then study your prospect—the employer.
- From the information gained, construct the résumé, cover message, and reference sheet.
- In writing the résumé (a listing of your major background facts), you can choose from two types.
- The *print résumé*—traditional and scannable.
- The *electronic résumé*—email, attached file, database, Web, scannable
- In preparing the traditional *résumé*, follow this procedure:
- List all the facts about you that an employer might want to know.
- Sort these facts into logical groups: experience, education, personal qualities, references, achievements, highlights.
- Put these facts in writing. As a minimum, include job experience (dates, places, firms, duties) and education (degrees, dates, fields of study). Use some personal information, but omit race, religion, gender, marital status, and age.
- If you include references, put them on a separate sheet entitled "References."
- Include other helpful information: address, phone number, email address, webpage address, and career objective.
- Write headings for the résumé, and for each group of information use either topic or talking headings.
- Organize for strength in reverse chronological, functional/skills, or accomplishments/highlights approach.
- Write the resume without personal pronouns, make the parts parallel grammatically, and use words that help sell your abilities.
- Present the information for good visual appeal, selecting fonts that show the importance of the headings and the information.
- In preparing the scannable résumé, follow these procedures:
- Include industry-specific keywords.
- Choose precise nouns over action verbs.
- Present the information in a form read accurately by scanners.
- In preparing the electronic résumé, follow these procedures:
- Create a plain text version for copying and pasting.
- Use standard fonts in formatted resumes sent by email.
- Save emails in a common file format such as .docx or .pdf.
- Use social media sites or Web templates to create your online presence.

LO 11-5
- As the cover message is a form of sales message, plan it as you would a sales message.
- Study your product (you) and your prospect (the employer) and determine the strategy for presentation.
- Begin with words that gain attention, begin applying for the job, and set up the presentation of your sales points.
- Adapt the tone and content to the job you seek.
- Present your qualifications, fitting them to the job you seek.
- Choose words that enhance the information presented.
- Drive for an appropriate action—an interview or further communication.

LO 11-6
- Your major contact with a prospective employer is the interview. For best results, you should do the following:
- Research the employer in advance so you can impress the interviewer.
- Present a good appearance through appropriate dress and grooming.
- Try to anticipate the interviewer's questions and plan your answers.
- Make a good impression by being at ease.
- Help the interviewer establish a dialogue with questions and comments that enable you to present the best information about you.

LO 11-7
- You may need to write other messages in your search for a job.
- Following the interview, a thank-you message is appropriate.
- Also appropriate is an inquiry about the status of your application.
- You also may need to write messages accepting, rejecting, or resigning a job.
- Write these messages much as you would the messages reviewed in the preceding chapters: direct order for good news, indirect order for bad.

LO 11-8
- To learn information about the changes occurring in your field and to be aware of better job opportunities, you should.
- Maintain your résumé.
- Read job ads and professional journals.

ANSWERS TO PRACTICAL APPLICATION 1. Tone (too writer centered). Make it more reader centered with you-viewpoint and reader benefit. 2. Vague wording (Information Systems is a broad area). Suggest some areas within the field. 3. Wordy (XYZ Company). It's likely obvious where you want the internship since you sent it to the reader there.

ANSWER: (1) Make the "thank you" specific: "Thank you for your consideration." (2) Ask for an interview: "I would be happy to schedule an interview to discuss my qualifications further." (3) Provide contact information: "You may contact me by email at _____ or by phone at _____."

Relevant Concepts: Résumé writing, descriptive language, active verbs

Carissa is seeking a summer internship. Her primary interest in business is human resources management. She has written a draft of her résumé but has not followed many of the tips in Chapter 11 regarding clear, descriptive language and the need to highlight her best skills and abilities. The résumé also lacks parallelism in many sections. In addition, numbers are not represented correctly, and capitalization is incorrect in many instances.

As your instructor directs,

a. Correct parallelism, number, and capitalization errors.

b. Applying what you learned in Chapter 11, indicate how you would revise the message to be targeted toward Carissa's interest in a human resources internship and to highlight her skills. Make the language more precise and descriptive. While you may need to delete some words or sentences or add content, be sure your edits preserve the type of information Carissa wants to communicate.

c. Retype the message to reflect the revisions you indicated in steps a and b.

Carissa Reyes

http://www.linkedin.com/pub/carissa-reyes/24/66/b92

4903 State Street, Apt. 4
Houston, TX 77009
Phone: 713-837-0313

OBJECTIVE: I wish to obtain an internship in a professional setting where I can use my skills to better the company and help the company achieve its mission.

EDUCATION:
University of Houston-Houston, Texas

Bachelor of Business Administration: Business Management, May 2016

Worked at RG Office Supplies thirty-forty hours per week during the school year; maintained a 3.2 GPA

Rock Creek High School-Rock Creek, Oklahoma
Diploma
May 2012
National Honor Society, Class President, basketball Team Captain

EMPLOYMENT HISTORY
Crew Leader
August 2013 – Present: RG Office Supplies, Houston, TX

- Sold office supplies to Customers
- Responsible for all store operations when a Manager is not scheduled
- Recruit, hire, and train employees
- Planning and coordinating Vendor Displays for product sales
- Marketed products and specials to new members using in-store advertisements
- Accountable for prioritizing tasks within a fast-paced work environment

Ad Sales Rep
September 2012 – August 2013: University of Houston *The daily cougar* student newspaper

- Sold ads to fifteen regular customers
- Marketer for Student Discount Cards to local businesses
- Promote ads on the newspapers Social Media sites
- I was responsible for coordinating ad territories for 6 sales representatives
- Developed a spreadsheet for tracking ad sales more efficiently

EXTRACURRICULAR ACTIVITIES
- Beta Upsilon Sigma (BUS)
- Tutor Business Statistics students in the Quantitative Analysis for Business Class
- Skilled in Computer Programs: Microsoft Excel, PowerPoint, and Word; Adobe Photoshop, InDesign, and Dreamweaver

ready, set, edit: exercise 7
Preparing Slides for an Oral Presentation

Relevant Concepts: Presentation delivery tools, formal presentations

The slide below comes from a presentation on team presentations. The construction of the content on the slide violates many of the principles discussed in Chapter 10 and contains parallelism and word choice errors.

As your instructor directs,

a.   Correct parallelism errors and word choice errors.

b.   Applying what you learned in Chapter 10, indicate how you would revise the slide. While you may need to delete some words or sentences or add content, be sure your edits preserve the writer's ideas.

c.   Create a PowerPoint slide (or slides) to reflect the revisions you indicated in steps a and b.

Things to Think About

- The team must decide where members will sit or stand, how visuals will be presented, how to adjust microphones, and team members should know about entering and exiting.

- Chose who will present the closing and what it will include. What will be summarized? Will the audience ask questions? Who will facilitate the question-and-answer session? How will the questions effect your conclusion?

- After rehearsal sessions, outsiders (non-team members) might be asked to view the team's presentation and provide there feedback. The team might also consider video taping a rehearsal so that all members can evaluate it.

- The principle step: Take special—to care determine the sequence of the presentation as well as the content of each team member's part and which sections are best lead by each group member.

- Plan for the physical delivery: type of delivery, use of notes, and you must dress appropriately.

- Several rehearsals should be planned. Practice the presentation in it's entirety several times as a group before the actual presentation. During rehearsals, individual members should critique each other's contribution and offer specific ways to improve.

- Remember: "Only the prepared speaker deserves to be confident."

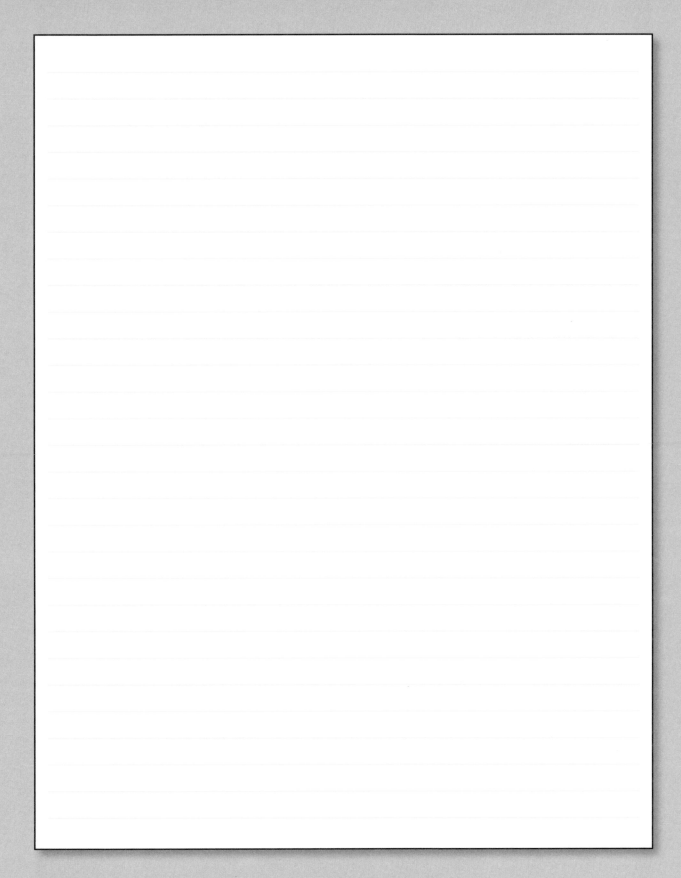

ready, set, edit: exercise 6
Reporting a Group's Progress on an Assignment

Relevant Concepts: Problem and purpose statements, logical report structure, report headings, appropriateness and sufficiency of data, progress report, commonly confused words

Assume that the report below was written by a student to report her group's progress on a report-writing assignment. The professor gave each group in the class a different client and project. This group was given the task of helping a local hotel decide whether to build a mid-sized conference/meeting facility on an empty lot adjacent to the hotel. The students' research goal is to identify the likely demand for such a facility. The goals for their progress report are to convince the professor that the group is functioning well, has a good research plan, has made progress on their research, and is on target to complete a high-quality product on deadline.

As your instructor directs,

a. Circle and correct all incorrectly used homophones (words that sound alike) and other commonly confused words.

b. Mark up the report to indicate its strengths and weaknesses as a progress report.

c. Suggest a more effective, logical structure for the report by writing topic headings. Assume that any additional content that is needed has been added to the report.

Subject: Progress Report

Our group is doing a grate job on our project. We get along well, and our strengths are very complimentary.

Dan and Joan have been researching the current competition in the area since this could significantly effect the demand for a meeting space on the Highgate Hotel's proposed cite. We have identified only two hotels near the Highgate that offer meeting space and cater to large groups. They're fees for space rental and catering are also quiet high.

I have been contacting various organizations in the area to find out if they would like to have a meeting space nearby. We decided to contact the five nearby health facilities, selected faculty at the University, 10 of the larger businesses in the area, and several non-profits. I am in the process of conducting phone or email interviews with contacts at 19 organizations. I'm asking them four principle questions. So far, the findings have been really been interesting.

Sara choose to aide me with the designing the interview questions. She is also helping me conduct the interviews. We hope to illicit detailed answers from the respondents that will enable us to imply the likely demand for a new meeting facility in the neighborhood.

Once we have all our data, I'll write a rough draft, which all of us will review and revise. Dan will also be preparing the transmittal message, Joan will be in charge of the graphics and the layout, and Sara will serve as our final quality-check person.

We expect to deliver a well-prepared, helpful report to the Highgate Hotel manager by the project deadline.

Persuading Potential Customers to Call You, continued

Relevant Concepts: You-viewpoint, elements of effective persuasion, parts of a sales message, effective use of formatting, use of the apostrophe, subject-verb agreement, pronoun-antecedent agreement

Assume that the message below was written by Traymore, a large national insurance company that specializes in car insurance. The purpose of the message is to introduce readers to their local Traymore agent and invite them to call or come in for a free quote.

As your instructor directs,

a. Correct any errors you see regarding use of apostrophes, subject-verb agreement, and pronoun-antecedent agreement.

b. Rewrite the message to make it more persuasive. The main facts you need are here. You should also make the format more appealing, but keep in mind that the letter is intended to seem like a personal letter, not a glitzy piece of corporate advertising.

Jeff Gardner

7023 Norwood Street

Cincinnati, OH 45220

Dear Car Owner,

As you've probably heard, we can provide insurance coverage for you're car for less than most other companies can. I'm writing to let you know that I'm your Traymore agent in Cincinnati.

I can give you a free quote if you will call me at 513-227-4250. Sometimes our rates are even 15% lower than other company's rates. Our coverage is great, and its' good to deal with an agent who is near you. And there's more reasons to switch to Traymore. New Traymore customers save an average of $500 per year. If you'll call me, I can find out what discounts I could offer you. Insuring more than one car, having certain safety features on your car, or wearing a seat belt make the rate lower. Anybody on active duty, in the National Guard/Reserve, or retired can also make their rate lower by using the Military Discount. Low rates doesn't mean that we compromise on customer service. We have an agent on call around the clock every day. If you're in an accident, the claim can be filed any time. We usually settle claims within 48 hours, and our settlements are fair. When you call, ask about insurance for homeowner's and renter's. I might be able to help. I can also provide a quote for motorcycle, ATV, or RV coverage.

Contact me today at 513-227-4250. I'd also be willing to meet face to face. Let me create a policy that will fit your needs and save you money.

Sincerely,

Joe Trahern

Traymore Agent

624 Marindale Road

Cincinnati, OH 45239

513-227-4250

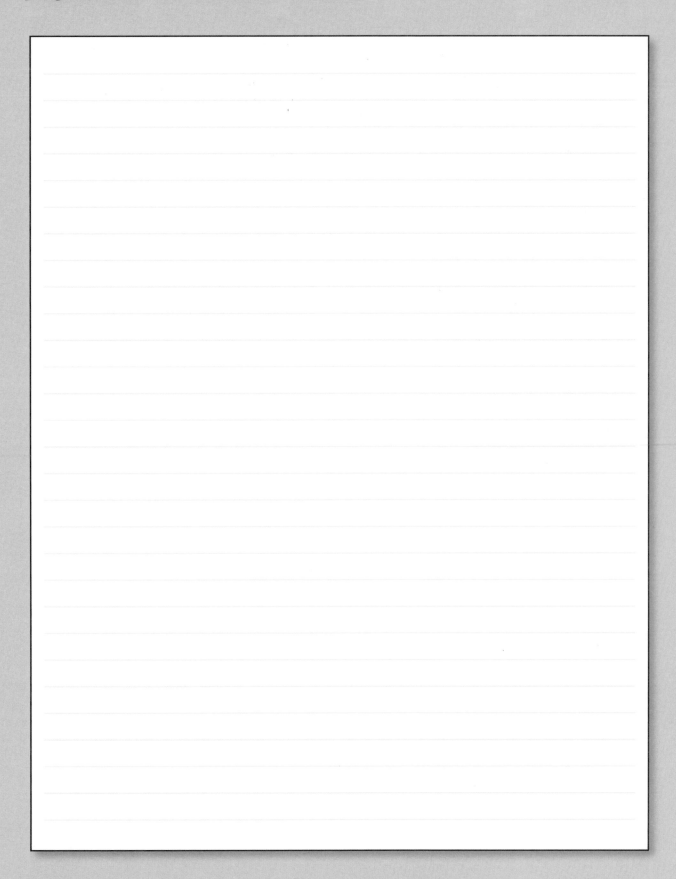

ready, set, edit: exercise 4

Denying an Insurance Claim

Relevant Concepts: Indirect order messages, adjustment refusals, punctuation

During a recent storm, several trees fell on Elizabeth Sampson's porch, damaging the porch and its roof. Ms. Sampson claimed the trees belonged to Nelson Insurance's client, so the claims field representative, Carl Anderson, inspected the damage. Mr. Anderson discovered that the trees were on Ms. Sampson's side of the lot line. Ms. Sampson contested Mr. Anderson's investigation and took her claim to the field claims manager, Mee Vue. She reviewed Ms. Sampson's claim and responded with the message below. The message he has written to Ms. Sampson, however, is too direct and poorly punctuated and contains unnecessarily negative and hostile language.

As your instructor directs,

a. Punctuate the message correctly using commas, colons, and semicolons. Cite the correctness standard from Bonus Chapter B that you applied.

b. Applying what you learned in Chapter 6, indicate how you would revise the message to follow the indirect order and improve the tone. While you may need to delete some words or sentences or add content, be sure your edits preserve the writer's ideas and relevant content.

c. Retype the message to reflect the revisions you indicated in steps a and b.

July 1, 2014

Ms. Elizabeth Sampson

10586 California Avenue

Hayward, WI 54843

Dear Ms. Sampson:

As you requested I have completed a review of this matter. I am sorry but I agree with our representative's assessment that we are not responsible for the claim. As you are aware Carl Anderson, field claims representative, completed a comprehensive investigation your claim. You claimed that you believe our insured, Mr. Conrad Dodge, is responsible for the cost of replacing the porch roof the dented siding and the torn screens because the trees that fell were his. Nelson Insurance has secured a copy of the most recent survey of the properties I have enclosed a copy of those documents.

My review of this matter has come to the same conclusion as Mr. Anderson's review. Unfortunately, we simply cannot honor your claim the current survey of your property and Mr. Dodge's property clearly shows that the trees are yours. While I can understand your disappointment you can obviously see these are your trees if you look at the survey. Therefore, Nelson Insurance has determined the following your claim lacks legal standing and our insured is not liable to you in any way.

If you have any further evidence that you wish for me to consider please provide it to me and I will review that information promptly. If you have any questions or concerns I can be contacted at 715-432-4321, ext. 4. I apologize for any inconvenience.

Sincerely,

Mee Vue, CPCU, AIC, AIS

Nelson Insurance: Field Claims Manager

Phone: 715-432-4321, ext. 4

Email: vuem@nelsoninsurance.com

Enclosure: Property Survey Results

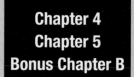

ready, set, edit: exercise 3
Announcing a Change in Procedure

Relevant Concepts: Direct-order messages, direct operational messages, sentence structure, hidden verbs, cluttering phrases

The following operational message informs employees of a new time-saving routing process for creating a bill of materials (BOM) for manufacturing your company's circuit boards. The BOM includes all resources for making the circuit boards, from purchasing raw materials and parts to assembling and selling the boards, so creating the BOM requires the approval of several divisions within the company. This message is indirect and unclear. In addition, the message contains several sentence errors (sentence fragments, run-on sentences, and comma splices), hidden verbs, and cluttering phrases.

As your instructor directs,

a. Correct the sentence errors and fix the hidden verbs and homophone errors.

b. Indicate how you would revise the message to make it more direct, better organized, and clearer. Be sure your edits preserve all of the ideas even as you revise or delete material.

c. Retype the message to reflect the revisions you indicated in steps a and b.

To: <MGMT@forwardcircuits.com>; <supstaff@forwardcircuits.com>

From: Ross Gilman <rgilman@forwardcircuits.com>

Subject: Circuit Boards

I initiated and led a process review for the reduction of the routing time our BOMs, two parts of the process were slowing it down. First, Purchasing's function is in the middle of the process. Which means the purchase order cannot be issued until the three departments before us in the process take up to three days to finish their assigned tasks. Also, once the Purchasing Department places the order, we have to wait five days for the supplier to ship the ISO 9001:2000/AS9100-compliant parts.

As you know, at the present time, the BOMs are routed as follows: Accounting, Sales, Scheduling, Purchasing, Accounting, Scheduling, Master Scheduling

In the new process we are able to get the consolidation of the number of steps from seven to five. Unnecessary processes were removed steps were eliminated. So that no department will be affected by long wait times. We have had two jobs go through the new process to test its effectiveness, average time from issuance of the purchase order is in the neighborhood of one day. The total time for creating the BOM averages 21 days instead of 29 days.

The new process routes the BOMs in the following order: Purchasing, Accounting, Scheduling, Sales, Master Scheduling

Please follow the new process outlined above to route any BOMs contact me with any questions or concerns.

Ross

Ross Gilman, CPM & Senior Buyer

Purchasing Department

Forward Circuits, Incorporated

Office: 715-555-5550 | Cell: 715-555-5551

Relevant Concepts: You-viewpoint, effective word choice, active verbs, economical wording

The following letter announces to customers the merger of two cellular service companies.

Applying what Chapter 4 says about using the you-viewpoint and a clear, interesting style, mark up the message to indicate where and why it needs improvement.

Dear Customer,

First, I would like to start by thanking you for being a customer of Speed mobile service. As you know, at Speed, we are committed to delivering exceptional value through our state-of-the-art products, services, and experiences. An important step was recently taken to leverage our assets and bolster our business.

As you may have heard, Speed will combine with NationalPS to create a bigger, better, bolder wireless provider. The combined entity will be a new, publicly traded company called Speed that will be the premier challenger in the marketplace by providing the best value. The merger will become effective on 1 July 2014.

Here is how we're going to the next level with this game-changing transaction:

• We will be rolling out a national NTE network with even greater speed and capacity. This will provide access to a denser, higher-capacity network, including deep LTE coverage in key metropolitan areas.

• We will be able to expand our broad device portfolio to offer more products, services, and devices.

• We're going to continue to change the way wireless is done with our creative, compelling, and unique 4G plans, including the industry's first Any Device plan.

This is a terrific opportunity for two companies that share a common commitment to innovation and customer service to come together to improve wireless communication.

We couldn't be more excited about what the future holds for Speed.

Regards,

Sam Williams, President & CEO

Speed

---

chapter 4

ready, set, edit: exercise 1
Instructing Employees on How to Ask for Tech Help

Relevant Concepts: You-viewpoint, positive emphasis, effective word choice, active verbs, logically worded sentences

The following email informs employees of a new system for requesting help from the company's technical support staff.

As your instructor directs,

a. Mark up the message to indicate its stylistic problems.

b. Revise the message so that it is more positive, reader friendly, and well written. Invent any plausible details necessary to make the message effective.

Subject: Asking for Tech Support Correctly

We in the IT department are doing our best to deal with your computer problems, but often you don't help us help you, making our jobs more difficult and a delay in correcting the problem.

Please be sure the following system is employed when you need tech help:

- Employees shouldn't stop us in the hall and ask us to take a look at your computer. If you need our help, you should either email us at techhelp@bestcompay.com or call us at 555-2420. The reason is because we need to be able to create a service record in our database. If we don't have a service record, we don't get full credit for our work, plus running the risk of losing track of your request.

- When you report a problem, don't be so vague. Describe the problem in detail along with including any error message you are getting and your phone number.

- If your problem is urgent, put the word "urgent" in your subject line or you can say it when you call. Urgent does not mean that you don't know how to do something in Excel or wanting a software upgrade or wishing you had a faster computer.

- As far as follow-up calls, don't keep calling to ask when your computer will be ready. We understand that you need your computer to do your work. Sometimes you have caused the problem by irresponsible use of the Internet and opening dangerous emails, and it can take a long time to clean viruses and other malware off a computer. Pestering us won't speed up our work. If you urgently need your computer, you can make a duplicate call or email, but you should let us know that you have already left a voice or email message about the problem. We have loaner computers we can give you if you ask for one.

Thank you for your cooperation on this matter.

Jeff
Jeff Jenkins, IT Manager
